PRINCIPLES of the MANUFACTURING OF COMPOSITE MATERIALS

HOW TO ORDER THIS BOOK

BY PHONE: 866-401-4337 or 717-290-1660, 9AM–5PM Eastern Time

BY FAX: 717-509-6100

BY MAIL: Order Department

DEStech Publications, Inc.

439 North Duke Street

Lancaster, PA 17601, U.S.A.

BY CREDIT CARD: American Express, VISA, MasterCard

BY WWW SITE: http://www.destechpub.com

PRINCIPLES of the MANUFACTURING OF COMPOSITE MATERIALS

Suong V. Hoa

Department of Mechanical and Industrial Engineering
Concordia University, Quebec, Canada

DES*tech* Publications, Inc.

Principles of the Manufacturing of Composite Materials

DEStech Publications, Inc.
439 North Duke Street
Lancaster, Pennsylvania 17602 U.S.A.

Printed in the United States of America
10 9 8 7 6 5 4 3 2 1

Main entry under title:
Principles of the Manufacturing of Composite Materials

A DEStech Publications book
Bibliography: p.
Includes index p. 337

ISBN: 978-1-932078-26-8

Contents

Preface

This book is a culmination of efforts and experience that I have gathered over the past 15 years from teaching the course of manufacturing of composites at Concordia University, from training engineers at local aircraft companies, and from research activities on analysis, design and manufacturing of composites over the past three decades.

I came to work on composites from a mechanical engineering background. This may be similar to many other people in the late 1970s and 1980s. I began work on mechanics of composites. For mechanics problems, we assumed that resin and fibers possess certain properties and that most resins would behave similarly as long as they are of the same category (for example, epoxies). As I got to know more about the field, I came into contact with people in various industries from which I found that there exists a world of knowledge and experience for composites manufacturing: in order to have sufficient expertise in composites, one must know not only the mechanics of the materials/structures, but also the intricacies of their manufacturing processes.

When I listened to presentations or attended industrial shows on composites manufacturing, the discussions and/or exhibits of composite parts were impressive. However, the question that always came to my mind was: what is the principle behind the manufacturing of these parts? How can one grasp the concepts and the principles behind them?

There are a few books on composites manufacturing that exist in the literature. One well-known book is that of George Lubin (*Handbook of Composites*), which is a comprehensive presentation of many aspects of composites and composites manufacturing. Another book is that of Brent Strong (*Fundamentals of Composites Manufacturing*). This book presents the principles of composite manufacturing, describing the ingredients: matrix, fibers, interface and the main processes for composite manufacturing. It deals more with the practical aspects of composites manufacturing. In 1997, T. G. Gutowski's *Manufacturing of Composites* was published. This book is a collection of works of well-known scientists in the field and focuses on the scientific aspects of composites manufacturing. Also in 1997, there was a book on processes entitled *Composite Materials, Processing, Fabrication and Applications,* by Mel M. Schwartz. In 2000, a book by Sanjay Mazumdar on practical aspects of composites manufacturing came out. These are excellent books in Composites Manufacturing and provide essential knowledge for the field.

However, during my 15 years of teaching I have always found that the existing literature presents composites manufacturing either from the practical point of view (Brent Strong), the collection of opinion of scientific experts (Gutowski), or the collection of different processes (Mazumdar). The existing literature describes more of *what* has been done, with little attention paid to *why* these processes have been done. As its name implies, composites is a field that requires knowledge from many fields. Manufacturing of composites involves a significant amount of knowledge in materials and chemistry. For mechanics people, trying to grasp the chemical aspects of composites manufacturing can be daunting. From the point of view of the learner (students), if they understand the "why," then it may be easier for them to grasp the essentials of composite manufacturing.

The objective of this book is to present composites manufacturing from the rationale of why things are done in a certain way. The book is intended for students from different backgrounds such as mechanical engineering, aerospace engineering, civil engineering, materials engineering, and chemical engineering. It covers the main principles governing manufacturing using composites.

The book is divided into two parts. The first part deals with the fundamental elements for composites manufacturing. This includes the discussion on the essential principles behind composites manufacturing in Chapter 1. This is followed by a discussion on matrix material in Chapter 2, and a discussion on reinforcements in Chapter 3. The second part presents the five most common techniques for composites manufacturing. These are: manufacturing using autoclave in Chapter 4, filament winding in Chapter 5, pultrusion in Chapter 6, liquid composite molding in Chapter 7 and, finally, thermoplastic composites in Chapter 8.

This book is a written version of the instructional materials that I have used in my manufacturing of composites course over the last 15 years. I hope that it will be of help to future students and also to future instructors.

SUONG V. HOA
Montreal
November 2008

Acknowledgements

The preparation of this book has received assistance from many people. I thank my students for asking questions, giving comments and feedback to the course delivery over the years. I thank Dr. Ming Xie for having done most of the drawings, Dr. Ngo Tri Dung for making the chemical formulae, Mr. Heng Wang for doing the calculations for the examples, and Dr. Minh Tan Ton That for reading Chapter 2.

I thank my mother Hoa thi Tho for giving me the appreciation of hard work and inspiration for moving ahead, my wife Do thi Dong for her love, encouragement and support, and my children Vincent, Glenn, Sabrina and Victoria for the joy they give me.

PART 1

Fundamentals of Constituents for Composites Manufacturing

CHAPTER 1

Introduction

Advanced composite materials have been used to fabricate many structural parts in engineering applications. This is due to their many attractive characteristics such as light weight, high strength, high stiffness, good fatigue resistance and good corrosion resistance. Also, the ability to manufacture parts with complicated geometry using fewer components enables manufacturers to save cost as compared with the same parts made of conventional metallic materials. Before presenting the fundamental aspects of manufacturing and different techniques used for composites manufacturing, it is appropriate to present composite structural parts currently in use and the main techniques that have been used to fabricate them.

1. EXAMPLES OF PRODUCTS MADE USING DIFFERENT MANUFACTURING TECHNIQUES

Figure 1.1(a) shows a schematic of an Airbus 380 airplane (the largest airplane in the world as of 2008). This airplane has more than 50% of its structure made of composite materials. These components include the flaps, ailerons, rudder, radome etc. Most of these components are flat in shape and they are usually made using hand-lay-up (HLU) and autoclave molding techniques. Figure 1.1(b) shows a schematic of the hand-lay-up fabrication technique and a representative lay-up sequence. Autoclave molding is a well-established method for composites used in the aerospace industry with certified resins and fibers. A photograph of an autoclave is shown in Figure 1.1(c). Autoclave Molding will be discussed in detail in Chapter 4.

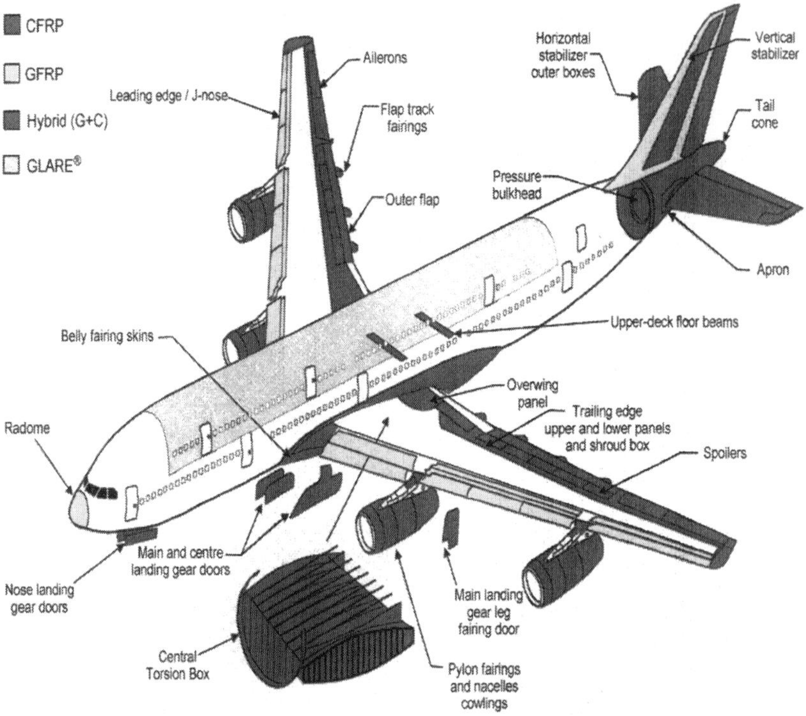

FIGURE 1.1(a) Airbus 380 with its composite component (from http://www.specialchem4adhesives.com/home/editorial.aspx?id=752).

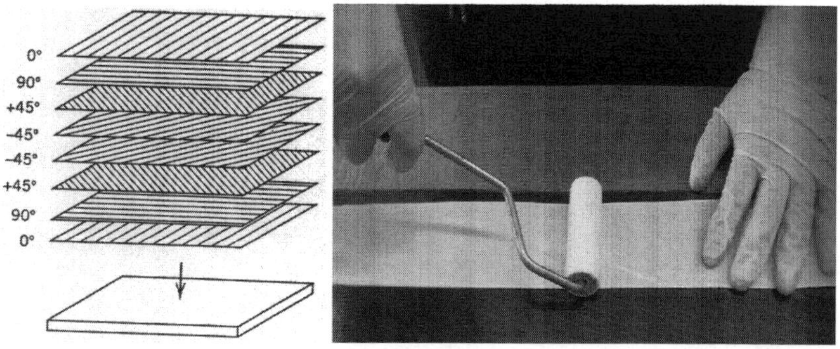

FIGURE 1.1(b) Schematic of the hand-lay-up fabrication method and a representative lay-up sequence. Individual layers can be cut by hand or by a computerized machine cutter. The layers can be stacked one on top of the other by hand or by a robot.

FIGURE 1.1(c) Photograph of an autoclave (courtesy of ASC Ltd.).

Figure 1.2(a) shows a pressure vessel made of composite materials using the combination of hand-lay-up and filament winding processes. Composite pressure vessels are light weight and can contain pressures higher than those contained by metallic vessels. These components are made using the filament winding process [Figure 1.2(b)]. Figure 1.2(c) shows a photograph of a filament winding machine. The filament winding process will be discussed in detail in Chapter 5.

FIGURE 1.2(a) Composite pressure vessel made by combination of hand-lay-up and filament winding.

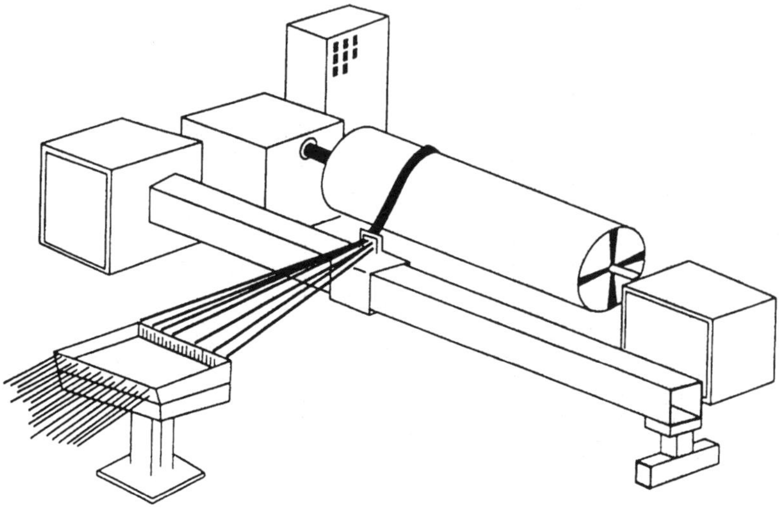

FIGURE 1.2(b) Schematic of the filament winding process (courtesy of Wiley Interscience).

FIGURE 1.2(c) A two-spindle winder with a carriage-mounted resin bath and a free-standing creel in the background (courtesy of *Composites Technology* magazine, August 2005).

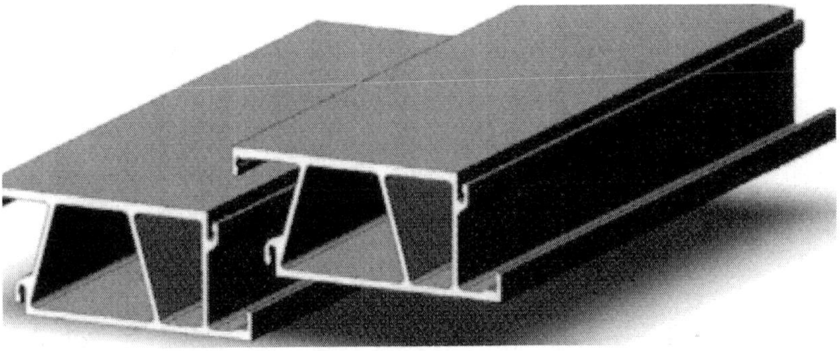

FIGURE 1.3(a) A composite pultruded connector.

Figure 1.3(a) shows a component made using pultrusion. Pultrusion is used to make many structures for civil engineering applications. Figure 1.3(b) shows the schematic of the pultrusion process, and Figure 1.3(c) shows a photograph of a lab scale pultrusion machine. Pultrusion will be discussed in Chapter 6.

Figure 1.4(a) shows a composite component made using the liquid composite molding (LCM) method (5 piece). LCM has been used to make automobile composite components. Figure 1.4(b) shows a schematic of the liquid composite molding process and Figure 1.4(c) shows a pump, a mold and accessories for the liquid composite molding hardware. Liquid composite molding will be discussed in Chapter 7.

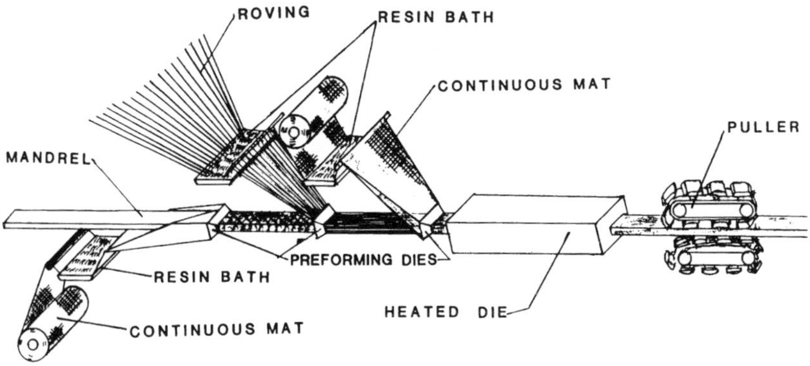

FIGURE 1.3(b) Schematic of the pultrusion process (courtesy of Springer).

FIGURE 1.3(c) A lab pultrusion machine.

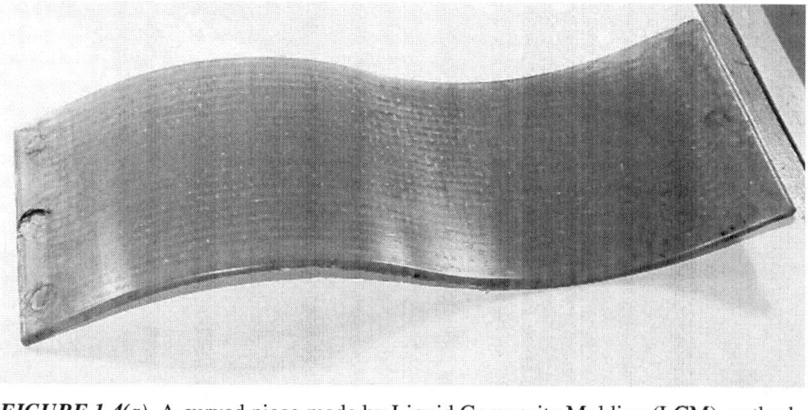

FIGURE 1.4(a) A curved piece made by Liquid Composite Molding (LCM) method.

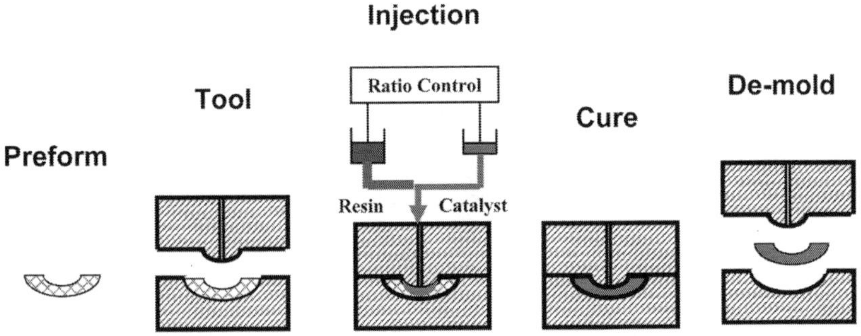

FIGURE 1.4(b) Schematic of the liquid composite molding process.

FIGURE 1.4(c) Instrumentation for LCM: pump, mold and accessories. Resin is filled into the vertical cylinder, then pumped into the mold cavity on the left-hand side.

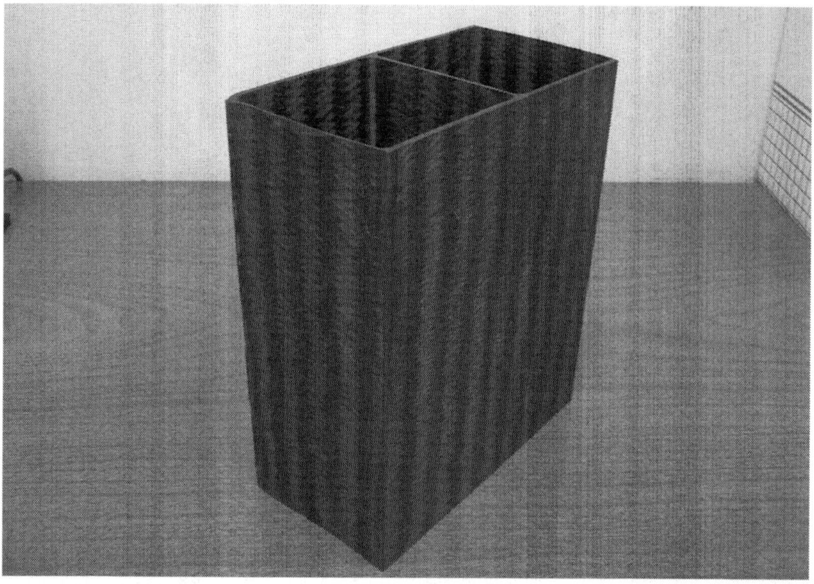

FIGURE 1.5(a) A thermoplastic composite wing box panel made by compression molding.

Figure 1.5(a) shows a composite wing box panel made using thermoplastic composites and compression molding method. Figure 1.5(b) shows the schematic for the thermoplastic composite molding process. Figure 1.5(c) shows a compression molding machine. Molding of thermoplastic composites will be discussed in Chapter 8.

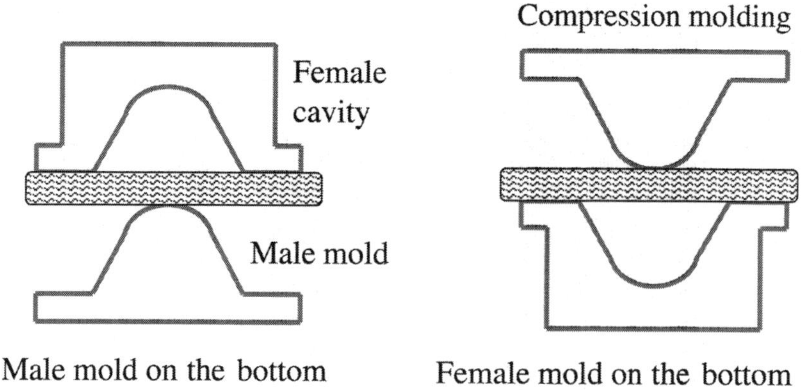

FIGURE 1.5(b) Schematic of the thermoplastic composite molding process.

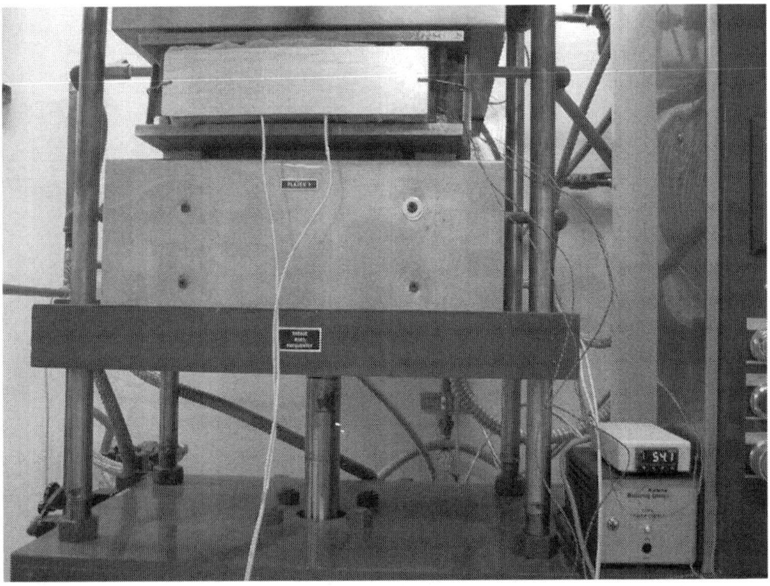

FIGURE 1.5(c) A compression molding machine.

Figure 1.6(a) shows a thermoplastic composite tube made by the fiber placement process. Figure 1.6(b) shows the schematic of the thermoplastic composite placement process, and Figure 1.6(c) shows a photograph of a fiber placement machine. Fiber placement of thermoplastic composites will be discussed in Chapter 8.

FIGURE 1.6(a) A thermoplastic composite tube made by the fiber placement process.

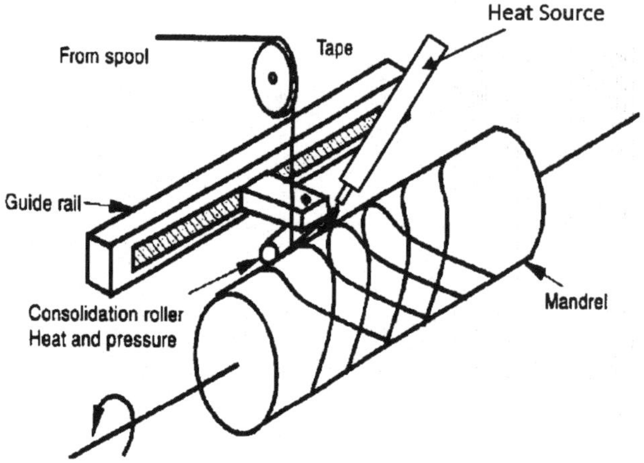

FIGURE 1.6(b) Schematic of the thermoplastic composite placement process.

FIGURE 1.6(c) A fiber placement machine (courtesy of Aerospace Manufacturing Technology Center, National Research Council of Canada).

A few specific features can be extracted from the above components and the different manufacturing techniques used to fabricate them.

Normally structural components can be classified according to their shape, and the manufacturing technique used depends significantly on the shape of the component as follows:

- *Relatively thin flat plate or shallow shell with free edges.* Normally aerospace components have these types of shapes. These are usually made using the hand-lay-up method. The autoclave is the common tool used for making aerospace composite components having these shapes.
- *Components of revolution, such as cylindrical or spherical pressure vessels and pipes.* These structures usually have no free edges (except for the end openings). These are usually made using the filament winding method.
- *Components having constant cross section such as tubes, rods, or even components with complex but constant cross section along the length* such as door frames. These are usually made using the pultrusion method.
- *Components having complex 3-D configurations.* These can be thick or thin. These are usually made using the liquid composite molding (LCM) method.
- *Large structures such as boat hulls, wind turbine blades* etc. These are made using a modified form of LCM such as vacuum-assisted LCM. A special process called SCRIMP (seaman composite resin infusion molding process) is usually used to make boat hulls.
- *Small and large components, either without free edges or with free edges.* These can be made by the fiber placement method. These machines are versatile but require a large amount of capital investment (on the order of several millions of dollars).

2. GENERAL CHARACTERISTICS OF MANUFACTURING USING COMPOSITES

Generally, manufacturing using composites involves the processing of two main ingredient materials to make a final product. The ingredients involve the matrix and fiber materials. This processing requires the following:

- Good bonding between matrix and fibers

- Proper orientation of the fibers
- Good amount of volume fraction of fibers
- Uniform distribution of fibers within the matrix material
- Proper curing or solidification of the resin
- Limited amount of voids and defects
- Good dimensional control for the final part

The implications of the above are as follows.

Good bonding between matrix and fibers. To provide reinforcement so that properties such as strength and stiffness can be enhanced, the fibers need to be bonded to the matrix. If at a certain location, the fibers are not properly bonded to the matrix, dry spots will occur. At this location, there is no proper shear transfer of load between fiber and matrix and the domino effect (as will be discussed in Section 3.1 of this chapter) will occur. These locations will also serve as nuclei for cracks to form. However, there are situations, such as the requirement to absorb impact energy, where partial dry spots may enhance the energy absorbing capability of the composite.

Proper orientation of the fibers. Proper orientation of the fibers is important since properties such as stiffness and strength are very sensitive to fiber orientation. If the fiber orientation deviates by about 10° from the 0° direction, the stiffness can drop by more than 30%. Fiber orientation may be deviated from the intended orientation due to improper position of the layer during the lay-up step, or due to the flow of liquid resin that pushes the fibers during the filling period in liquid composite molding.

Good amount of volume fraction of fibers. In composite materials, the fibers provide stiffness and strength. Therefore the greater the amount of fibers, the better will be these properties. The amount of fibers is usually expressed in terms of volume fraction, v_f which is defined as:

$$v_f = \frac{V_f}{V_c} \tag{1.1}$$

where V_f is volume of fibers and V_c is volume of the composite material.

Properties such as stiffness of a unidirectional composite along the axial direction is given by the rule of mixtures:

$$E_c = E_f v_f + v_m E_m \tag{1.2}$$

Where subscript f refers to fiber, and m refers to matrix.

The fiber volume fraction and matrix volume fraction are related by:

$$1 = v_f + v_m + v_v \qquad (1.3)$$

Where the last term refers to volume fraction of voids.

For good quality composites, the amount of voids should be minimum (less than 1%) and Equation (1.3) can be approximated to be

$$1 = v_f + v_m \qquad (1.4)$$

Using Equation (1.4) in Equation (1.2), the modulus is expressed as:

$$E_c = (E_f - E_m)v_f + E_m \qquad (1.5)$$

The modulus is linearly proportional to the fiber volume fraction. Therefore the larger the fiber volume fraction, the better the mechanical properties. It should be noted, however that the fiber volume fraction cannot be 1 because this would mean that there is no matrix material which, in turn, would mean dry bundles of fibers and the domino effect as mentioned in Section 3.1 will prevail.

Uniform distribution of fibers within the matrix material. Figure 1.14 shows a cross section of a unidirectional composite layer. The white dots show the cross section of the fibers and the dark area represents the matrix. One can see that at the fine scale, the distribution of the fibers is uniform in some region but not in other regions. The region where there is more matrix than fiber is called a *resin rich area*. It is not a good idea to have large or many resin rich regions because there will also be weak areas. Under loading, these areas can serve as locations for crack to nucleate.

Proper curing of the resin. In the manufacturing of polymer matrix composites, the resin first occurs in the form of low viscosity liquid so that it can flow and wet the surface of the fibers. After wetting has been completed, the resin needs to solidify and harden. For thermoset resin, this is called *curing*; and for thermoplastic resin, this is called *solidifica-*

tion. In both cases, the resin needs to be hard and stiff for the reinforcement effect to take place. If there are regions where the resin is not hard enough, they will be weak and can serve as crack nucleation areas.

Limited amount of voids and defects. Voids and defects may be formed during the manufacturing of composites. Voids can arise due to lack of compaction of many layers together, or due to low pressure in the resin during curing. The amount of voids needs to be a minimum to be acceptable. A limit of about 1% is commonly used. Defects such as delamination between layers, cracks, fiber mis-orientation, or nonuniform fiber distribution may not be acceptable.

Good dimensional control for the final part. Polymeric resins shrink when they change from liquid state to solid state. The degree of shrinkage can be between about 5%–8% depending on the type of materials. This shrinkage of the material may cause residual stresses in the part, and also out-of dimensions or warping. For a large structure such as the wing of an aircraft, a few percentages of shrinkage of the material can translate into significant deformation of the structure. Another problem that may occur is the surface finish of parts such as automobile panels which may be adversely affected by this shrinkage. Resins with Low Profile Additives are usually used to control shrinkage.

2.1 Metal versus Composite Manufacturing

Manufacturing using composites has differences from manufacturing using metals:

- In metals such as steel or aluminum, materials with finished form such as rods, slabs, or sheets are available. The making of a finished product such as a car body or the box frame for a computer only requires working on these finished forms. Processes such as cutting, bending, forming, welding, or drilling are used on these finished forms to make the finished product.
- In composites, the steps that transform the finished form to the final structure are usually bypassed. A manufacturer using composite materials has to work directly from the ingredients of fiber and matrix to make the finished product itself. Figure 1.7(a) shows the different stages of existence of composite constituents up to the final product:

—*Stage a:* At this stage, the materials appear in raw basic form. For fibers, these consist of fiber either in the form of filaments or fi-

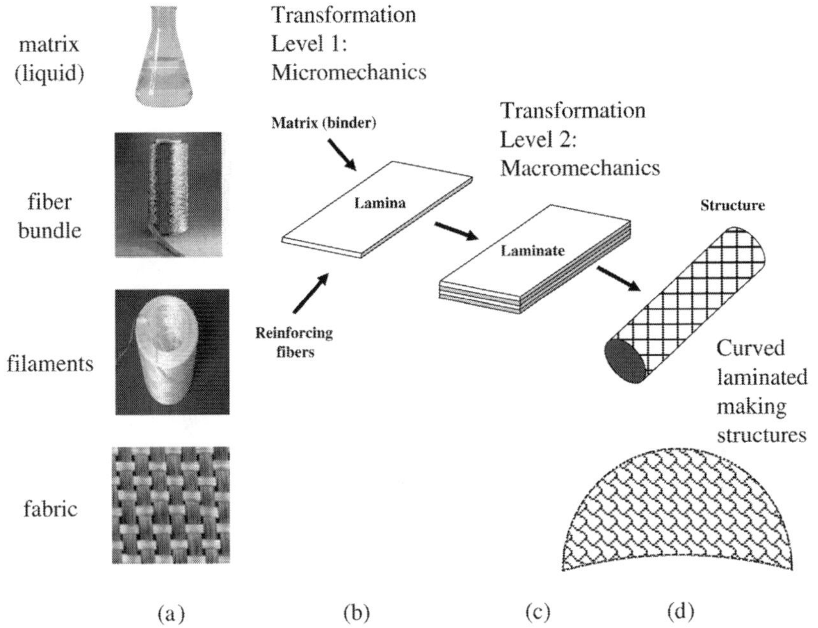

FIGURE 1.7(a) Stages of existence of constituents in the manufacturing of composites.

ber bundles. Fibers may also be woven into fabrics or braided into braided perform. For matrix, the material usually appears in liquid form for thermoset resin or in granular form in the case of thermoplastics.

—*Stage b:* At this stage, the fibers and matrix may be combined into a single layer. For the case of thermoset matrix composite, the matrix may appear in a semi-liquid, semi-solid form so that the sheet can hold its shape. For the case of thermoplastic composite, the matrix is solidified. This form for thermoset matrix composites is called prepreg. For thermoplastic composites, it is called towpreg.

—*Stage c:* At this stage, the layers in stage *b* are stacked on top of each other to make flat plate laminates. This intermediate step is important for the analysis where material properties are tested or calculated. However this step is usually bypassed in the manufacturing process of practical composite parts.

—*Stage d:* This is the final stage where the final product configuration is formed.

The involvement of these stages in the different manufacturing processes is as follows:

- Hand-lay-up (with or without autoclave): Stages *a*, *b* and *d* are involved. Stage *c* is bypassed.
- Filament winding: Stages *a* and *d* are involved. Stages *b* and *c* are bypassed.
- Pultrusion: Stages *a* and *d* are involved. Stages *b* and *c* are bypassed.
- Liquid composite molding: Stages *a* and *d* are involved. Stages *b* and *c* are bypassed [Figure 1.7(b) shows stages for Liquid Composite Molding].
- Thermoplastic composites: Stages *a* and *d* are involved. Sometimes stage *b* and even stage *c* may be involved.

The mentality of working with metals therefore cannot be applied when manufacturing using composites.

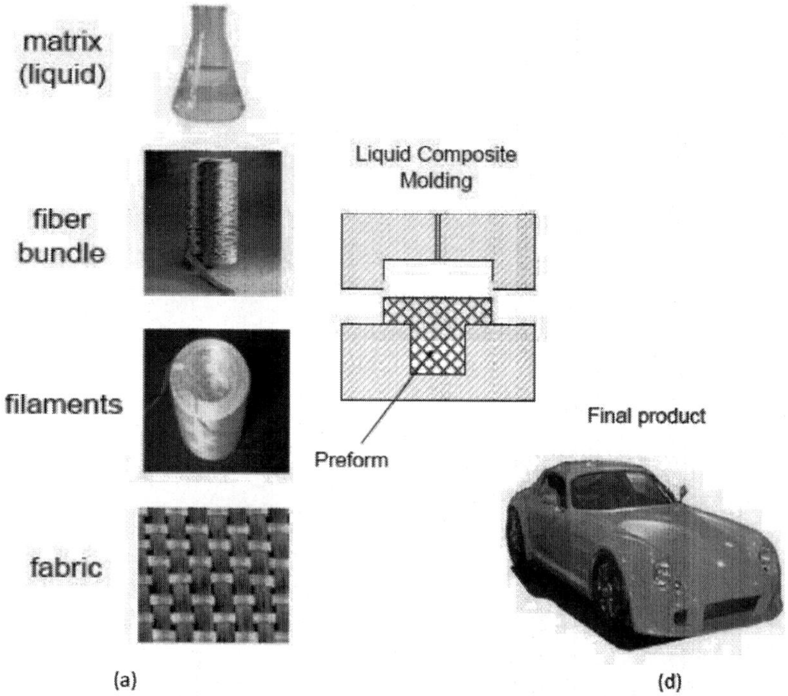

FIGURE 1.7(b) Stages of existence of constituents in the LCM process.

3. FUNCTIONS OF THE CONSTITUENTS OF COMPOSITES

There are two main constituents making up advanced composites. These are fibers and matrix. The interface between the fiber and matrix is critical for the function of the composite material. The interface may be considered as a third constituent of the material. Each of these constituents will be presented in the following.

3.1. Fibers

Fibers provide strength and stiffness to the composite materials. Fiber materials are usually glass, carbon or Kevlar. One may ask the question why do composites appear in fiber form. There are many reasons for this as follows.

3.1.1. Advantages of the Fiber Form

3.1.1.1. Strength of Material in Fiber Format is Better as Compared to Bulk Format

Materials can appear in different forms. These can be bulk form (relatively large volume), fiber form (diameter of about 10 μm and length from a few millimeters to a few meters) powder form (more spherical shape with diameter on the order of micrometers), or flake form (thin sheets). As a rule, the smaller the volume of a certain piece of material, the less defects there are in that volume, because there is less chance for defects to occur when a smaller volume of material is made. As such, bulk-form pieces have smaller strength than fiber-form pieces. The difference in strength of materials in bulk form and in fiber form is illustrated by the comparison between the properties of glass in fiber form and in plate (bulk) form. While the moduli of E glass (72 GPa) and plate glass (70 GPa) are about the same, their strengths are very different. E glass has a strength of 3448 MPa (and S glass has a strength of 4585 MPa) while plate glass has a strength of only 70 MPa.

Materials in powder form do have small volume, however, the reinforcement effect is not as good as that in fiber form. This is because the reinforcement effect depends on the aspect ratio (ratio of l/d) where l is the length of the reinforcement and d is its diameter. If the aspect ratio is smaller than a certain critical value (under uniform shear stress distribution assumption, equal to $\sigma/2\tau$ where σ is the tensile strength of the reinforcement material and τ is the shear bond strength between the reinforcement and the matrix), failure will occur due to slipping between the reinforcement and the matrix, making the reinforcement ineffective. For the case of glass/epoxy where $\sigma = 3448$ MPa and $\tau = 20.5$ MPa, the critical aspect ratio is 84. Reinforcements in powder form having aspect ratios on the order of 2 or 3 do not give the same reinforcement effect as fibers with small diameter and long length. If the diameter of a fiber is about 10 μm, then a length of about 1 mm would be sufficient. However due to stress magnification at the end of the fibers, the smaller the number of ends of the fibers, the better the reinforcement effect. As such, long continuous fibers give better reinforcement than short fibers.

Materials in flake form are also available (such as mica or clay sheets). However these usually occur naturally and are limited in their variety.

3.1.1.2. Availability of More Fabrication Techniques

The fiber format allows fiber processing steps that are difficult or impossible in bulk. Examples of this are stretching and orientation (carbon, polyaramide, Kevlar, and polyethylene fibers), vapor deposition (boron fibers), solvent removal (polyaramide and kevlar-type fibers), and rapid oxidation (carbon fibers). Hence, the fibers used in advanced composite structures frequently represent unique materials that are not possible or at least difficult to achieve in bulk. The same explanation goes for the reduction of the strength of the fiber as the diameter increases (as shown in Figure 1.8).

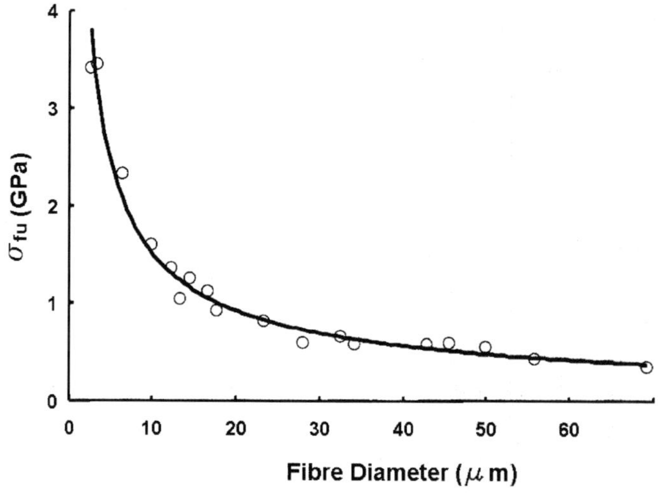

FIGURE 1.8 Effect of fiber diameter on strength [1] (courtesy of ASTM).

3.1.1.3. *Flexibility in Forming*

The fiber format allows formation of very complex shapes out of strong and stiff materials at very low forces and without breaking the fibers. This is because at these very small diameters, the fibers may conform to complex shapes by essentially elastic bending. For example, the maximum axial strain in a fiber of diameter *d* bent to a radius ρ under typical elastic assumptions is:

$$|\varepsilon_{max}| = \frac{d}{2\rho} \tag{1.6}$$

Hence, if a fiber of diameter 10 μm is bent to a radius of 2.54 mm (0.10 in), the axial strain will be 1.97×10^{-3}, or about an order of magnitude smaller than the typical strain at breaking for a glass fiber (5% or 0.05). Hence very small features may be molded into advanced composite parts without damaging the fibers.

The fiber format provides many advantages as mentioned above. However, the fiber format also presents difficulties and disadvantages that need to be addressed. These are described below.

3.1.2. *Disadvantages of the Fiber Form*

3.1.2.1. *Requirement of a Large Number of Fibers*

Fibers have very small diameter (about 10 μm, while the diameter of one hair is about 100 μm). In order to make something of a good dimension for engineering applications, one needs to make components with thickness on the order of millimeters or centimeters (about 1000 times the diameter of a fiber, and of width in the order of decimeters or meters). Therefore, one needs millions and millions of these fibers to make an engineering component of significant size. Individual fibers by themselves are very flexible and fragile. The fibers tend to curl and form entangled pieces if not aligned. Figure 1.9 shows a photograph of three tows of entangled fibers.

The fibers need to be aligned, and slightly tensioned in order for their properties to be effectively utilized. In order to withstand loads of significant magnitude, millions of fibers need to be aligned and work simultaneously. Not only that, fibers need to be straight and a small amount of tension may be required to keep them straight. Special techniques and care are required to attain this configuration.

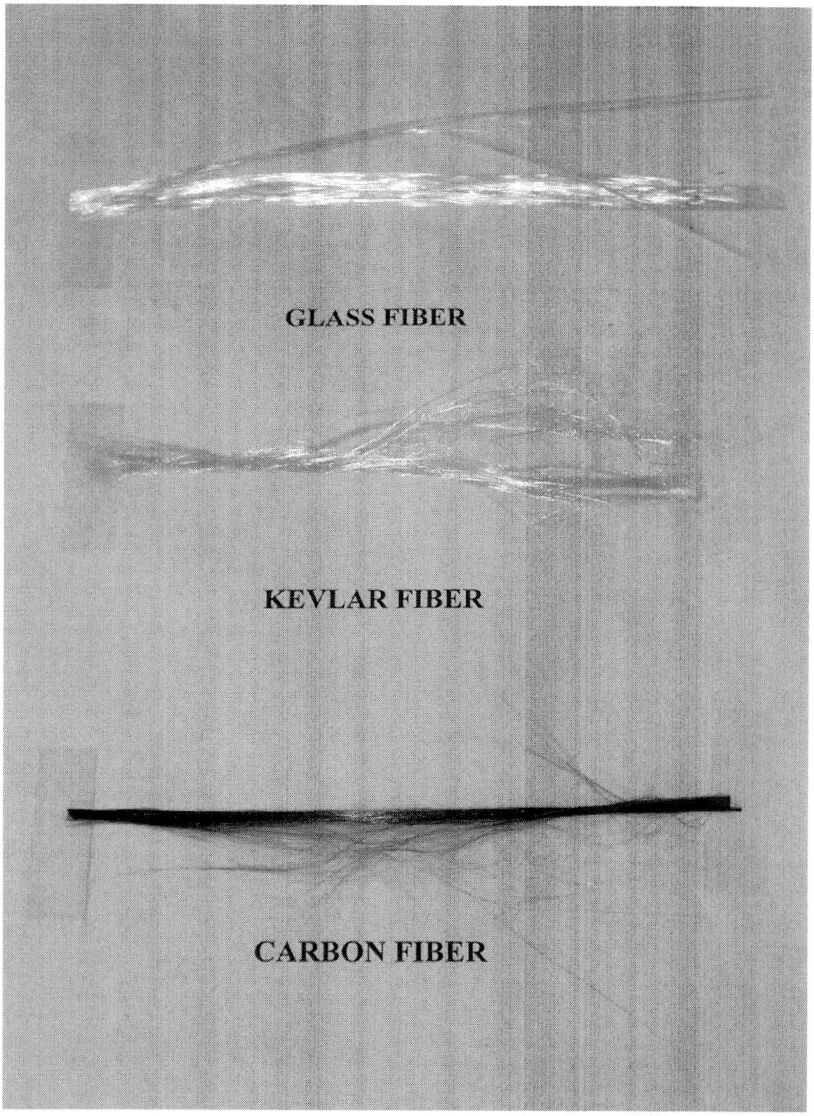

FIGURE 1.9 Photograph of entangled bunches of fibers. Each shows a tow with fibers aligned and fibers that are frayed.

3.1.2.2. Fibers Need to be Bonded Together to Provide Good Mechanical Properties

Fibers used in composite materials can have a significant variation in their strength because strength depends on the microstructure of the material and is very sensitive to the presence of defects. Fibers are brittle materials and their strength exhibits a significant amount of variation. Figure 1.10 shows the variation in the strength of graphite fibers. This figure shows that graphite fibers can have strength that varies from 0.14 MPa–0.4 MPa. The strongest fiber can have strength that is about three times more than that of the weakest fiber. If the fibers are aligned but not bonded together, the strength of the whole bundle of fibers would be governed by the strength of the weakest fibers.

The strength of a dry bundle of fibers can be much less than the average strength of a bundle of fibers. A dry bundle of fibers means that the fibers are not bonded together by the matrix material. The following example illustrates why a dry bundle of fibers has much lower strength than that of the average strength of the fibers and why the use of adhesive bond between the fibers can improve the strength of the bundles.

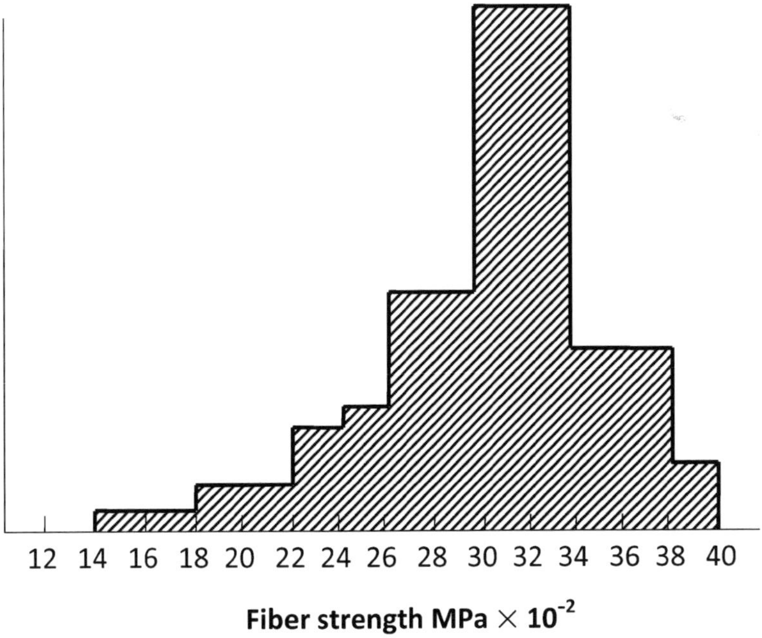

FIGURE 1.10 Typical strength distribution for graphite fibers.

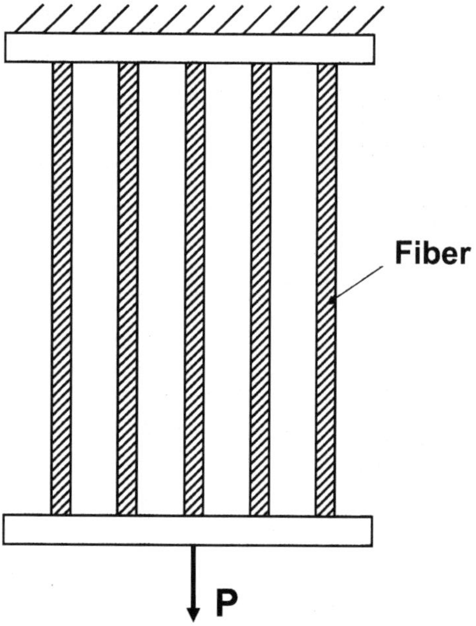

FIGURE 1.11 A dry bundles of 5 fibers.

Example of Domino Effect of Strength of a Dry Bundle Of Fibers

For illustration purposes, a dry bundle of five fibers is shown in Figure 1.11. Assume that this bundle of fibers is held fixed at the top and all five fibers are joined together at the bottom by a common bar. This bar is in turn subjected to a load P. Due to the variation in the properties of the fibers, assume that the strength of the five fibers is as follows:

$$\begin{aligned}&\text{Load required to break fiber 1} = 0.30 \text{ N}\\&\text{Load required to break fiber 2} = 0.35 \text{ N}\\&\text{Load required to break fiber 3} = 0.25 \text{ N}\\&\text{Load required to break fiber 4} = 0.40 \text{ N}\\&\text{Load required to break fiber 5} = 0.50 \text{ N}\end{aligned} \tag{1.7}$$

The sum total of the above five loads is 1.80 N. However, if the load P were to increase slowly from 0 N, the whole bundle of fibers will break when the load reaches 1.25 N, much less than the total value of 1.80 N. The reason for this is as follows:

a. When the total load *P* reaches 1.25 N, the load in each of the fibers is 0.25 N (1.25 N/5). This is the load at which fiber 3 breaks.
b. After fiber 3 breaks, only 4 fibers remain to sustain the load of 1.25 N. The average of 1.25 N over 4 fibers is 0.31 N. This load in turn is more than the breaking load of 0.30 N of fiber 1. So fiber 1 breaks and leaves only 3 fibers to sustain the load of 1.25 N.
c. The average of 1.25 N over 3 fibers is 0.42 N. This is more than the breaking load of 0.35 N of fiber 2 so this fiber breaks.
d. The average of 1.25 N over 2 fibers is 0.63 N. This is more than the breaking loads of 0.40 N and 0.50 N of fibers 4 and 5 and therefore these fibers also break.

The domino effect above results in the total bearing load of the bundle being controlled by the strength of the weakest fiber in the bundle. For this situation, the stronger fibers cannot contribute much to the enhancement of the strength of the material.

On the other hand, if the fibers were bonded together by some adhesive matrix such that the load from one fiber can be transferred to another through the adhesive mechanism, the situation is different. To illustrate this point, Figure 1.12 shows again the fiber bundle as in Figure 1.11 but in this case the fibers are bonded together via matrix adhesive.

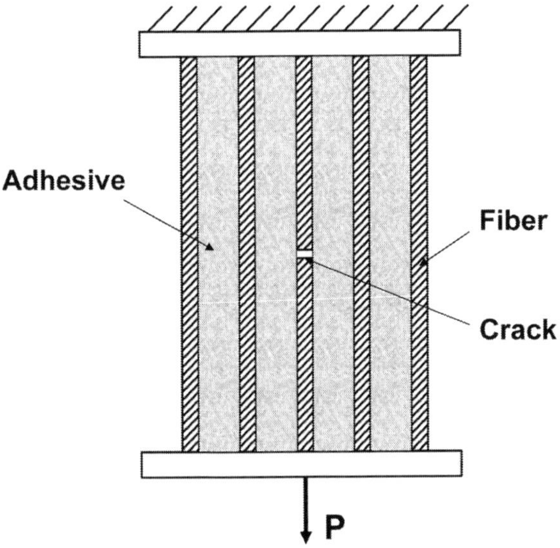

FIGURE 1.12 Bundles of five fibers bonded together via adhesive.

Assume that we also have the same individual fiber failure loads as in Equation (1.7). Now assume that the load is increased slowly from 0–1.25 N. Assume also that the matrix has negligible (zero) tensile strength (matrix can have good adhesive shear strength but low tensile strength). As such, fiber 3 will break at a location within its length. A small crack is shown in Figure 1.12. For fiber 3, load can no longer be transferred between the two pieces of the fiber above and below the crack. However, load can still be transferred by shear action between fiber 3 and fiber 4 and between fiber 2 and fiber 3. The two segments of fiber 3 above and below the small crack therefore do not become totally useless. In fact they still contribute to the bearing of the load and the remaining structure that supports the load P is more than just the four fibers 1, 2, 4, and 5. The stress in the surrounding fibers may be more than before, as shown in Figure 1.13. In Figure 1.13(a), the adhesive is shown as the binder. The middle fiber has a crack. The shear stress τ at the interface between fiber and binder shows a maximum close to the cracked end of the fiber. The normal stress σ is zero at the fiber end and increases as one moves away from the end. In Figure 1.13(b), the presence of the crack in the middle fiber causes the normal stress σ in the fiber on the right to increase a little in the vicinity of the crack. Depending on the strength of the fiber on the right at that location, that fiber may or may not break. If one assumes that the normal stress in the fiber to the right and left of the cracked fiber increases by 8% (0.02 N) due to occurrence of the crack in the middle fiber, the maximum stress in the remaining fibers can be as follows:

Fiber 1: 1.25/5 = 0.25 N
Fiber 2: 0.25 + 0.02 = 0.27 N
Remaining of Fiber 3: Less than 1.25/5 = 0.25 N (assumed to be 0.21 N)
Fiber 4: 0.25 + 0.02 = 0.27 N
Fiber 5: 0.25 N

Total sum = 1.25 N

(The lower stress in fiber 3 is due to the presence of crack which relaxes the stress in this fiber and the load is shifted to the other fibers.)

Comparing the strengths of the fibers as shown in Equation (1.7), it can be seen that the crack will not propagate at 1.25 N load. Load needs to be increased if further cracks are to happen. The presence of the adhesive therefore allows the stronger fibers to participate in the load bearing action. This is because after the occurrence of the first crack, fiber 3 does

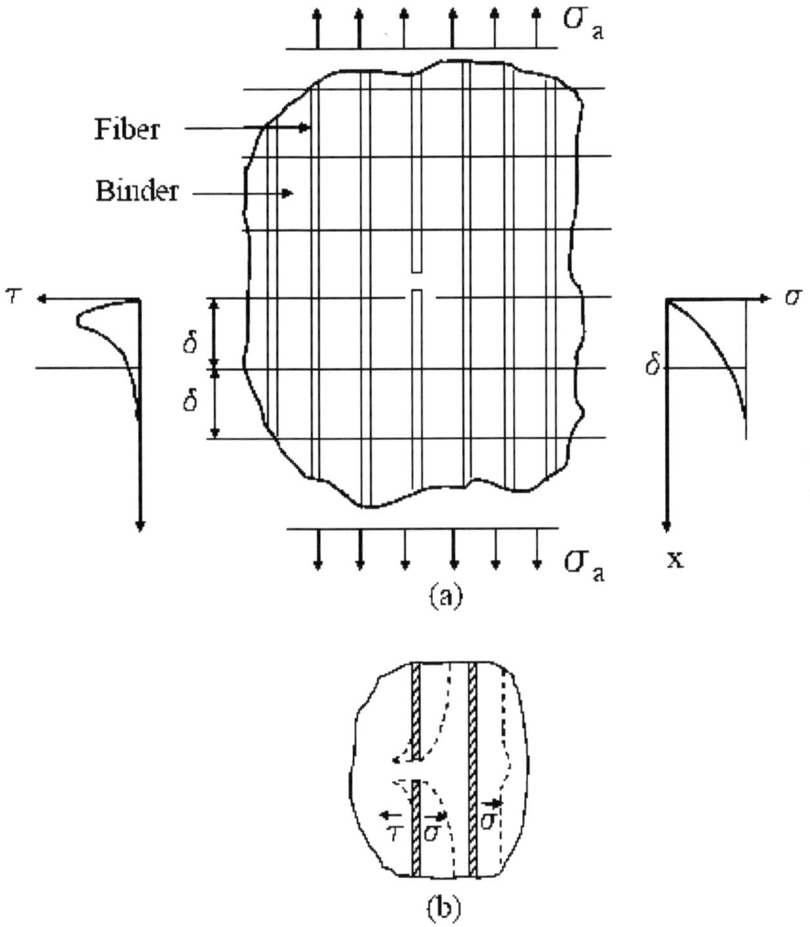

FIGURE 1.13 Stress redistribution after a fiber cracks.

not become totally useless. It can be seen in Figure 1.13(b) that even though the normal stress σ becomes 0 at the crack, it picks up as one moves away from the location of the crack. This occurs on both portions of fiber 3 above and below the crack. As such the loss of load bearing of fiber 3 is not 100%.

Note that the assumption of 8% increase in the load in surrounding fibers and 16% reduction in the load in the broken fiber are assigned for this illustration only. The real values of the modified stresses (increase or decrease) depend on the particular arrangement of fiber and matrix materials and need more rigorous analysis to be accurate.

The adhesive is therefore essential for the strength of the composites. At locations where there is no adhesive (the so called dry spots), cracks may appear and propagate and premature failure may happen. One of the challenges for manufacturing using composites is to assure that the matrix adhesive surrounds each and every one of the fibers (this is the so-called wetting action).

3.1.2.3. The Need for a High Fiber Volume Fraction, v_f

Since the fibers provide strength and stiffness for the composite materials, it is essential that one has as much fiber as possible in a composite material. The modulus of the composite along the fiber direction is proportional to the fiber volume fraction as expressed in Equation (1.2).

In Figure 1.14, the white dots represent the fiber cross section and the dark area represents the matrix material. The volume fraction of the fibers can be obtained from the micrograph by determining the ratio of area of fiber over area of the material. It can also be calculated based on some idealized arrangement. Figure 1.15 shows a square array of fibers.

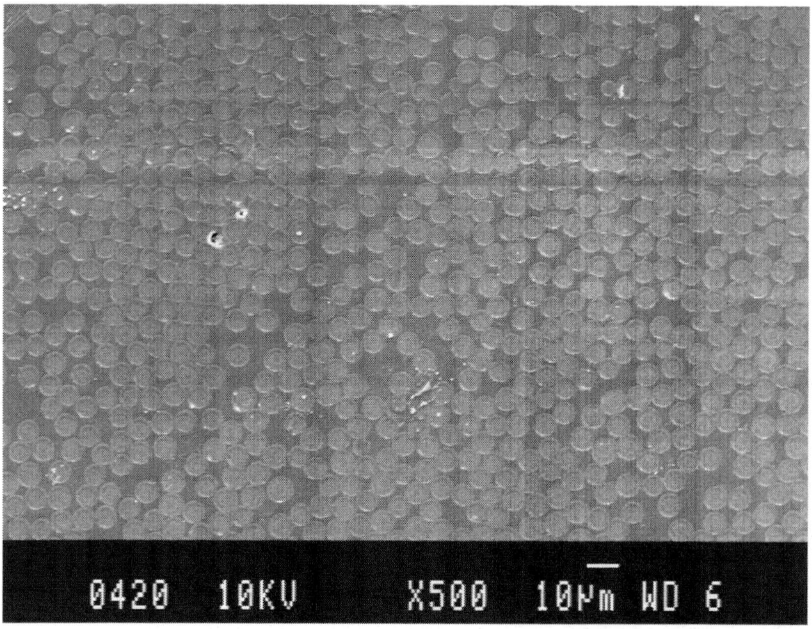

FIGURE 1.14 Micrograph of a cross section of a unidirectional composite sample.

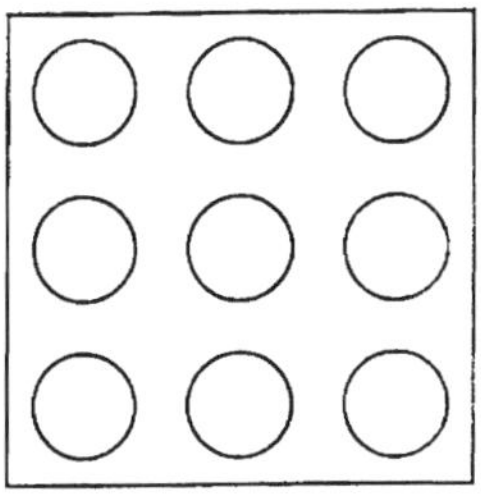

FIGURE 1.15 Square array of fibers.

Based on this arrangement, the fiber volume fraction can be calculated as:

$$v_f = \frac{\text{Volume of fiber}}{\text{Total volume}} = \frac{n^2 \pi d^2}{4(nd + n\delta)^2} \tag{1.8}$$

where d is the fiber diameter and δ is the shortest space between two fibers. In the limit when the fibers are touching ($\delta = 0$), one has:

$$v_f = \frac{\pi}{4} = 0.785 \tag{1.9}$$

So the limit for fiber volume fraction in the case of square array is 0.785.

One can also assume the fiber arrangement to be hexagonal such as those shown in Figure 1.16. The maximum fiber volume fraction is 0.875 for open packing and 0.907 for closed packing.

Normally the fiber volume fraction may not reach the high levels cal-

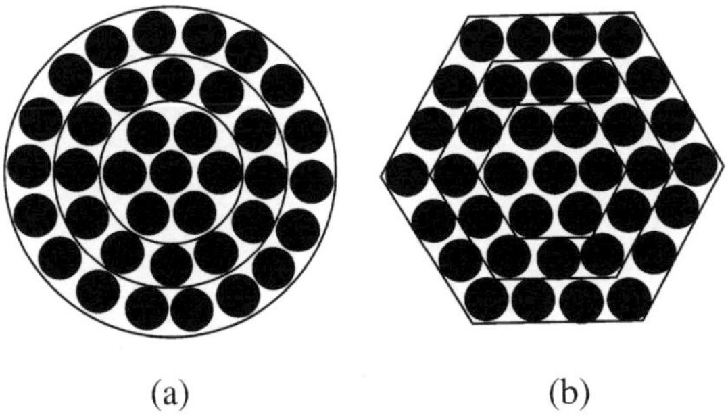

FIGURE 1.16 Hexagonal packing arrangement of fibers: (a) open packing, (b) closed packing.

culated using the ideal arrangement. Fiber volume fractions achieved in practice is around 68% for hand-lay-up using autoclave molding and may be 70% for pultrusion. Note that there should always be a layer of resin in between two fibers, otherwise dry spots will occur and dry spots are points of weakness. Therefore one important thing to remember in the case of composite materials is:

For composite materials, one wants to have as much fiber content as possible as long as the fibers do not touch each other.

This can be put another way as:

For composite materials, one wants to have as little matrix material between two fibers as possible, but not zero.

3.1.2.4. Small Interfiber Spacing

One important consequence of the high fiber volume fractions for advanced composites is a small interfiber spacing. For example, if the actual microstructure is approximated as a square array (Figure 1.15) where the maximum allowable fiber volume fraction is $\pi/4$, the average interfiber spacing δ can be calculated as:

$$\delta = d\left[\sqrt{\frac{\pi / 4}{v_f}} - 1\right] \qquad (1.10)$$

where d is the fiber diameter.

Hence for a typical case (graphite/epoxy) with $d = 10\ \mu\text{m}$ and $v_f = 0.68$, one gets $\delta = 0.74\ \mu\text{m}$.

Important Consequences of this Small interfiber Spacing

1. *Stress concentration.* In the solid composite, the resin is highly constrained in small volumes between the fibers, which results in stress concentrations and reduced strength in the matrix-dominated directions.

   During processing, the small interfiber spacing also has important consequences:

2. *Fiber-to-fiber contact.* Because there is much variation in fiber spacing for real composites, a small average δ suggests considerable fiber-to-fiber contact. This can make the fiber bundle

load-bearing when compressed in the transverse direction. This means that when a load is applied on a bundle of fibers containing liquid resin, fibers may support the load through their contacts as shown in Figure 1.17. A direct consequence of this can be reduced resin pressure during cure, which can lead to potential voids in the matrix.

3. *Large shear resistance of prepregs.* Another important effect is the resultant large shear resistance of the prepregs (Prepregs are fibers impregnated with partially cured resin). This affects properties such as drape, which translates into poor handling properties during manufacture. *Drape* is a term used to denote the ability of the fiber fabric to conform to the shape of the tool.
4. *Small permeability values.* Permeability is a characteristic of the bed of fibers that indicates the ease (or difficulty) for the resin to penetrate into the bed of fibers. Permeability depends on the space between the fibers and can be shown to scale roughly as δ^2. As such, the small interfiber spacings result in very small permeability values. The effects of increasing liquid volume fraction v_R (or decreasing interfiber spacing and therefore decreasing v_f) on the axial permeability S_{11} and the transverse load-carrying capacity of the aligned fiber beds are shown in Figure 1.18. It can be seen that the higher is the liquid volume fraction (or the lower is the fiber volume fraction), the higher is the values of the permeability.

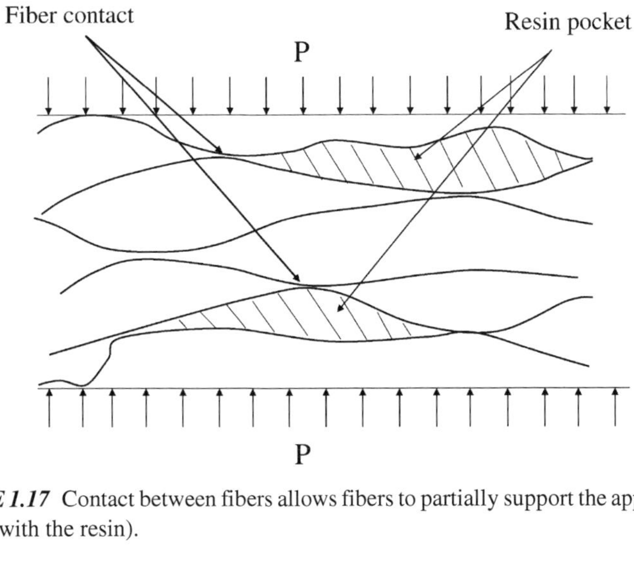

FIGURE 1.17 Contact between fibers allows fibers to partially support the applied load (sharing with the resin).

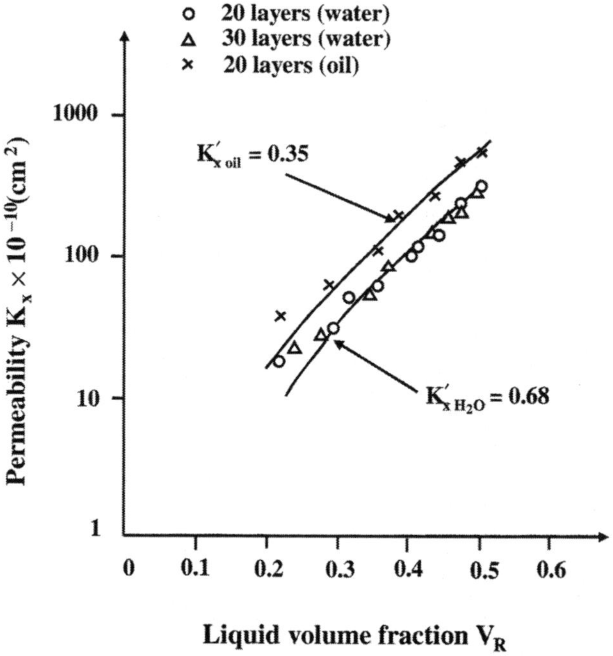

FIGURE 1.18 Effect of liquid volume fraction on the axial permeability of an aligned fiber bed, S_{11} (reproduced from Reference [2], with permission from John Wiley and Sons).

The above four factors act to limit the maximum obtainable fiber volume fraction, generally making it much below the theoretical maximum values of 0.785 for square packing, and 0.907 for hexagonal packing.

3.1.2.5. *Anisotropic Behavior*

Figure 1.19 shows a representative element of an aligned fiber bundle. It exhibits anisotropic behavior (properties depending on direction). Apart from mechanical properties such as stiffness and strength, there are also implications of anisotropic behavior for manufacturing. One particular influence is the anisotropy of the permeability of liquid resin into the interstices between the fibers. For the graphite/epoxy system shown in the figure, the ratio between elastic modulus along the fiber direction over elastic modulus transverse to the fiber direction E_{11}/E_{22} is about 16. Similarly, the ratio of axial to transverse resin permeabilities

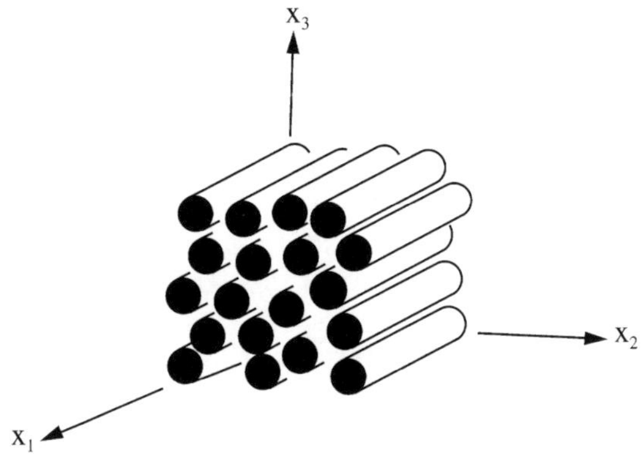

FIGURE 1.19 Representative element for an aligned fiber bundle.

for the fiber bundle, S_{11}/S_{22}, is on the same order. Figure 1.20 shows the transverse permeability of an aligned fiber bundle, which is about one twentieth the values as those along the axis of the fibers, as shown in Figure 1.18.

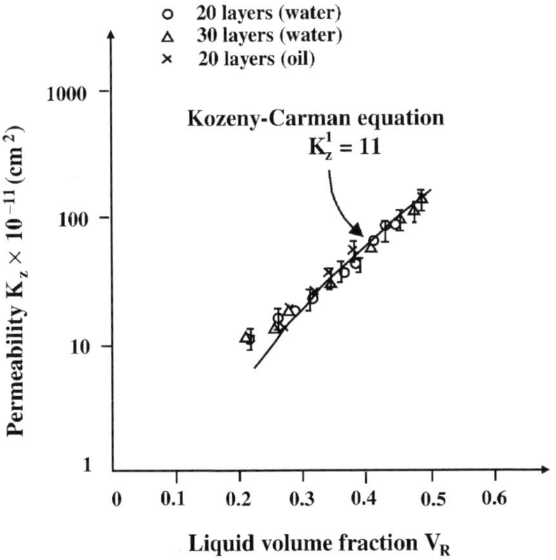

FIGURE 1.20 Transverse permeability of aligned fiber bed. Note that the unit for permeability is multiplied by 10^{-11} cm^2, as compared to Figure 1.18 where the unit is multiplied by 10^{-10} cm^2. The ratio between axial permeability and transverse permeability is about 19 (reproduced from Reference [2], with permission from John Wiley and Sons).

3.2. Matrix Materials

It was mentioned at the beginning of Section 3.1 that fibers are made of strong and stiff materials and that they can provide strength and stiffness for the composite materials. However fibers by themselves cannot provide these properties to the composites. This is because fibers exist in tiny quantities (the diameter of a fiber is about 7 mm). In order to make composite structures of a dimension of engineering significance, parts of dimension in the order of centimeters (0.01 m) or decimeters (0.1 m) need to be made. Matrix materials serve the function of making this possible. Matrix materials usually have low normal strength (tensile or compressive) but they can provide good adhesive shear strength. The tensile strength of epoxy resin is about 35–130 MPa (as compared to the tensile strength of carbon fiber of about 3000 MPa). The shear strength of epoxy adhesive is about 20 MPa. Even though this number (20 MPa) seems to be small as compared to the tensile strength of carbon fiber (3000 MPa), the aspect ratio of the fibers (length over diameter) is usually large. This provides comparative shear load as compared to tensile load (more analysis to illustrate this effect is given later in this chapter). As such, matrix materials in composites are utilized such that shear is the main mode of loading. The matrix serves the following functions.

3.2.1. Aligning the Fibers

It can be seen in Figure 1.9 above that a bundle of dry fibers consists of fibers that can faze. Individual fibers can take random orientation and may not align with each other well. In order to make an engineering component out of composite materials with a certain significant dimension, the fibers need to be aligned. One can grasp the fibers and align them, however in order to keep them aligned, some form of glue (adhesive) needs to be used. The matrix material serves the function of the glue. It is essential that the matrix resin (glue) surrounds the total surface of each individual fiber.

3.2.2. Transfer the Load Between the Fibers

In Section 3.1, it was mentioned that fibers need to be bonded together so that their strengths can be utilized effectively. Otherwise the domino effect will take place and the strength of the composite material is governed by the strength of the weakest fibers. The bond is provided by the matrix material, since the matrix material serves as a glue. This bonding action serves to transfer the load from one fiber to the matrix and then

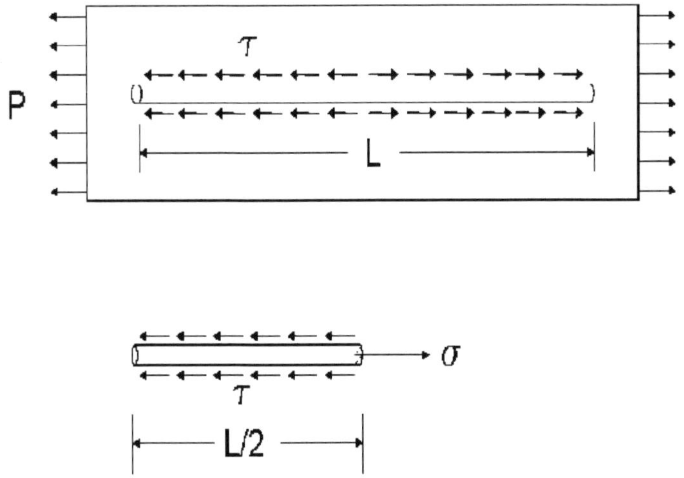

FIGURE 1.21 Shear load transfer between fiber and matrix.

from the matrix to the next fiber. While the fiber supports the load via its tensile strength, the matrix provides the load transfer via shear strength (Figure 1.21).

When a load is imposed on the matrix portion of the material, this load is transferred to the fiber in the form of shear. Usually the shear stress at the interface is maximum at the end of the fiber and minimum toward the center of the fiber [Figure 1.13(a)]. If one assumes that the shear stress is constant to simplify the calculation, and assuming that the diameter of the fiber is d and its length is l, equilibrium of the broken segment gives:

$$\frac{\pi}{4} d^2 \sigma_f = \pi d \frac{l}{2} \tau_i \tag{1.11}$$

yielding

$$\frac{l}{d} = \frac{1}{2} \frac{\sigma_f}{\tau_i} \tag{1.12}$$

Equation (1.11) shows the balance between load provided from tensile resistance of the fiber and load provided from shear resistance at the interface, where σ_f represents the tensile stress in the fiber and τ_i represents the shear stress at the interface. Equation (1.12) shows the aspect ratio of the fiber as a function of the two strengths. If the aspect ratio is as given in Equation (1.12), then failure will occur by both fiber breaking and interface slipping simultaneously. If the aspect ratio is larger than that given in Equation (1.12), then the fiber is longer than the critical length and failure will occur by fiber breaking. If the aspect ratio is less than that given in Equation (1.12), then failure will occur by slipping at the interface. In this case, the fiber is not well utilized. In order to fully utilize the strength of the fiber, it is important that the fiber be longer than the critical length given as:

$$l_c = \frac{1}{2}\frac{\sigma_f}{\tau_i}d \tag{1.13}$$

3.2.3. *Assisting the Fibers in Providing Compression Strength and Modulus to the Composites*

Fibers are long and slender pieces that behave like strings. Individual fibers resist tension well but they cannot resist compression. In order to possess good compressive properties, fibers need to be aligned. Matrix materials assist in aligning the fibers and providing the composite materials with compressive properties.

3.2.4. *Assisting the Fibers in Providing Shear Strength and Modulus to the Composites*

Similar to the above, a bunch of individual fibers cannot provide good shear properties because the fibers can slide relative to each other. The presence of the matrix material provides the shear transfer between the fibers and this also provides good shear properties for the composite material.

3.2.5. *Protecting the Fibers from Environmental Attack*

Fibers such as carbon and glass usually have high surface energies. Moisture can easily adsorb on the surface of these fibers. With adsorption of water, it is difficult for the matrix material to adhere to the fiber to make a good bond. The presence of the matrix on the surface of the fiber prevents moisture from adhering to the fiber surface. Also water attacks and creates cracks in glass fiber over a long time. The presence of the resin on the surface of the fiber prevents glass fibers from being attacked by moisture in the surrounding environment.

3.3. Interface

It was mentioned in the previous section that the matrix needs to be bonded to the fibers. The bond between the fiber and the matrix constitutes the interface between them. The interface area within a certain volume of a composite material made up of aligned fibers of diameter d can be estimated as:

$$\frac{\text{Interface area}}{\text{Composite volume}} \approx \frac{4v_f}{d} \qquad (1.14)$$

Equation (1.14) is obtained by considering a unit cell containing one single fiber (which can be extracted from Figure 1.15). The interface area is πd while the composite volume is $(\pi d^2/4)v_f$.

For a composite 1 m × 1 m × 0.02 m with $v_f = 1/2$ and $d = 8 \times 10^{-6}$ m, the interface area is: $(0.02\ \text{m}^3)(4 \times 0.5)/(8 \times 10^{-6}\ \text{m}) = 5000\ \text{m}^2$! To develop this interface, the resin must come into intimate contact with the fibers. A good interface is needed for a coherent structure that transfers loads around broken fiber ends and carries transverse loads. In general, one does not want the strongest possible bond. Separation between the fiber and the resin can be an important energy absorbing mechanism during the failure of the composite. This idea is used to great advantage, for example, when making composites that stop ballistic projectiles. In order for this interface to develop, there are two requirements that a manufacturing process needs to satisfy: availability and compatibility.

3.3.1. Availability of the Resin at the Surface of the Fibers

For the interface to develop between the resin and the fiber, the resin needs to be available at the surface of the fiber. This may seem obvious, but the concept is important for manufacturing. Recall that the diameter of a fiber is about 10 μm. Assuming a fiber volume fraction of 0.6, the interfiber distance can be calculated from Equation (1.10) to be 0.14 μm. The dimension of the unit cell would be 10.14 μm. If one were to make a laminate for an aircraft wing about 3 mm thick and 500 mm wide, there are about 15 million fibers over the cross section of the part. One needs to get the resin onto the surface of each and every one of these fibers to assure a good quality part.

To bring the matrix to the surface of the fiber, the matrix needs to be in the form of a liquid with low viscosity. The bulk flow is generally pressure driven by an external pressure source, with the final degree of wetting accomplished by capillarity. Often the bulk flow may be complicated by the inhomogeneity of the fiber distribution. For example, in woven and braided fabrics, the different length scales that characterize the fiber structure result in different time scale for resin flow. The basic event of flow of resin through a fiber bundle can be modeled as flow through an anisotropic porous medium. The flow of the liquid through a bed of these fibers normally is estimated using Darcys law of flow through a porous medium as:

$$q_i = \frac{S_{ij}}{\mu} \frac{\partial p}{\partial x_j} \tag{1.15}$$

Here the vector q_i is the so-called filter velocity, which consists of the three components of the volumetric flow rate divided by the cross-sectional area. Where μ is the Newtonian viscosity and $\partial p/\partial x_j$ is the pressure gradient vector. The term S_{ij} constitutes the permeability matrix, with units of $(\text{length})^2$.

When Darcys law is written in the material coordinate system, such as the direction shown in Figure 1.19 for a bundle, the permeability matrix is diagonalized and Sij takes the form:

$$S_{if} = \begin{bmatrix} S_{11} & 0 & 0 \\ 0 & S_{22} & 0 \\ 0 & 0 & S_{33} \end{bmatrix} \tag{1.16}$$

where S_{ii} are the principal values corresponding to each of the three principal directions. It is usually further assumed that the fiber bundle is transversely isotropic which leads to the simplification that $S_{22} = S_{33}$.

For the majority of cases, the flow is along one direction (normal to the plane of the composite component), for this case the equation simplifies to:

$$q_z = \frac{S_T}{\mu} \frac{dp}{dz} \tag{1.17}$$

where z represents the direction along the normal to the plane and S_T represents the transverse permeability of the fiber bed.

A faster flow rate [large value of q_z in Equation (1.17)] can bring the liquid resin matrix to the surface of the fiber faster. If the liquid resin has viscosity like water (1 centipoise) then it runs well. If the liquid resin has viscosity like honey at room temperature (10^5 centipoise), then the flow is not as rapid.

The viscosity of the matrix has important implications on the type of matrix materials that are commonly used. Thermoset resins such as polyesters, vinyl esters and epoxies have reasonably low viscosities at reasonably low temperatures (from room temperature up to about 200°C). Viscosities of polyesters are about 1 Pa-sec (1000 centipoise) at room temperature whereas those of epoxies are about 50 centipoise (0.05 Pa-sec) at 115°C. On the other hand, viscosities of thermoplastic resins such as polyetheretherketone (PEEK) (1000 Pa-sec or 10^6 cP at 400°C) are very high. As result, there are more composite materials made of thermoset matrix than with thermoplastic matrix in spite of the fact that composites with thermoplastic matrix can offer better properties such as fracture toughness and environmental resistance.

3.3.2. Compatibility Between the Fiber and the Matrix Materials

The second important requirement for good bonding is that the matrix and fiber be thermodynamically compatible. The required intimate contact can take place spontaneously by a wetting process. The thermodynamics of wetting can be illustrated by calculating the surface and interface energies for a wetting system as illustrated in Figure 1.22.

$$\text{Surface energy} = A\gamma_j \tag{1.18}$$

where A is the area, γ_j is the specific surface energy, and the subscript $j = s$ (solid), l (liquid) and ls (liquid-solid) interface.

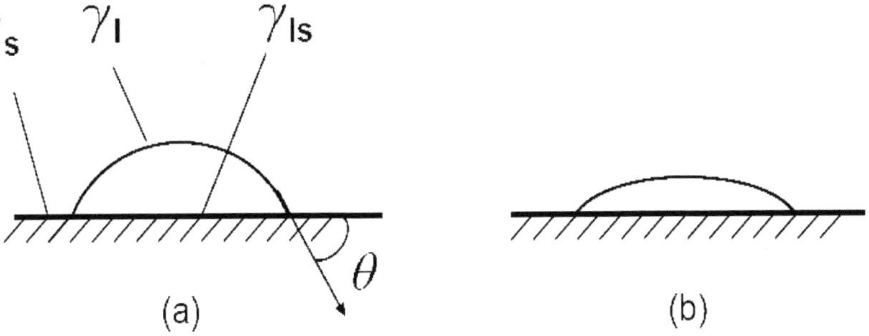

FIGURE 1.22 Illustration of (a) non-wetting and (b) wetting.

The equation of equilibrium for the situation shown in Figure 1.22 is:

$$\gamma_{sv} = \gamma_{lv} + \gamma_{ls} \cos\theta \tag{1.19}$$

where γ_{sv} is the surface tension of the solid surface, γ_{lv} is the surface tension of the liquid surface, γ_{ls} is the surface tension of the liquid/solid interface and θ is the contact angle. Situation (a) represents the case where the liquid does not want to spread over the solid surface and situation (b) represents the case where the liquid wants to spread on the solid surface. For the case of complete wetting, $\theta = 0$ and $\gamma_{sv} = \gamma_{lv} + \gamma_{ls}$.

For wetting to occur, one needs to have:

$$\gamma_{sv} \succ \gamma_{lv} \tag{1.20}$$

Simply put, low-energy liquids wet high-energy solids. For many polymer composite systems, the liquid resin will naturally wet the solid fibers. This is because the fibers are high surface energy elements, whereas the liquid polymers have much lower surface energy. Exceptions to this rule can occur when we try to wet solid polymeric polymers with other liquid polymers, or dirty and/or wet solid fibers with liquid polymers. With metal matrix composites, it is difficult for the liquid metal to wet the fibers, since the molten metal may have a higher surface energy than do the solid fibers. The surface energy of metals varies from 400–2000 dyne/cm, that of clean glass is about 500 dyne/cm, graphite fiber is about 50, and that of polymers is about 30–45 dyne/cm. As such, polymeric liquids have lower surface tensions than most fibers, making the polymeric liquids compatible with the fibers. This facilitates bonding between the polymeric liquids and the solid fiber materials. This explains why, in practice, there are more polymer matrix composites than metal matrix composites even though metal matrix composites may provide higher temperature resistance than polymer matrix composites.

In summary, in order for the liquid matrix materials to bond to the solid fibers, two conditions need to be satisfied:

1. The liquid matrix has to be available at the surface of the solid fibers. Various means have been designed to do this and this has resulted in different forms of manufacturing techniques. This is governed by the viscosity of the liquid matrix material μ, the permeability of the fiber beds S (or K), the amount of pressure applied

to move the matrix Δp and the distance that the liquid matrix has to flow Δx. Even though there are many forms of manufacturing techniques, these can all be understood from this simple principle.

2. The liquid matrix and solid fiber materials have to be thermodynamically compatible with each other. To put in simple terms, the surface energy of the liquid matrix has to be less than that of the solid fiber. This principle limits the number of material systems that can be used to make composites and it also explains why there are more polymer matrix composites as compared to metal matrix composites.

Composite materials consist of two groups, based on the length of the fibers. These are short fiber composites and long fiber composites. Short fiber composites consist of those where fibers are about 1.25 cm or less, whereas long fiber composites are those consisting of fibers that are continuous and long. Short fibers are usually incorporated into the polymer system. Traditional processing techniques used for regular polymeric materials such as injection molding or compression molding are also used to process polymers reinforced with short fibers. The principle governing the processing of regular polymers also applies to the processing of polymer reinforced with short fibers. The focus of this book is on long fiber composites.

4.0 REFERENCES

1. Metcalf A.G. and Schmitz K.G. *ASTM Proc.* 64, 1974, p. 1075.
2. Lam R.C. and Kardos J.L. The permeability and compressibility of aligned and cross-plied carbon fiber beds during processing of composites, *Polymer Engineering and Science,* 1991, Vol. 31, No. 14, pp. 1064–1070.

5.0 APPENDIX

Volume Fraction and Weight Fraction

In the manufacturing of composites, weights and weight fractions are usually used. However, in the determination of different properties for the composite materials in terms of the properties of the fiber and the matrix, volume fraction is usually used. The following gives relations between volume fractions and weight fractions.

The following symbols will be used:

V_f = Volume of fibers
V_m = Volume of matrix
V = Volume of the composite
v_f = Fiber volume fraction
v_m = Matrix volume fraction
w_f = Fiber weight fraction
w_m = Matrix weight fraction
W = Weight of the composite
W_f = Weight of fibers
ρ_f = density of fibers
ρ_m = density of matrix
ρ = density of composite materials

The volume fraction of fibers can be written as:

$$v_f = \frac{V_f}{V}$$

This can be expanded as:

$$v_f = \frac{\dfrac{W_f}{\rho_f}}{\dfrac{W_f}{\rho_f} + \dfrac{W_m}{\rho_m}} = \frac{\dfrac{w_f}{\rho_f}}{\dfrac{w_f}{\rho_f} + \dfrac{w_m}{\rho_m}}$$

This can be written as:

$$\frac{1}{v_f} = 1 + \frac{\rho_f w_m}{\rho_m w_f} \quad \text{or} \quad v_f = \frac{\rho_m w_f}{\rho_m w_f + \rho_f w_m} \tag{1.21}$$

The weight fraction of the fiber can be written as:

$$w_f = \frac{W_f}{W}$$

This can be expanded as follows:

$$w_f = \frac{V_f \rho_f}{V_f \rho_f + V_m \rho_m} \quad \text{and} \quad w_f = \frac{v_f \rho_f}{v_f \rho_f + \rho_m (1 - v_f)} \tag{1.22}$$

Percent (%) and Parts per Hundred (phr)

In the discussion for relative amounts of different constituents (particularly for the case of matrix material where curing agents are involved), two types of quantities are used: percent and parts per hundred (phr). The following example illustrates the difference between the two quantities.

A mixture containing 100 g of epoxy and 40 g of amine curing agents is to be made. The percent of amine curing agent in the mixture is 40/140 = 29% and the parts of amine per hundred of epoxy is 40 phr.

If 10 g of certain fillers are added into the mixture, the percent of the amine now is 40/150 = 27% while the parts of amine per hundred epoxy remains 40 phr.

The use of phr is convenient in that the phr for a particular constituent does not change when one adds some other components into the mixture.

6.0 HOMEWORK

1. Indicate three main functions provided by the matrix in a composite material. Indicate the main functions provided by the fiber in a composite material.

2. Indicate the four most common manufacturing processes for making components using advanced thermoset composite materials. For each process, indicate in which sector of the industry the process is used the most. Give a reason why this is so.

3. Indicate in what way manufacturing using advanced composites is different from manufacturing using metals.

4. Why is wetting of the fibers important? What are the two most important aspects that affect wetting of the fibers?

5. Using the square packing array, show that the interfiber spacing is given by:

$$\delta = d_f\left(\sqrt{\frac{\pi}{4v_f}} - 1\right)$$

   where δ is the interfiber spacing, d_f is the fiber diameter, and V_f is the fiber volume fraction.

6. What is the general difference between the viscosity of a thermoset composite and a thermoplastic composite. What is the implication of this phenomenon on the manufacturing strategy?

7. What are the three important advantages of a material (like glass material) existing in fiber form over the same material in bulk form?

CHAPTER 2

Matrix Materials

1. INTRODUCTION

As presented in Chapter 1, composite materials consist of three main parts: fibers, matrix and interface between fibers and matrix. The functions and importance of the matrix were described in Chapter 1. The principles involving the matrix in the manufacturing of composites are covered in this chapter.

It is not the intention of this book to cover all matrices that are available for the manufacturing of composites. There are many excellent books that cover the topics and these should be consulted for the properties of these matrix materials. One example is the *Handbook of Composites* by Lubin [1].

The emphasis of this book is on the principles that govern the behavior of the matrix material during the manufacturing of composites. This involves an understanding of the basic chemical structure of the material, how this structure evolves during the manufacturing process, and how this evolution will influence the quality of the resulting composite structures. Both main categories of polymer matrix materials will be discussed: thermoset matrix composites and thermoplastic matrix composites.

2. DIFFERENT TYPES OF MATRIX MATERIALS AND THEIR PROMINENCE

Matrix materials are generally polymers, metals, or ceramics. In a way this sentence says that any material can serve as a matrix material. In reality, however, the majority of composites that exist on the market are made of polymer matrix composites. Among these, thermoset matrix composites are more predominant than thermoplastic composites.

The reason why there are more polymer matrix composites than metal matrix composites and ceramic matrix composites is due to the second requirement in the interface discussion in Section 3.3 of Chapter 1: Compatibility between the matrix and the fibers. It was shown in Chapter 1 that the surface energy of metals is on the order of 400–2000 dyne/cm while the surface energy of polymers is on the order of 30–45 dyne/cm. The surface energy of glass fibers is about 500 dyne/cm, that of graphite is about 50 dyne/cm and about 44 dyne/cm for Kevlar fibers. Thermodynamic requirements call for the surface tension of the matrix (in liquid form) to be less than that of the fibers to ensure bonding. Since the surface energy of liquid metals is much larger than that of the solid fibers, it is very difficult for liquid metals to bond onto the surface of solid fibers. As such, it is difficult to make metal matrix composites, in spite of the fact that metal matrix composites can offer desirable properties such as high temperature resistance. The same explanation can be used for ceramic matrix composites. This does not mean that metal matrix and ceramic matrix composites do not exist. They do, but only in rare cases of high temperature applications.

Among the polymer matrix composites, there are thermoset matrix composites and thermoplastic matrix composites. The differences between thermoset and thermoplastic resins are explained below.

2.1 Thermoset and Thermoplastic Matrix Materials

The similarities and differences between thermoset and thermoplastic composites can be understood if one compares the processes by which these two types of materials are made.

First consider a typical thermoset polymer such as epoxy. To make this material, one first starts with the epoxy molecules. The epoxy molecules are relatively small [on the order of about 20–30 carbon-carbon (C–C) links]. This is relatively short as compared to the order of a few hundreds or thousands of C–C links for thermoplastic molecules. Since the length of the thermoset molecules is short, the material consisting of them usually has low viscosity and appears in the form of liquid at room temperature or moderately high temperature (about 100°C). Figure 2.1(a) shows a schematic of the molecules in a thermoset resin. Since the material appears in liquid form, in order to make a solid out of it, the molecules must be tied together with molecules of some other type. The tying molecules are called the *linkers* or *curing agents*. Figure 2.1(b) shows a schematic of the linker molecules.

In some cases (such as polyester) the linker molecules may not react easily with the resin molecules when they come into contact. For these cases, the linker molecules can be mixed together with the resin molecule in a container for shipping purposes [Figure 2.1(c)]. When the linking is desired, one needs to add into the mixture an initiator (an unstable type of molecule) which will start the reaction.

In other cases (such as epoxies) the linker molecules may react easily with the resin molecules. For these cases, the linker molecules can not be mixed together with the resin molecules until the time the manufacturer is ready to incorporate the resin systems together with the fibers.

When the proper conditions for linking occur (discussed later in this chapter), the tying molecules will link the resin molecules together as shown in Figure 2.1(d). This 3-D linking network is a solid and it represents the solid thermoset resin. Since the ties (links) are made by chemi-

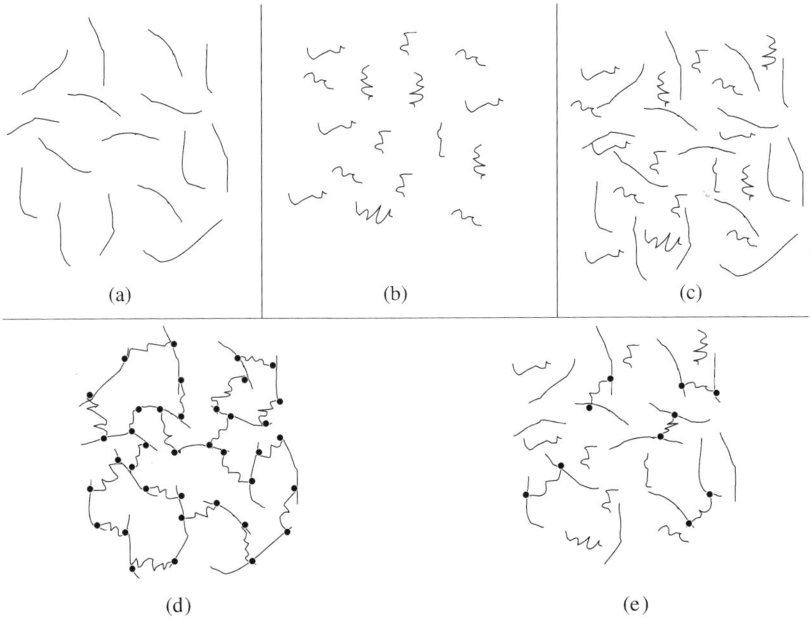

FIGURE 2.1 Schematic of (a) the molecules in a thermoset resin, (b) the linking molecules, (c) the resin molecules and the linking molecules in a container before linking reactions, (d) the thermoset resin network after linking reactions, and (e) a partially linked network.

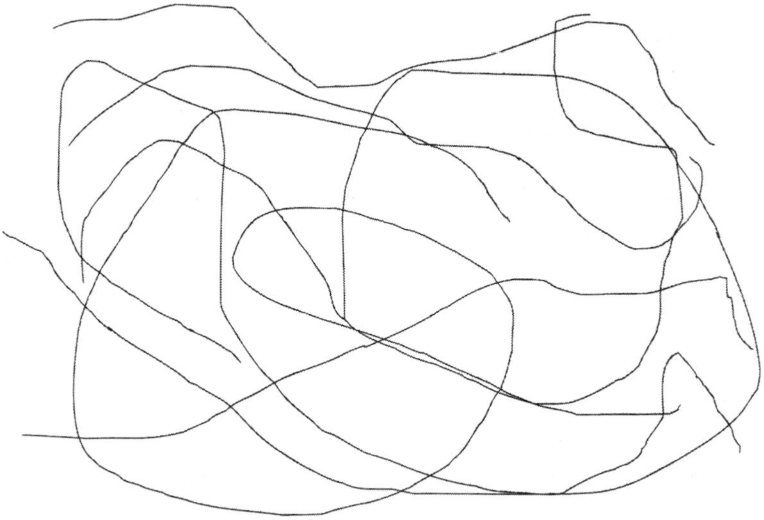

FIGURE 2.2 Schematic of the molecules in a thermoplastic resin.

cal bonding, once set, the shape of a component made of thermoset resin cannot be changed by heating.

The linking between the linker molecules and the resin molecules takes place whenever an active end of the resin molecule is in the vicinity of an active region of the linker molecule. All links (millions and millions of them) need to be complete in order to create a solid 3D network. This process takes time (several hours, sometimes several days). One can intervene in the process by allowing only a portion of the links to be formed and retarding the remaining reactions. This can be done either by lowering the temperature or adding in retarding molecules (called inhibitors) somewhere during the process. The result of this is a partially linked network [Figure 2.1(e)] which exhibits itself as a viscous liquid (or flexible solid) which can be handled like a liquid but remains tacky for bonding purposes. This is the process for making preimpregnated layers (prepregs).

Next consider a typical thermoplastic resin for composite applications such as polyetheretherketone (PEEK). Thermoplastic molecules can be very long. Each molecule may contain up to several hundreds or thousands of C–C links. Figure 2.2 shows a schematic of these very large molecules. Due to high molecular length, it is difficult for these molecules to move around at room or moderate temperature. In order for these molecules to be able to move relative to each other, high temperature needs to be applied. The viscosity of these resins is large even at high

TABLE 2.1 Viscosity (in centipoise) of a Few Thermoset and Thermoplastic Materials (1 Pa-sec = 10 Poise = 1000 centipoise).

Material	20°C	25°C	T°C
Air	0.0187		
Water	1		
Polyester	100–300		
Vinyl ester	100–300		
#10 Motor oil	500		
Golden syrup	2,500		
Epoxy (Shell Epon 828-14 phrMPDA, 15 phr BGE)	600		
Epoxy (Shell 826 16 phr MPDA, 10phr BGE)	750		
Epoxy (Dow 332-16 phr MPDA, 10 phr BGE)	500		
Molasses		10^5	
Epoxy 5208			100 @ 177°C
BMI			1000 @ 150°C
Ryton (thermoplastic)			10^7 @ 313°C
PEEK (thermoplastic)			10^6 @ 400°C
Utem (thermoplastic)			10^8 @ 305°C
Torlon (thermoplastic)			10^9 @ 350°C

temperature (Table 2.1). However when the material is cooled down, it becomes solid fairly quickly. The processing time therefore can be much shorter (on the order of minutes) as compared to thermoset resins (on the order of several hours or days) where time needs to be allowed for all the linking reactions to complete.

There are more thermoset matrix composites than thermoplastic composites. The explanation can be referred to the first condition discussed in the interface section (3.3) in Chapter 1, i.e., availability of the resin at the surface of the fibers. In order for the matrix to bond to the surface of the fibers, the resin has to be available at the surface of the fibers. This seems to be an obvious requirement but it has strong implications. For the resin to be available at the surface of the fibers, the manufacturer has to put it there. In the case of prepregs, the resin is already placed on the surface of the fibers and so this does not seem to be critical during the fabrication of the part (it is critical to assure the availability of resin at the surface of the fiber during the fabrication of the prepregs). For a process such as resin transfer molding, however, resin needs to be pumped so that it can flow to the surface of the fibers. The flow of resin depends on the permeability of the fiber networks, and also on the viscosity of the resin. At reasonably low temperatures (less than 100°C), the viscosity of thermoset matrix is much lower than that of thermoplastic matrix. Table 2.1 shows the viscosity of a few thermoset and thermoplastic matrix materials. Due to their low viscosity, thermoset matrix materials can flow to the surface of fibers more easily as compared to thermoplastic matrix. The difference in viscosity between thermoset matrix and thermoplastic matrix is the main reason for the predominance of thermoset matrix composites as compared to thermoplastic matrix composites.

3. THERMOSET MATRIX MATERIALS

The presentation on thermoset matrix materials will concentrate on two materials: polyester and epoxy, with some brief presentation on other types of materials. This is because the mechanism of operation of these two materials is representative for other materials.

3.1. Polyester Resins

3.1.1. General

Compared with epoxy resins, polyester resins are lower in cost but limited in use due to less adaptable properties. Polyesters have been used mainly with glass fibers (normally E glass) to make many commercial products such as pipes, boats, corrosion resistant equipment, automotive components, and fiber reinforced rods for concrete reinforcement.

3.1.2. Polyester Chemical Structure and Polymer Formation

Polyesters are formed by the condensation polymerization of a diacid and a dialcohol (a diacid means two organic acid groups are present in a molecule, and a dialcohol, sometimes called a diol, has two alcohol groups in the molecule). A typical reaction is shown in Figure 2.3, in which maleic acid is made to react with ethylene glycol to form polyester. In this reaction, the acid group (O=C–OH) on one end of the diacid reacts with the alcohol group (CH_2OH) on one end of the diol to form a bond linking the two molecules and to give out water as a byproduct. The linking group which is formed is called an *ester* (C–O–C=O). This step is called a condensation reaction.

The resulting product still has another acid group on one end and another alcohol group on the other. Both of these ends are still capable of undergoing further condensation reactions and then to repeat again and again. Therefore, with sufficient reactant materials, chains of alternating acid and alcohol groups will form and will have regularly repeated units as shown in the polymerization step of Figure 2.3. One unit of the repeating units is shown within the bracket at the bottom of the figure. Subscript *n* represents the number of repeated units. A large *n* value indicates a longer (or larger) molecule. Many polymers would have an *n* of several hundred, although some polymers exist with *n* of less than 20.

REACTANTS

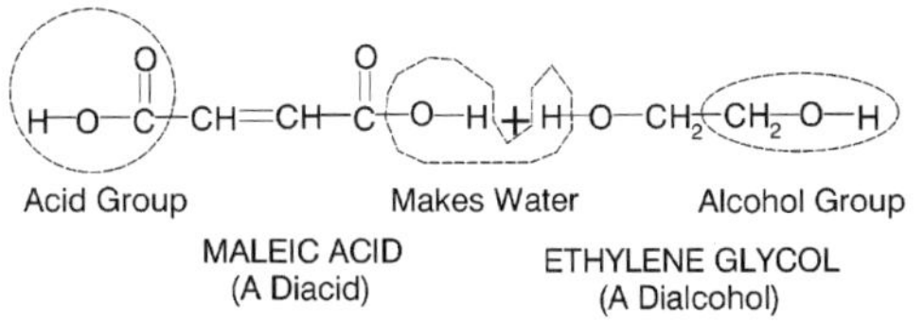

In the above, the –O–H end of the maleic acid molecule is reaction with the –H end of the glycol molecule.

FIRST CONDENSATION
REACTION PRODUCTS

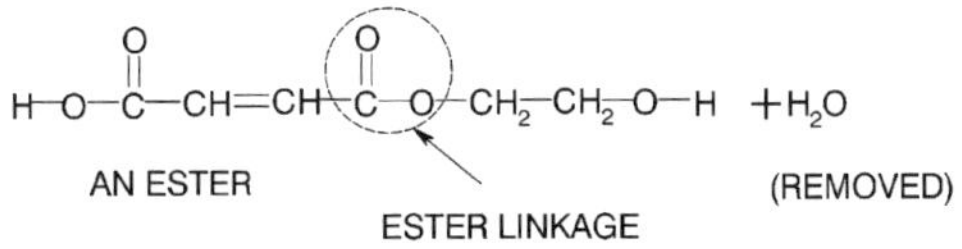

In the above, combination of –O–H and –H forms water H_2O. The remaining parts of the two types of molecules connects together to form an ester linkage.

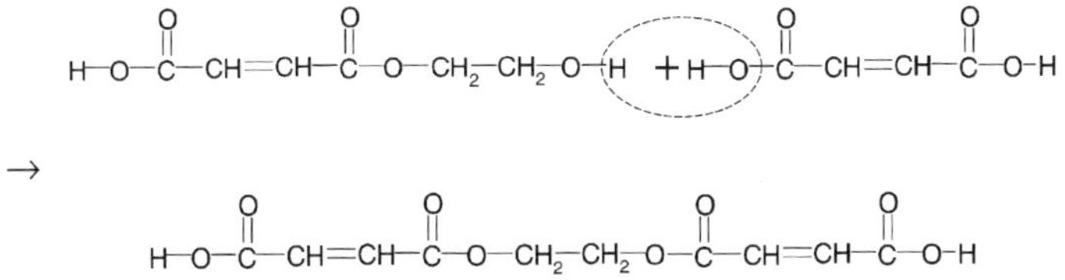

In the above, similar reaction at other ends of the acid molecule and the glycol molecule can take place. The result is an ester molecule with two carboxylic ends (COOH).

POLYMERIZATION

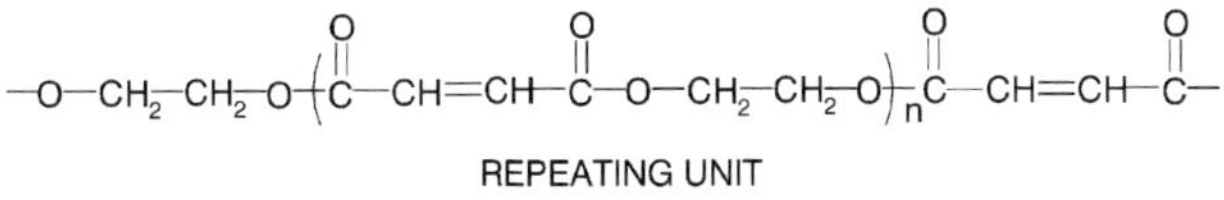

When many units of ester connects together due to the reactions, polyester molecules will be formed. The part in the bracket shows one unit of the repeating units.

FIGURE 2.3 Condensation polymerization of a polyester.

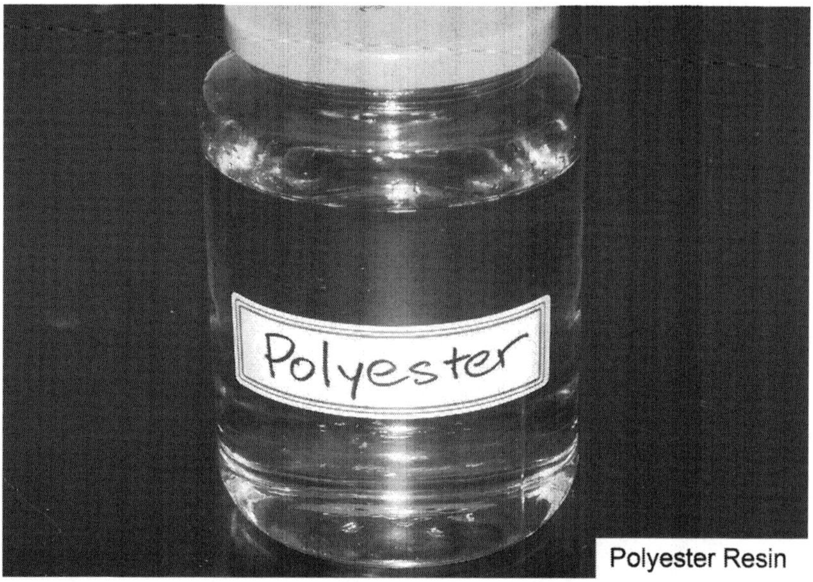

FIGURE 2.4 Glass container with liquid polyester.

These chains are called *polymers* (from the word for many parts). Because the linking group formed by acids and alcohols are esters, the term given to this type of resulting polymer is *polyester.* A quantity of these polyester polymer chains is collectively called *polyester resin*. (Polymers, in general, in the uncured state, are called *resins*.) For a certain amount of resin, there is a distribution of sizes of molecules, i.e. the resin is an ensemble of molecules of different lengths. This is because the formation of the molecules is due to the availability of the reactants. The molecular weight of the material is the average value of all molecules in the material. At room temperature, polyester appears as a liquid. Figure 2.4 shows a sample of polymer in a glass container.

Example 2.1

It is desired to make a polyester resin using 100 g of maleic acid and a corresponding amount of ethylene glycol. A stoichiometric amount of ethylene glycol is used. After the condensate is removed, how many grams of polyester are obtained?

Mass of different atoms:

C = 12 g/mol, H = 1 g/mol, O = 16 g/mol, N = 14g/mol

Solution

The molecular structures of maleic acid and of ethylene glycol are shown below. From this, the molecular mass of each material is calculated as:

Maleic acid

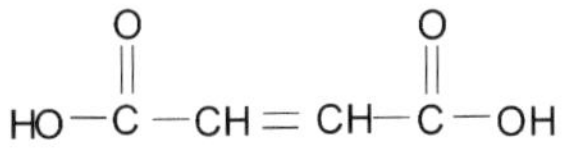

Ethylene glycol

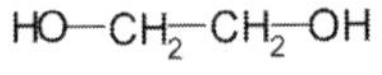

Maleic acid: 4C + 4H + 4O = 4 (12 + 1 + 16) =116 g/mole

Ethylene glycol: 2C + 6H + 2O = 24 + 6 + 32 = 62 g/mole

The reaction takes place as shown in the following:

It can be seen that one molecule of maleic acid reacts with one molecule of ethylene glycol to make a unit of ester and two water molecules.

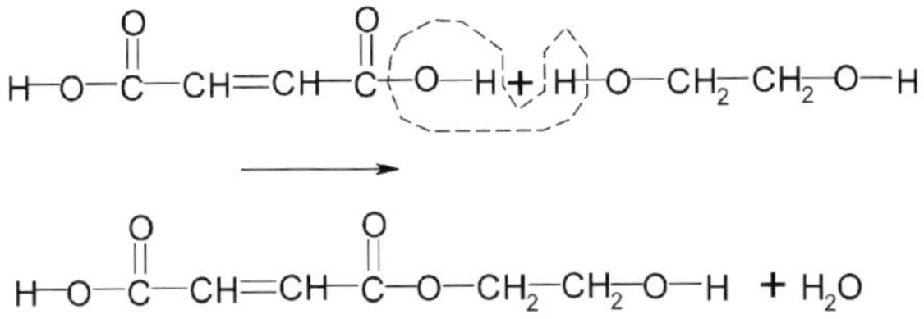

and

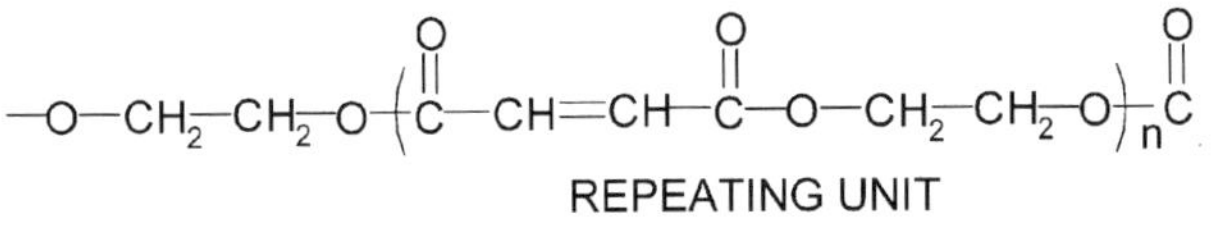

Mass of the water molecule: 2H + O = 18 g/mole.

Let M_p be the mass of the polyester made using 100 g of maleic acid, we have:

$(M_p/100) = (116 + 62 - 36)/116 = 1.224$

$M_p = 122.4$ g

3.1.3. *Polyester Crosslinking*

The polymers formed in the reaction illustrated in Figure 2.3 are not crosslinked since no chemical bond has been formed between the various chains. (The chains are often mechanically intertwined, but that is not crosslinking.) However, the diacid chosen in this case (maleic acid) con-

TABLE 2.2 Structures of Commercial Organic Peroxides [2].

Name of Peroxide	Chemical Structure
Hydrogen peroxide	H–OO–H
Hydroperoxides	R–OO–H
Dialkyl peroxides	R–OO–R
Diacyl peroxides	R–C(O)–OO–C(O)–R
Peroxyesters	R–C(O)–OO–R
Peroxy acids	R–C(O)–OO–H
Peroxy ketals	R_2–C–OO–R_2
Peroxy dicarbonates	R–OC(O)–OO–C(O)O–R

tained a carbon-carbon double bond which survived the polymerization reaction and is contained in every repeating unit of the polymer. (When a reactant or polymer contains a carbon-carbon double bond, the term unsaturated is often applied to it. Therefore maleic acid is an unsaturated diacid and the resulting polymer is an unsaturated polyester.) This unsaturation is critical since the carbon-carbon double bond is the location where crosslinking occurs.

The crosslinking occurs by the addition polymerization reaction as shown in Figure 2.5. In this figure, the RO– tends to react with another active site from another molecule [a dash (–) on the right hand side of the oxygen atom O indicates that it is reactive]. This reaction utilizes a crosslinking agent (styrene, in the example) that reacts with the polyester polymer chains to provide the crosslinks. (The styrene also lowers the initial viscosity to improve processing.)

Normally the styrene molecules are mixed together with the polyester molecules and shipped in a container. Under normal conditions [room temperature for a limited time (months)], the styrene molecules do not react with the polyester molecules. Initiators are usually required to start the reaction. The reaction steps shown in Figure 2.5 are explained below.

3.1.3.1. Initiation Step

The crosslinking reaction is initiated by a molecule that readily produces free radicals (a chemical species with unpaired electrons). The most common group of such molecules is the organic peroxides (for example, methyl ethyl ketone peroxide—MEKP).

Organic peroxides are useful as initiators or crosslinking agents because of the thermally unstable O–O bond which decomposes to form free radicals. Organic peroxides may be viewed as derivatives of hydrogen peroxides in which one or both hydrogens are replaced by organic radicals. Table 2.2 lists the structures of several commercially available organic peroxides.

The rate of free radical production is highly temperature dependent, therefore, the reaction to produce free radicals, which in turn initiates the crosslinking reaction, can be accelerated by increasing the temperature.

REACTANTS

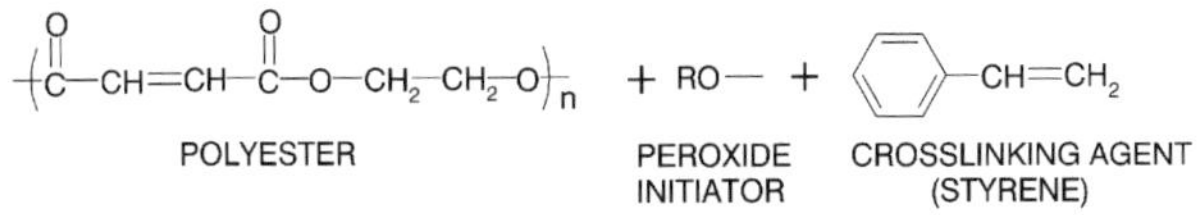

In the above reaction, a polyester material system consists of the polyester molecules, the peroxide initiator molecules (such as MEKP) and the curing (or cross linking) agents are mixed together inside a pot.

INITIATION STEP

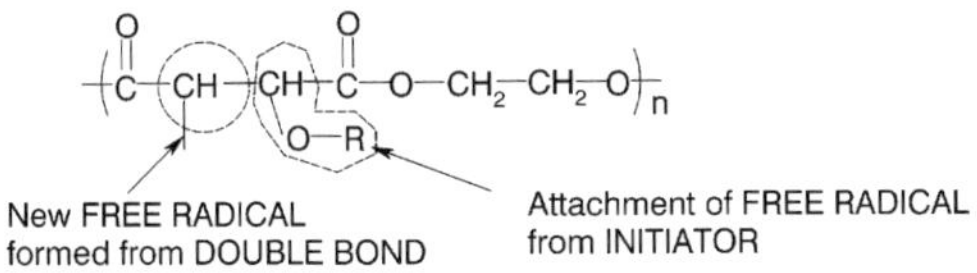

In the above, the initiator (–O–R) is making a reaction with one of the carbon atoms that had a double bond (=). After the reaction, the double bond in this carbon atom (and its partner carbon atom) is broken and it is replaced by a single bond (–). The partner carbon atom also has an active arm which is ready to make reaction with another adjacent molecule).

BRIDGING STEP

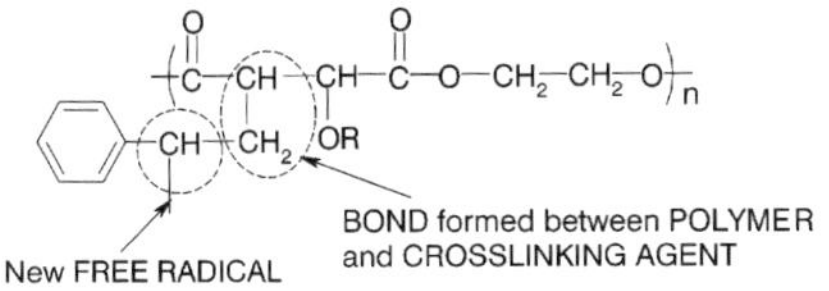

In the above, the active arm of the partner carbon atom makes a link with an adjacent styrene monomer. Due to this reaction, the double bond between the two carbon atoms in the styrene monomer is broken and is replaced by a single bond between the two carbon atoms. The carbon atom in the styrene monomer that is not connected to the ester molecule now has an active arm that is ready to react with another adjacent molecule.

CROSSLINKED POLYMERS

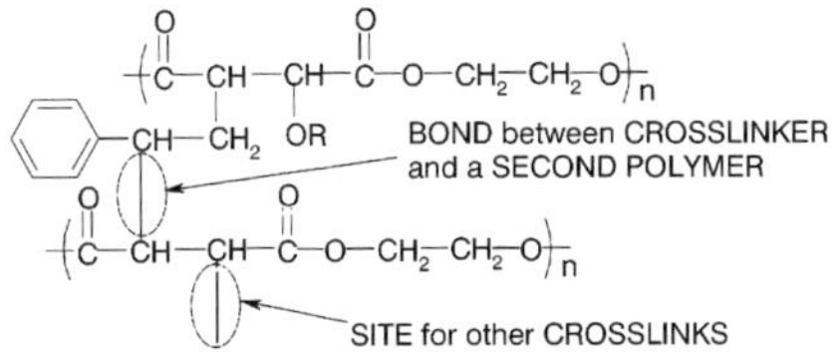

In the above, repeats of the reaction between the polyester molecules and styrene molecules gives rise to a crosslinking network between polyester molecule and styrene molecule. This is the polyester resin system.

FIGURE 2.5 Addition (or free-radical) cross-linking of polyester [3].

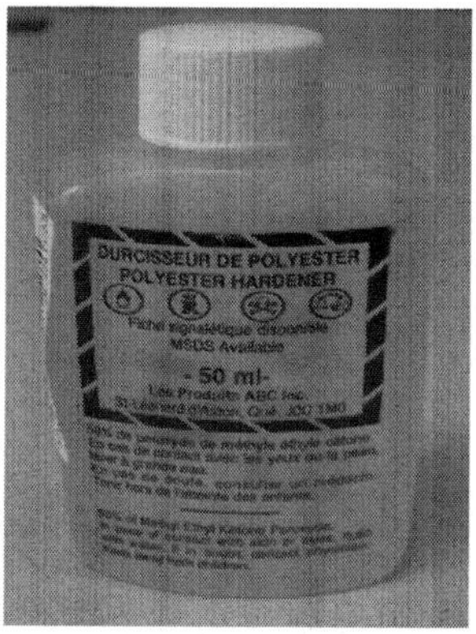

FIGURE 2.6 Small container of MEKP.

Since polymerization reactions take place at various temperatures, organic peroxides have been developed with different decomposition rates and these can be conveniently expressed by half-lives (the time required to decompose 50% of the peroxide in a diluent solution at a given temperature). The half-lives of most peroxides at about 100°C vary from 0.5 hours to 50 hours [2]. Initiator compounds are sometimes called *catalysts*. Figure 2.6 shows a small container of MEKP. These free radicals attack the carbon-carbon double bonds and bond to one of the double-bond carbons, reducing the double bond to a single bond and producing a new free radical on the other formerly double-bonded carbon. This new free radical is then free to react with any other carbon-carbon double bond. Since the styrene is more mobile than the polyester molecules, the most likely new double bond to react with would be in the styrene. Usually only a small amount of initiators (about 1%) is added to the polyester/styrene system to start the reaction. Adding large amounts of initiators is dangerous due to the simultaneous reactions of many links and this can cause a large amount of heat generated over a short duration of time.

3.1.3.2. Bridging Step

At the end of the initiation step, many radicals (the free bonds from the carbon atom C–) are produced. The radicals are produced from the breaking of the double C=C bonds in either the styrene molecules or in the polyester molecules. When these free radicals are in the vicinity of each other, they connect. The results are the single C–C bonds between the radicals. This forms the link between the styrene molecules and the polyester molecules. The formation of the single C–C bonds in turn creates new free radicals. These free radicals are again available for reaction with other carbon-carbon double bonds, and reactions with other polyester molecules would create further crosslinks. The reaction continues to propagate the crosslinks.

3.1.3.3. Crosslinking Step

The net result is a large network of interconnected polymers in which styrene serves as the crosslink (bridge) between the polymers. Since the peroxide is generally present in small quantities (typically 1%) the final structure does not show the peroxide attached to the polystyrene molecule as this would generally not be present except on the endmost crosslink. This crosslinking mechanism is called addition or free-radical crosslinking. This entire crosslinking reaction process is called *curing*.

Styrene is not the only crosslinker (sometimes crosslinkers are called *curing systems*) for polyester systems, although it is the most widely used. Other chemicals in use include vinyl toluene, chlorostyrene, methyl methacrylate, and diallyl phthalate. The physical and chemical properties of the crosslinked polymer are affected by the curing system. For instance, the use of chlorostyrene imparts flame resistance to the polymer.

Example 2.2: Crosslinking of Polyester

It is desired to make a polyester using 100 g of maleic acid and ethylene glycol. A stoichiometric amount of ethylene glycol is used. Cross linking is done using styrene. Assume that one styrene monomer corresponds to one oligoester (this assumption is to simplify the calculation to illustrate the principle; in reality the crosslink between oliester molecules can be in the range from 1 to 4 styren monomers). From the cross linking process, how many C=C bonds are broken, how many C–C bonds are formed?

Solution

Continuing from the same problem in Example 2.1:

Number of bonds broken and formed

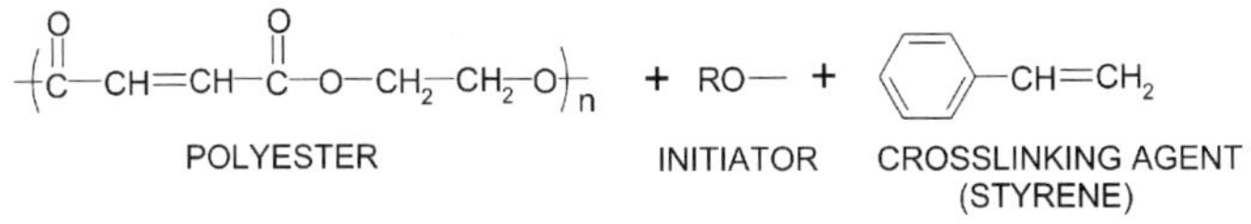

Initiating Step

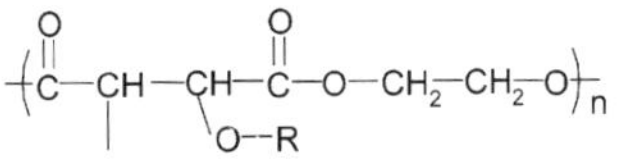

Bridging Step

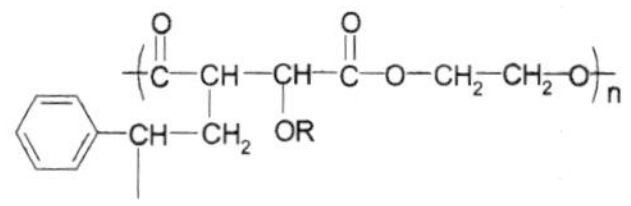

Crosslinking Step

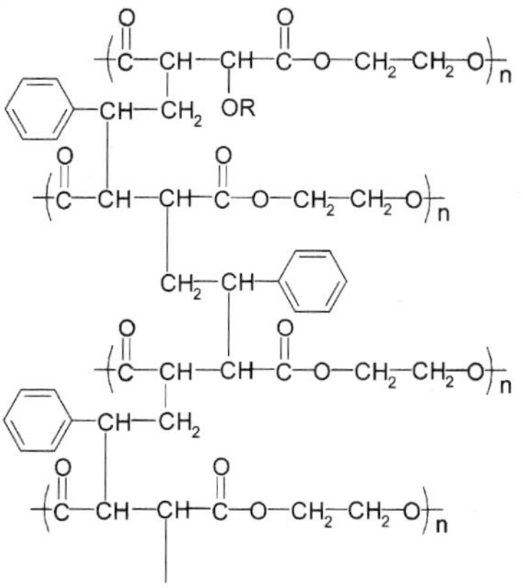

The crosslinking process is illustrated above.

For each pair of polyester units, it can be seen that four C=C bonds are broken and eight C–C bonds are formed. For the C=C bonds, two are from the two polyester units, and the other two are within the structure of the two styrene molecules. For the eight C–C bonds, two are replacing the two C=C bonds within the ester units, the other two are replacing the two C=C bonds within the two styrene molecules, the other four are connecting the two styrene molecules with the two ester units.

Therefore for each polyester unit, there are two C=C bonds broken and four C–C bonds formed.

The chemical formula for a polyester unit is:

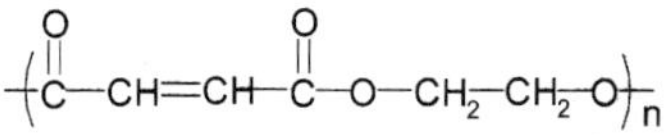

The mass of a polyester unit is:

6C + 6H + 4O = 72 + 6 + 64 = 142 g/mole

Since there are 122.4 g of polyester made, the number of bonds involved is:

C=C bonds: (122.4/142) 2 = 1.724 mole or 1.038×10^{24} bonds. (Note that 1 mole = 0.602×10^{24}).

C–C bonds: $2 \times 1.724 = 3.446$ moles or 2.076×10^{24} bonds

3.1.4. Crosslinking (Curing) is an Exothermic Reaction (Heat is Generated)

The crosslinking of polymer from individual molecules into a polymer 3-D network is an exothermic reaction (i.e., heat is generated from the process). What this means is that the temperature of the material will increase as a result of curing. For the curing of thin laminates (about a few millimeters thick), this increase in temperature may not create a problem; but for the curing of thicker laminates the heat generated from inside the laminate may not be conducted away easily, due to the fact that polymers such as polyesters (or epoxies) do not conduct heat. Over a short period of time, the temperature increase at a given location can be quite large, resulting in the burning or degradation of the matrix material.

From the thermodynamic point of view, to break a bond, one needs to add energy into the system. For example, to break a piece of chalk with one's fingers, it has to be pulled apart or bent. Energy is added to the piece of chalk by the pulling or bending motion of the fingers. On the other hand, if the bonds between the broken pieces of chalk were to be bonded together again, energy would be generated from the bonds by the law of conservation of energy. This energy will increase the temperature of the chalk unless the heat is conducted away.

Example 2.3: Heat Generation and Temperature Increase

It is desired to make a polyester using 100 g of maleic acid and ethylene glycol. A stoichiometric amount of ethylene glycol is used. Crosslinking is done using styrene.

Continuing from the same problem in Examples 2.1 and 2.2:

a. The energy required to break a C=C bond is 680 kJ/mole and the energy created by forming a C–C bond is 370 kJ/mole. How much energy is generated during the polymerization process? 1 mole = 0.602×10^{24}
b. The heat capacity of polyester is 0.25 cal/g/°C. Assuming no heat loss, what is the increase in temperature of the polyester?

Mass of carbon C= 12 g/mole, H = 1 g/mole, O = 16 g/mole
1 calorie = 4.18 Joule

Solution

a. Energy generated:

Considering the energy in the bonds, one has:

Energy inputted into the system to break the double bonds:

(1.724 mole) (680 kJ/mole) = 1172.32 kJ

Energy generated from the system to form the single bonds:

(3.446 moles)(370 kJ/mole) = 1275.02 kJ

Net energy generated:

1275.2?1172.32 = 102.7 kJ

b. Temperature increase:

Heat stored in the material:

$Q = mc\Delta T$ or $\Delta T = Q/(mc)$

For the mass of the material, apart from the polyester, there is also the styrene. Each unit of the polyester corresponds to each unit of the styrene for complete crosslinking. The chemical formula for styrene is shown in the following:

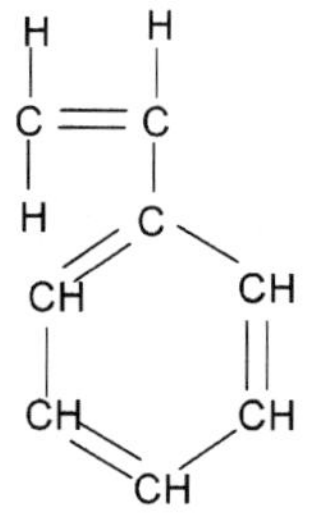

The mass of a styrene molecule is therefore:

8C + 8H = 96 + 8 = 104 g/mole

Mass of styrene corresponding to 122.4 g of polyester is:

(122.4/142)104 = 89.65 g

Total mass of polyester and styrene:

122.4 + 89.65 = 212.05 g

The temperature increase can now be calculated using the above equation:

$$\Delta T = \frac{(102.7 \text{ kJ})(1cal / 4.18 \text{ J})}{(212.05 \text{ g})(0.25\text{cal} / \text{g}^{\circ}\text{C})} = 463^{\circ}\text{C}$$

Note that the temperature increase is very high. The reason for this is because of the assumption of no heat loss. In reality, there is heat loss to the surrounding environment. Also not all the heat is generated instantly. The process of bonding takes time. The curing process may take hours and even days, depending on the surrounding temperature. As such in real applications the temperature increase will be a lot less than the number above.

TABLE 2.3 Energy Associated with the Atomic Bonds.

Bond	Energy (kJ/mole)
C–C	370
C=C	680
C–H	435
O–H	500
N–H	430
C–O	360
C=O	535

On the atomic level, there is an amount of energy associated with the breaking or forming of each of the bonds. The energy associated with some of the bonds is shown in Table 2.3.

3.1.5. Photo-Activated Curing Systems

The initiators for curing described above (such as MEKP) are readily activated agents. They start acting as soon as they are mixed together with the resin systems. There are also initiators that are activated by ultraviolet rays. These photo-initiators can be mixed in the resin system but they are not activated until photon energy provided by ultraviolet light is introduced. Once activated they start reactions similar to the case of conventional initiators. The use of photoinitiators allows some control over the time for the reaction to start but requires that the resin be translucent enough for light to be transmitted.

3.1.6. Polyester Properties

One of the advantages of polyester resins is that the reactants, which are called monomers (Table 2.4), can be chosen from a wide assortment of diacids and diols to meet specific physical and chemical properties desired. Apart from maleic acids, one can also use maleic anhydrides, orthophthalic acids, isophthalic acids or orthophthalic anhydride. The difference between orthophthalic acids and isophthalic acids is that due to the location of the bonds on the benzene ring, the molecule of the isophthalic acids is straighter. Straighter molecules provide better packing of the material, and this in turn gives higher density, better strength, stiffness and environmental resistance. The anhydride on the other hand has a close ring which must be broken by the application of more energy (heat) before further reaction can occur. As a rule of thumb, when a molecule contains benzene rings rather than just straight C–C bonds it dis-

TABLE 2.4 Choices of Reactants for Polyester Resins [3].

Reactants	Chemical Structure	Advantages
Ethylene Glycol	HO—CH_2—CH_2—OH *This molecule is aliphatic since it contains mainly linear chains.*	Basic reactant
Propylene Glycol	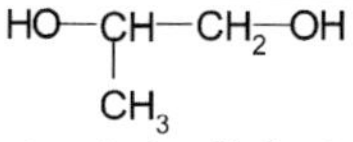 *This molecule is aliphatic and branched.*	More compatible with styrene than ethylene glycol
Maleic Acid (also Fumaric Acid)	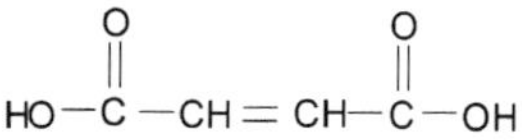 *This molecule is aliphatic.*	Flexible, Low cost
Maleic Anhydride	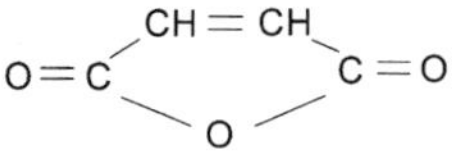 *The anhydride has a ring which needs to be opened before reaction can occur.*	—
Orthophthalic Acid (Ortho)	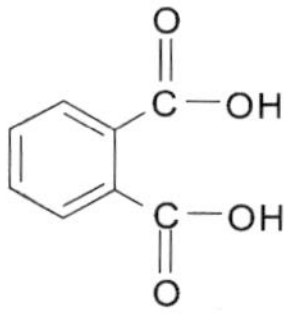 *This molecule is aromatic since it contains the benzene ring.*	Rigid Relatively low cost
Orthophthalic Anhydride	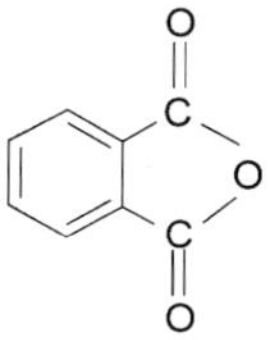 *This aromatic molecule has the ring which needs to be opened before reaction can occur.*	Rigid
Isophthalic Acid (ISO)	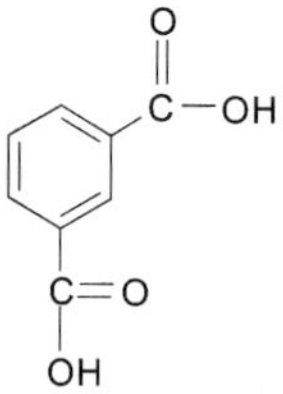 *The isophthalic molecule gets its name from the relative position of the connection between the COOH group with the benzene ring.*	Resilient (tough) Thermal stability

plays better mechanical properties and environmental resistance. Molecules that contain benzene rings are usually called aromatic while those that do not are called aliphatic.

Apart from the variation in the type of matrix material to make a composite, the manufacturer may use different types of resins at different locations in the structure for various reasons. For example, in the case of a fiberglass/polyester pipe used for transporting corrosive liquids, where the inside surface of the pipe is exposed to the corrosive environment, it is necessary to use a resin that is more corrosion resistant such as isophthalic resin (as compared to orthophthalic resin). However isophthalic resin is more expensive than orthophthalic resin. In order to save cost, the manufacturer may use a thin layer of isophthalic resin only on the inside of the pipe while the remaining part is constructed of orthophthalic resin. A layer of resin meeting particular requirements applied on the surface of a structure is usually called the *gel coat.*

There are disadvantages and advantages in using different types of resins at different locations within a structure. One disadvantage is that checking the quality of the product becomes difficult. Normally one can check the quality of the material by testing the surface for properties such as hardness or chemical resistance. But when the material at the surface is not the same as the material below, one cannot translate the finding at the surface to the material inside. One advantage is cost saving. Another possible advantage is that one can use resins with different thermal conductivities (such as the increase in thermal conductivity by adding carbon nanotubes) at different locations across the thickness of the structure to help dissipate the heat generated during the manufacturing of thick composite parts.

3.1.7. Polyester Use and Storage

Polyester resins are generally shipped and stored as a resin system (consisting of the polyester resin, curing agent, fillers and inhibitors) in a large container (drums), and the initiator in a separate container. Even though the polyester resin and the curing agent (such as styrene) are mixed in the same container, reaction does not take place if no activation occurs. Activation can occur in the form of addition of initiators, raising of temperature, or strong mixing activities. Figure 2.1(c) shows a schematic of the co-existence of the polyester molecule and curing agent molecule without reaction. The contents of the containers are mixed in appropriate proportions (usually about 40 phr of styrene and small amounts of inhibitors and fillers) depending on the applications. The styrene molecules not only serve as linking agents but they also reduce the viscosity of the polyester. For processes like liquid composite molding, low viscosity (less than 1000 cP) is required while for processes such as hand-lay-up for filament winding, higher viscosities may be required.

Occasionally the resin mixture will begin to crosslink spontaneously (without the addition of the initiator). This may happen when the temperature is raised or some violent sloshing occurs in the liquid resin. Since the crosslinking reaction is exothermic, once begun the reaction can proceed quite rapidly. To prevent this from occurring, molecules that absorb free radicals are added to the resin mixture. These molecules are called *inhibitors* and a typical example is hydroquinone. Even with inhibitors, however, after a sufficiently long time, the resins can crosslink, so the resin mixtures should be discarded after a specified length of time or at least tested to see if too much crosslinking has occurred. The length of time that they can be stored and still remain useful is called the *shelf life* or *storage life*. The test that is often performed to see if the resin is still acceptable for use is to check the viscosity. Viscosity of uncured polyester resin at room temperature is about 100–300 centipoise. If the viscosity is too high the resin mixture should be discarded.

Other chemicals that might be added to the container are *coupling agents*. These chemicals enhance the bond between the cured polymer resin and the fiberglass reinforcement. Typically these chemicals have one end that is a silane derivative and is, therefore, compatible with the fiberglass, and another end that is compatible with the resin. Hence, they act as bridges between matrix and reinforcement. Coupling agents will be discussed more in Chapter 3.

The incorporation of resins with fibers to make composites is done by either the wet process or dry process. Wet process means that liquid resin is introduced to the dry fibers at the time when the composite part is to be made. For the dry process the composite part is made in two stages. First, the liquid resin is introduced to the dry fibers and the resin is allowed to be partially cured. The result of the first stage is called *prepreg*. Prepregs are flexible sheets. When the manufacturer is ready to make the final composite parts, the prepregs are placed on the mold in different layers to build up thickness. The resin is then cured completely.

Even though there are recent efforts to make prepregs using polyester resin, most of the processes using polyester resins are wet. These include wet hand-lay-up, wet filament winding, and liquid composite molding. When the manufacturer is ready to make the part, a small amount of the resin system is transferred out of the large container and placed in a smaller container called a *pot*. A small amount of initiator (about 1%) is added and the crosslinking reaction then begins. The manufacturer needs to dispense all the resin in the pot before the viscosity becomes too high.

The duration that the resin stays in the pot without going bad is called the *pot life*. Different applications such as hand-lay-up, filament winding, pultrusion, or liquid composite molding may require different pot lives. The pot life can be modified by adjusting the amount of catalyst, inhibitor, or accelerator. Typical accelerators are cobalt naphthenate, diethyl aniline, and dimethyl aniline. They can be added by the resin system formulator or by the resin system user.

Curing agents such as styrene are unpleasant to the smell and are hazardous. There are government regulations that determine maximum amounts of styrene that may exist in the atmosphere of a working environment. Shops where styrene evaporation into the air exists should be well ventilated with high ceilings. Because of this problem, processes using polyester tend to go more toward the closed-mold approach (such as liquid composite molding) rather than the open-mold approach (such as hand-lay-up or filament winding).

3.2. Epoxy Resins

The most common matrix for advanced composites and for a variety of demanding applications is epoxy. Epoxies have been used extensively for these applications due to their excellent properties such as excellent adhesion, high strength, good corrosion resistance, low shrinkage (as compared to polyester resin), and processing versatility. In addition, the processing of epoxy resins does not involve toxic gases like styrene. Uncured epoxies emit some gas but these are not as unpleasant as styrene. The range of operating temperatures of epoxy resins can be up to 140°C. This is better than polyesters but lower than that of polyimides. On the other hand, epoxies are more expensive than polyesters and have higher viscosities than polyesters.

Figure 2.7 shows photographs of containers of Shell Epon 828 and

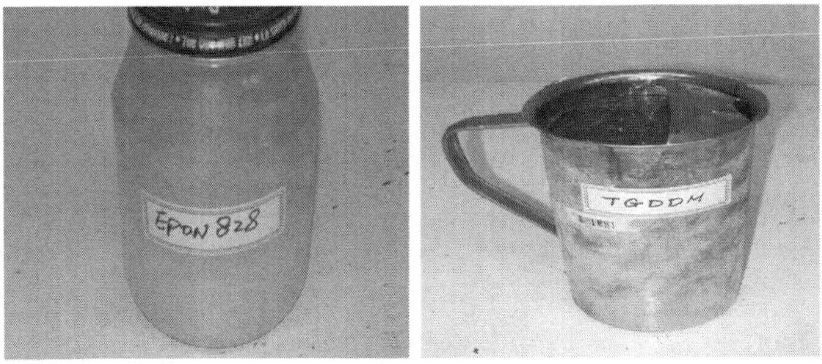

FIGURE 2.7 Photograph of a sample of Shell Epon 828 and TGDDM epoxy resin.

Vantico TGDDM epoxy resin. Shell Epon 828 is a low temperature cure resin (up to about 80°C). It has low viscosity at room temperature (about 600 cP) and appears in liquid form at room temperature. The viscosity of low temperature epoxy resins is about 500–2000 cP at room temperature. TGDDM is a high temperature cure resin (curing at 177°C). It is used as a base for many aircraft grade epoxy composites. It has high viscosity at room temperature and appears as a very viscous liquid. Its viscosity is about 20,000 centipoise at 50°C. Heat is usually required to reduce the viscosity of this resin in order to wet the fibers.

3.2.1. Epoxy Chemical Structure and Polymer Formation

Epoxies are characterized by the presence of an epoxy group—a three-membered ring with two carbons and an oxygen—as shown in Figure 2.8. The epoxy group usually occurs at the ends of the molecule. This epoxy group is the site of the crosslinking reaction in roughly the same way as the carbon-carbon double bond in polyesters, although the crosslinking reactions are quite different.

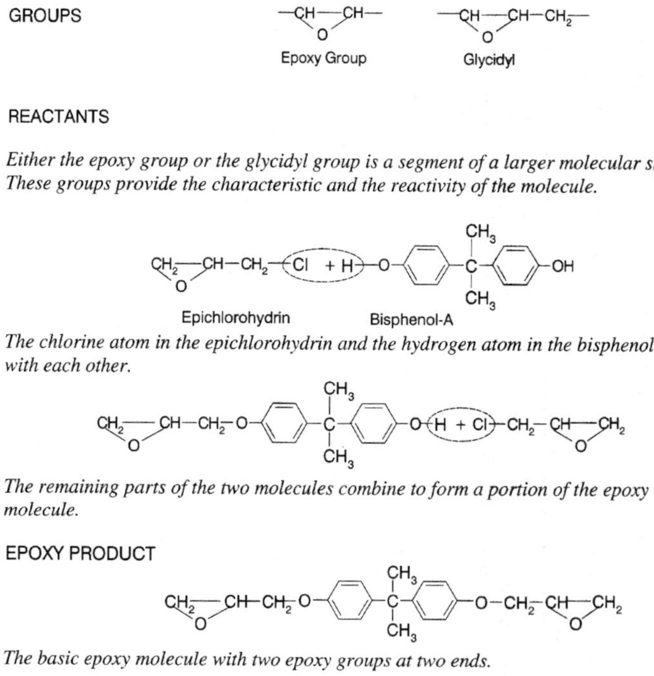

GROUPS
Epoxy Group
Glycidyl
REACTANTS
Either the epoxy group or the glycidyl group is a segment of a larger molecular structure. These groups provide the characteristic and the reactivity of the molecule.
Epichlorohydrin
Bisphenol-A
The chlorine atom in the epichlorohydrin and the hydrogen atom in the bisphenol-A react with each other.
The remaining parts of the two molecules combine to form a portion of the epoxy molecule.
EPOXY PRODUCT
The basic epoxy molecule with two epoxy groups at two ends.

FIGURE 2.8 Epoxy groups and reaction to form epoxy [3].

The properties of the crosslinked polymer are more dependent upon the choice of curing systems in epoxies than in polyesters, so both the nature of the epoxy and of the curing system must be understood. Many epoxies use the slightly modified epoxy group with one additional carbon. This group is called the glycidyl group and is illustrated in Figure 2.8.

In addition to being the site that is used for crosslinking, the epoxy group is the reactive site and provides for good adhesion with the reinforcement or with the surface of another material.

3.2.1.1. Diglycidyl Ether of Bisphenol A (DGEBPA)

The most common of the epoxy resin systems is based upon the condensation polymer formed from epichlorohydrin and bisphenol A and is called *diglycidyl ether of bisphenol A* (DGEBPA) (see Figures 2.8 and 2.9). The reaction to form the resin from the reactants is shown in Figure 2.9. In this figure, the epichlorohydrin molecule is made to react with a bisphenol A molecule. Hydrogen will combine with chlorine to form hydrochloric acid. The epoxidyl group will connect with the bisphenol. Reacting with another epichlohydrin will form another molecule of hydrochloric acid and a basic segment of the epoxy is formed. This epoxy molecule has two epoxy ends. The reaction will continue to form epoxy molecules with longer length.

The number of repeat units in epoxy is generally small (less than 20) and the average molecular weight can therefore be fractional. For molecules of short chain lengths, the resin is a liquid. Generally, processing becomes difficult as the molecular length is increased although the physical and mechanical properties also increase. Therefore, a compromise must be made between processability and properties.

3.2.1.2. Specialty Epoxy Resins

Apart from the base resin DGEBPA, there are other epoxies that have been developed for specialty applications. A few of these are shown in Figure 2.10 along with their trade names. If one wants to have better mechanical strength, better chemical resistance and higher operating temperature, one can have more benzene rings in the molecule, as in the case of epoxidized phenolic novolacs (one trade name is DOW DEN 438), or tetraglycidylmethylene dianiline (THMDA) (trade name: CIBA MY-720). The presence of halogen atoms such as chlorine (Cl) or Bromine (Br) usually imparts flammability resistance. However these molecules emit toxic gases upon burning.

3.2.1.3. Diluents

In addition to the epoxy resins, sometimes diluents are also added. These serve to reduce the viscosity of the resin, to improve shelf or pot life, to lower the exotherm (exotherm is the peak heat release during the curing process; large exotherm can bring the peak temperature of the

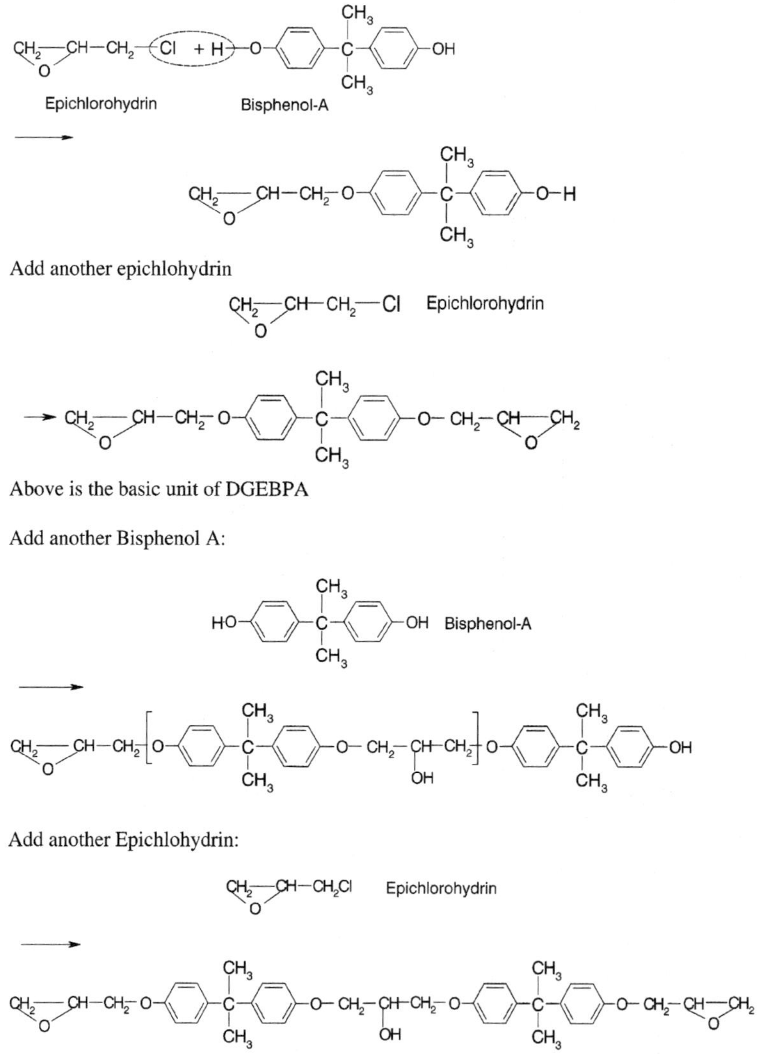

FIGURE 2.9 Epoxy groups and reactants for epoxy polymers.

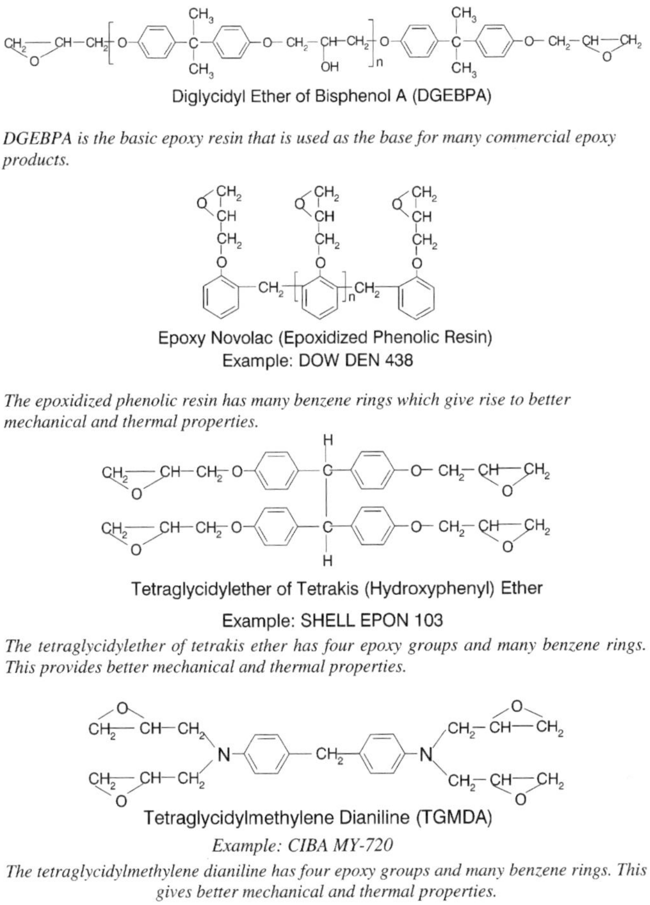

Diglycidyl Ether of Bisphenol A (DGEBPA)
DGEBPA is the basic epoxy resin that is used as the base for many commercial epoxy products.
Epoxy Novolac (Epoxidized Phenolic Resin)
Example: DOW DEN 438
The epoxidized phenolic resin has many benzene rings which give rise to better mechanical and thermal properties.
Tetraglycidylether of Tetrakis (Hydroxyphenyl) Ether
Example: SHELL EPON 103
The tetraglycidylether of tetrakis ether has four epoxy groups and many benzene rings. This provides better mechanical and thermal properties.
Tetraglycidylmethylene Dianiline (TGMDA)
Example: CIBA MY-720
The tetraglycidylmethylene dianiline has four epoxy groups and many benzene rings. This gives better mechanical and thermal properties.

FIGURE 2.10 Major epoxy resin systems [3].

resin to dangerously high level such that damages due to burning can occur), and to reduce shrinkage.

For most applications, diluents that will react with the resin and become an integral part of the cured system are preferred. Such reactive diluents include butyl glycidyl ether, cresyl glycidyl ether, phenyl glycidyl ether, and aliphatic alcohol glycidyl ethers.

3.2.2. Curing Systems for Epoxies (Hardeners)

Three curing systems are of primary importance for epoxies: amines; anhydrides; and tertiary amines and accelerators.

3.2.2.1. Amines

The most common curing system is the amine, in which one of the amine hydrogens (an amine hydrogen is a hydrogen atom that is attached to a nitrogen atom) reacts with the epoxy ring to form a hydroxyl group which can then react with another group to crosslink the chains. These reactions are addition reactions, so no byproduct is formed as would be formed in a condensation reaction. Primary amines (RNH_2), which contain two active amine hydrogens, are therefore capable of reacting with two epoxy groups and achieving a greater crosslink density (number of crosslinks per polymer) than single functional hardeners. This greater crosslink density generally improves physical properties, although at the expense of flexibility. If the amine is aromatic, the overall stiffness, low shrinkage, and temperature capability is improved, although toughness is sacrificed. The chemical structures, and viscosities of the commonly used amine curing agents are shown in Table 2.5.

An amine is a substance that contains nitrogen, of which there are three types. A molecule containing a reactive group consisting of nitrogen that is attached to two hydrogens, is called a primary amine. When the reactive group consists of a nitrogen that is attached to one hydrogen, this is called a secondary amine. When there is no hydrogen attached to the nitrogen, this is called a tertiary amine.

It can be seen that many of the structures given in Figure 2.10 and Table 2.5 include reactive groups at each end. These permit the formation of crosslinks between epoxy molecules. For example, an amine end group with two hydrogens on the nitrogen (a primary amine) reacts with the epoxy molecule as follows:

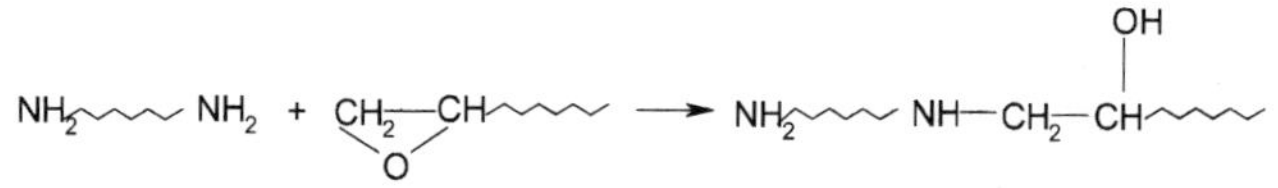

When another amine hydrogen combines with a second epoxy molecule, a crosslink is formed. The curing agents in Table 2.5 that contain secondary amine end groups (one hydrogen on the nitrogen) react in much the same way.

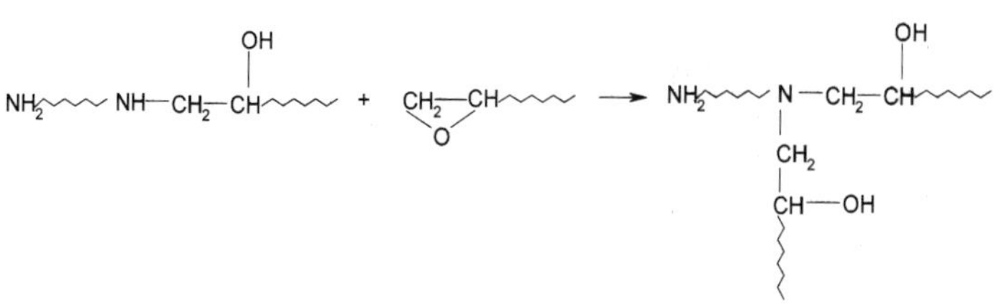

TABLE 2.5 Structures and Characteristics of Commonly Used Amine Curing Agents [1].

Amine Curing Agents	Amine Hydrogen Equivalent Weight (g/eq)	Viscosity @ 25°C (77°F), Pa.sec (cP)
Diethylenetriamine (DETA) $H_2N-CH_2-CH_2-NH-CH_2-CH_2-NH_2$	20	0.0055–0.0085 (5.5–8.5)
Triethylenetetramine (TETA) $H_2N-((CH_2)_2-NH)_2-CH_2-CH_2-NH_2$	24	0.020–0.023 (20–23)
Diethylaminepropylamine (DEAPA) $(CH_3-CH_2)_2N-(CH_2)_3-NH_2$	65	< 5.0

Amine Hydrogen Equivalent Weight (g/eq) and Epoxy Equivalent Weight

In Table 2.5, there is an expression of amine hydrogen equivalent weight (g/eq). What this means is the weight of the amine molecule over the number of hydrogen atoms that are attached to the nitrogen atoms. For example the DETA molecule has five hydrogen atoms that are attached to the nitrogen. These hydrogen atoms are called the amine hydrogens. The mass of the DETA molecule consists of 4C + 3N + 13H = 103 g/mol. Dividing this number by five gives the amine hydrogen equivalent weight of 20.6. The number reported in Table 2.5 is rounded off to 20.

The hydrogen equivalent weight is useful to determine the mass of amine required to react with a certain mass of epoxy for full reaction, as will be discussed later in Example 2.4.

A similar definition can be used for epoxy equivalent weight. This is obtained by dividing the molecular mass of the epoxy molecule by the number of epoxy groups in the molecule.

For thorough crosslinking, the hydrogens of the primary and secondary amines should be matched 1:1 with the epoxy groups. The amounts of curing agent and epoxy resin needed in order to obtain the 1:1 stoichiometric quantity are calculated as follows:

$$\text{Parts by weight of amine to be used with 100 parts by weight of epoxy resin (phr)} = \frac{\left(\dfrac{\text{Molecular weight of amine}}{\text{Number of available amine hydrogens per molecule}}\right)}{\text{Epoxy equivalent weight}} \times 100$$

FIGURE 2.11 Appearance of DDS amine curing agent.

Figure 2.11 shows the appearance of the DDS (one type of amine) curing agent. It appears in powder form at room temperature.

3.2.2.2. Anhydride Curing Agents

The amine curing agents work well with epoxy resins. However, these amines may not be environmentally friendly. The alternative to amine curing agents is the anhydride curing agent, which has a closed ring. One example is phthalic anhydride (PA)with the chemical formula as shown below. The characteristic of the anhydride is the closed ring O=C–O–C=O. This ring needs to be opened (usually by catalysts) for the

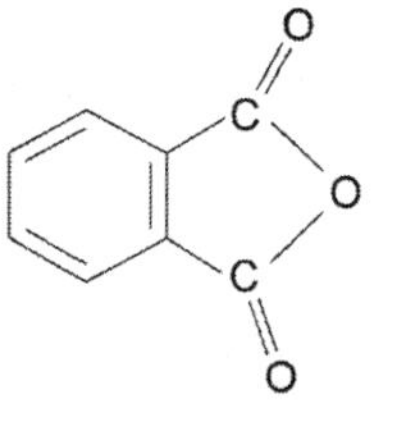

phthalic anhydride (PA)

linking reaction to occur. Phthalic anhydride has an anhydride equivalent weight of 148 g/eq and a melting point of 130°C. The crosslinking reaction of anhydride with epoxy is shown schematically below.

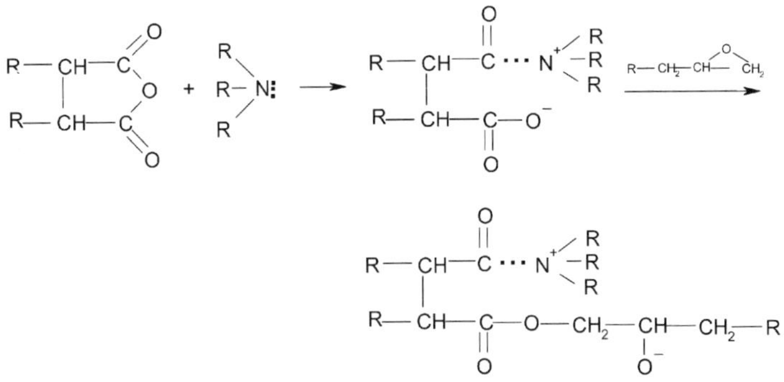

Theoretically, one anhydride group reacts with one epoxy group. The amounts of resin and curing agent that contain identical quantities of the two functional groups (i.e., 1:1 stoichiometrically) can be determined as follows:

$$\text{Parts by weight of anhydride to be used with 100 parts by weight of epoxy resin (phr)} = \frac{\left(\dfrac{\text{Molecular weight of anhydride}}{\text{Number of anhydride groups}}\right)}{\text{Epoxy equivalent weight}} \times 100$$

3.2.2.3. Tertiary Amines and Accelerators

Tertiary amines (no hydrogen on the nitrogen) are Lewis bases that cure epoxy resins in an entirely different manner than the primary and the secondary amines. The curing agent operates as a true catalyst by initiating a self-perpetuating anionic polymerization. They are added to an epoxy resin in small non-stoichiometric amounts that have been empirically determined to give the best properties.

This type of reaction is also termed *homopolymerization* because the epoxy molecules react among themselves and there is no curing agent that forms part of the final structure.

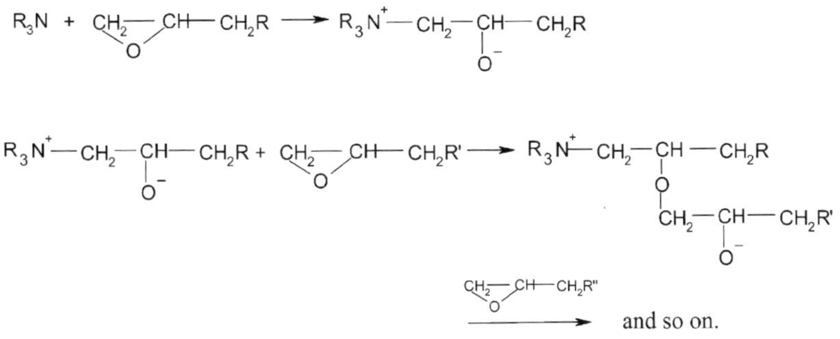

This homopolymerization of epoxy molecule to epoxy molecule results in a polyether. As such, this reaction is also called *etherification*. The ether linkages (C–O–C) are fairly stable against most acids (both organic and inorganic) and alkalis. Further, like ester linkages, they are more thermally stable than the carbon-nitrogen linkages formed by an amine cure.

One example of the accelerators is 2-ethyl-4-methyimidazole (EMI) with an amine hydrogen equivalent weight of 110 g/eq, a melting point of 25°C and a viscosity of 4000–8000 cP. EMI is a very efficient accelerator. It produces the highest degree of crosslinking and the highest heat distortion temperature.

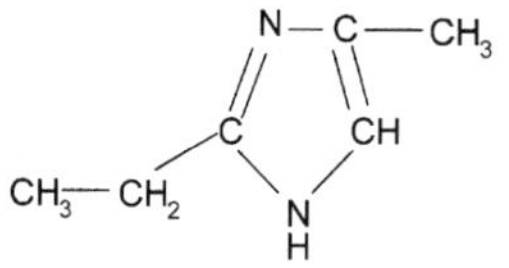

EMI accelerator

Another example is boron trifuoride methyl amine (BF_3MEA) with an amine equivalent weight of 28 g/eq and a melting point of about 208°C. This Lewis acid has achieved great popularity as a curing agent for epoxy resins in composites. Added in small amounts to the epoxy resin alone, it functions as a catalyst by cationically homopolymerizing the epoxy molecules into a polyether. Boron trifluoride causes very rapid (occurring in minutes) and very exothermic polymerization of the epoxy resin, and blocking techniques to halt the room temperature reaction must be used when other than very small amounts of the resin are being used. When blocked with monoethylene amine to form the complex BF3MEA, boron trifluoride is a latent curing agent at room temperature, but becomes active above 90°C causing rapid cure of the epoxy resin with a controllable release of heat. For prepregs, which are often stored for weeks before fabrication into a component, the use of latent curing agent is an absolute necessity. Epoxy resin systems containing BF_3MEA are popular for use in potting, tooling, laminating, and filament winding. It has been found, however, that prepregs and cured systems containing BF_3MEA generally have poor resistance to humidity [3].

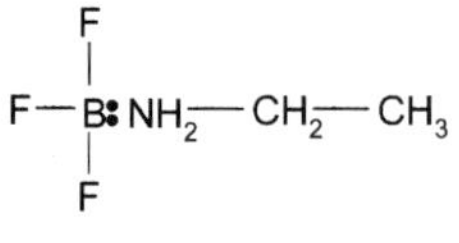

boron trifluoride-methyl amine

Normally the primary amine is preferred for the reaction over the secondary amine. However this also depends on the temperature. High temperature can activate the reaction with the secondary amines. Also the close proximity between the epoxy groups and the amines also has some influence. Usually at the beginning of the reaction the primary amines are involved. As the reaction proceeds, temperature is increased and the secondary amines may start to react.

3.2.3. Relative Concentrations [3]

The relative concentration of epoxy groups in the resin (active epoxy sites) and reactive sites on the hardener is very important in determining the physical and mechanical properties of the epoxy resin system. These effects can be summarized into four categories:

1. Large excess of epoxy. With a large excess of epoxy groups over reactive sites on the curing agent, a non-crosslinked epoxy-amine adduct is the predominant product. The physical properties of such a product are generally not as good as if it were highly crosslinked.
2. One reactive epoxy site for each reactive hardener site. At one epoxy group for each hardener reactive site, a crosslinked, thermoset polymer is obtained with most physical properties at a maximum. This gives a multidimensional crosslinked matrix because of the multiple crosslinking sites available.
3. One epoxy molecule for each hardener molecule. Because there are usually more reactive sites for each hardener molecule (typically four) than for each epoxy molecule (typically two), this case has slightly excessive hardener. When the concentration of curing agent reactive sites exceeds the number of epoxy groups, the material forms a thermoplastic resin and the physical properties again reduce and linear polymer results.
4. Large excess of hardener. The result is an amine epoxy adduct and the properties are those of thermoplastic polymer of low molecular weight. That is, poor physical properties compared to the other concentration options.

Example 2.4

Question

It is desired to cross link a DGEBPA epoxy resin using an amine curing agent called DETA. The formula for the two materials are as shown in Figure 2.8 and Table 2.5. How many grams of DETA should be used if 100g of the epoxy resin are used?

Solution

One repeat unit of the DGEBPA epoxy is shown at the bottom of Figure 2.8. Note that the symbol

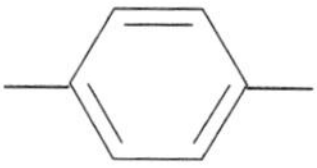

means

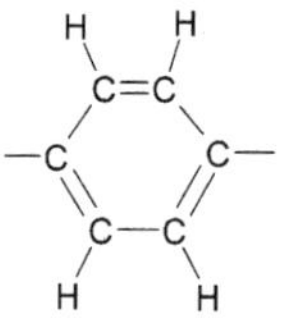

Counting the number of atoms, each unit of DGEBPA has 21 carbon atoms, 24 hydrogen atoms, and 4 oxygen atoms. The mass of the epoxy unit is therefore:

$m_1 = 21(12) + 24(1) + 4(16) = 340$ g/mole

The DETA molecule (Table 2.5) has 4 carbon atoms, 3 nitrogen atoms and 13 hydrogen atoms. The mass of DETA is:

$m_2 = 4(12) + 3(14) + 13(1) = 103$ g/mole

In the epoxy molecule, there are two epoxy groups, the epoxy equivalent weight is therefore:

$m_3 = 340/2 = 170$ g/mole

There are 5 available hydrogens in the amine molecule. The value of molecular weight of amine over the number of available hydrogens per molecule is:

$m_4 = 103/5 = 20.6$ g/mole

Parts by weight of amine to be used with 100 parts by weight of epoxy resin are:

$\% = 20.6/170 \times 100 = 12.1\%$

For 100 g of epoxy, the amount of amine curing agent to be used is: 12.1 g.

3.2.3. Cured Epoxy Resin Systems

Some generalizations can be made concerning the relationship between the chemical structure and properties of a cured epoxy resin:

Aromatic Rings:

- The greater the amount of aromatic rings 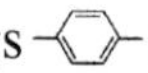 a cured epoxy resin contains, the greater its thermal stability and chemical resistance.
- Epoxy resins cured with aromatic curing agents are likely to be more rigid and often make a stronger cured product than those cured with aliphatic curing agents. However, these epoxy resin systems require higher cure temperatures because their very

rigidity reduces the molecular mobility needed to properly position two reactive end groups for reaction.

Crosslink Density

Crosslink density is defined as the number of crosslinks per volume of the material. The crosslink density depends on the mass of the molecular segments between the cross links and on the ratio of mass of curing agent over that of the epoxy. It has the following effects on resin system properties:

- A lower crosslink density can improve toughness (if strength is not significantly lowered) by permitting greater elongation before breakage.
- A lower crosslink density can also result in reduced shrinkage during cure.
- A higher crosslink density yields an improved resistance to chemical attack.
- A higher crosslink density also leads to an increase in the heat distortion temperature (and glass transition temperature), but too high a crosslink density lowers the strain to failure (increased brittleness).

3.2.4. Pot Life and Prepregs

The pot life (working life) also must be long enough to allow fabrication of the desired component without the complications of rapidly advancing cure.

Aliphatic amine curing agents react faster with epoxy resins than do

> ***In the manufacturing of composites, the resin usually is provided from the manufacturer in containers in the form of large drums. In the manufacturing process, it is necessary to transfer resin from the drum into smaller containers called pots. Curing agents are then added into the pot. While the resin system (resin and curing agent) is inside the pot, it should stay in liquid form for a certain amount of time. The manufacturing process should take place within this time. One should not wait too long so that the reaction has already taken place sufficiently and the viscosity becomes high. If the viscosity becomes high, the resin system is no longer of good usable liquidity. The time the resin stays in the pot and still retains its good liquidity is called pot life.***

aromatic amines. The former have pot lives ranging from minutes to a few hours, whereas the latter have long pot livesoften 24 hours or more.

Anhydride curing agents can produce long pot lives (e.g., 2 months for NMA). However when an accelerator is used, the pot lives of the epoxy-anhydrated systems can be as short as a few hours, depending on the amount and type of accelerator used.

Prepregs are normally stored at low temperatures to inhibit further reaction of the partially cured (B staged) resin until desired.

The term prepreg stands for pre-impregnated. Prepregs are made by running fibers that are uniformly separated by tows into a resin bath for wetting. The wetted fibers are then conditioned so that a small portion of the bonds in the resin is formed. In this state, the resin is a viscous liquid. This makes the prepreg a flexible sheet of fibers and resin. The resin is neither liquid nor solid [schematically represented in Figure 2.1(e)]. This state allows the prepregs to be rigid enough to be handled and yet liquid enough so that they can be draped onto a complex shaped mold and be bondable. One analogy to the prepregs is wet wallpaper, except that prepregs are sticky on both sides. In order to provide handlability, usually a piece of non-stick paper is placed on one side of the prepregs to prevent them from sticking to each other. As such, prepregs can be provided in roll form. Usually prepregs are shipped in refrigerated bags to slow down the reaction. Under storage, the resin in the prepregs may continue to crosslink. As such prepregs should be stored at low temperature (about −3°C). If enough bonding has occurred in the resin in the prepreg, the prepreg becomes too stiff for forming, and also the tackiness (stickiness) of the prepreg is no longer sufficient for further processing. The time that the prepregs can stay in storage before becoming stiff or non-sticky is called the shelf life.

For the manufacturing of aircraft composites, prepregs are usually used. A controllable and reproducible cure advancement that allows both sufficient pot life as well as shaping and arrangement of the prepreg is critically important.

3.2.5. Quality Control

The quality control of the resin takes place at two stages: the liquid stage and the solid stage.

3.2.5.1. Quality Control at the Liquid Stage

Before the liquid resin is used for making composite components, its quality should be checked. This is because resins or fillers may be blended with resulting mixtures of different qualities. Also if there is a significant amount of cure already taking place in the resin, then the resin does not have sufficient liquidity to flow to wet the fibers. Normally chromatography is used and viscosity of the resin is measured to determine the molecular weight/molecular weight distribution of the molecules in the resin and the state of liquidity of the resin.

3.2.5.2. Degree of Cure of the Solid Resin

In the manufacturing of polymer composites, it is very important to ensure that the resin cures properly. Proper cure (almost 100% cure) will give the proper mechanical properties such as strength, stiffness and hardness. On the molecular level, curing means that crosslinking has taken place. Complete curing means that all the crosslinks within the microstructure of the resin have already taken place. It is difficult to achieve 100% cure inside the composite material but a degree of cure approaching this limit (in the order of 95% or more) is desirable. The reason why it is difficult or it takes a long time to attain 100% cure is due to lack of accessibility of the species required for curing. If one epoxy end is at one location in the container while the amine end is at a location far away, being separated by other cured material, it would be very difficult for that epoxy end and amine end to react.

There are many methods used to monitor the degree of cure. These can be classified into methods using chemical, thermal, electrical or mechanical principles:

a. Method using chemical principle
  - Wet chemical or physical analysis method (solvent swell) is often used to directly measure the chemical reaction during cure. The molecules of a solvent for an epoxy can diffuse into the space between the epoxy molecules when the epoxy is still not completely cured. By applying the solvent onto the surface of a piece of the solid epoxy, if some portion of that epoxy is not yet completely cured, the diffusing action will take place and this usually gives rise to discoloration or gumminess of the material.
  - Infrared spectroscopy has been used to determine the degree of cure. Figure 2.12 demonstrates the use of near-infrared

spectroscopy for monitoring the appearance or disappearance of the epoxy groups with time. The infrared microscopy measures the vibration resonance of a certain bond inside the material. If a bond exists, it will show more resonance. If a bond disappears over time, the resonance will diminish.

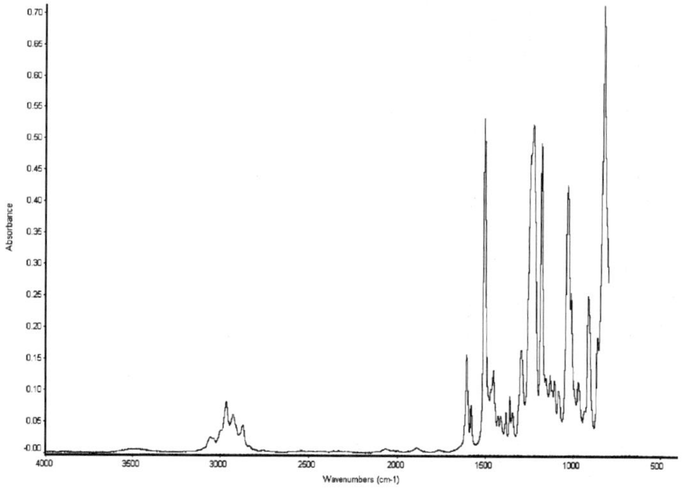

(a)

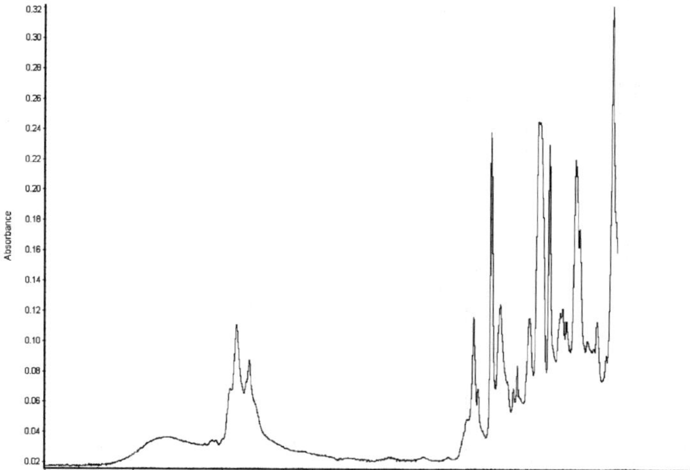

(b)

FIGURE 2.12 FTIR curve for Shell Epon 828. (a) Resin only, (b) Resin and curing agent 3046.

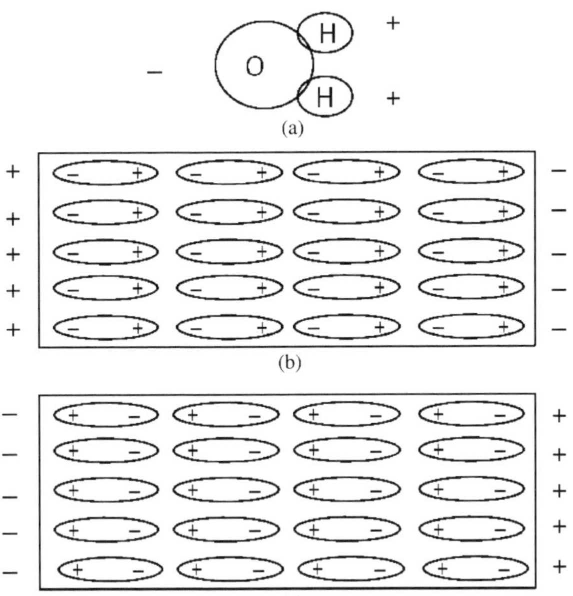

FIGURE 2.13 Switching of dipoles due to alternating electric field: (a) dipoles in water molecule, (b) two polar ends of the molecule in an electric field, and (c) switched dipoles due to switching of electric field.

b. Method using electrical principle

Dielectric measurement has become an increasingly popular cure monitoring technique. The mobility of the molecular segments and therefore the extent of cure is inferred by the response of molecular dipoles to an oscillating electric field. For polar molecules such as water, one end of the molecule has a positive charge and another end has a negative charge (Figure 2.13). If the molecules are subjected to an alternating electric field, the poles of the molecules will be switched back and forth. If the molecules are in the liquid state, it takes less energy to switch the dipoles. However, if the molecules are in a more solid state, it takes more energy for switching. By measuring the loss of energy due to the switching, one can infer the state of cure (or solidifying) of the molecules. Epoxy molecules also have positive locations (around the H atoms) and negative locations (around the oxygen atoms) and these dipoles are switched when the molecules are subjected to an alternating electric field. Figure 2.13 shows the switching of dipoles due to alternating electric field.

c. Method using mechanical principle

The degree of cure can be estimated from the mechanical properties of the resin. The available methods can be:

- Simple hardness test (Barcol Hardness test). In this technique, a simple hardness tester projects a sharp needle against the surface of the cured epoxy. If the epoxy is well cured, its surface is hard and the penetration of the needle is shallow. If the epoxy is not well cured, the penetration is deep. The machine gives the measure of the penetration. This method is good qualitatively and is useful for quick, quality control purposes.
- More complicated mechanical tests or sensitive dynamic mechanical tests such as the dynamic mechanical analyser (DMA). DMA is a machine that imposes mechanical and thermal loads onto the sample. The resistance of the sample against these loads provides the indication of stiffness which, in turn is a result of the degree of cure.
- Detection of the change of modulus of the material (E') or the damping capacity of the material (η) by the speed of propagation of the ultrasonic signals (ultrasonic cure monitoring). In this technique, ultrasonic signals are sent through the thickness of the sample. The time of flight of the signal through the sample is a function of its stiffness and its damping capacity. These in turn are indicators of the degree of cure of the sample.

d. Method using shrinkage measurement

As the resin cures, its volume decreases. There is a correlation between the degree of cure and the degree of shrinkage [4]. Monitoring shrinkage during cure can give an indication of the degree of cure.

e. Method using thermal principle

The thermal properties (such as thermal conductivity, specific heat, and heat content) of an epoxy resin depend on the extent of cure and can be used to assess cure advancement. Heat content measured by differential scanning calorimetry (DSC) has been used successfully to this end. The DSC machine enters a certain amount of heat into the sample and measures the net amount of heat coming out of the sample (heat out minus heat in). Figure 2.14 compares the DSC traces of a sample after being subjected to a few cycles of heating. The ordinate of the curve gives the net heat. Exotherm means net heat is coming out while endotherm means net heat is coming in. The area under the peak in the curve indicates the amount of heat. If there are uncured portions in the sample during the test, these portions will un-

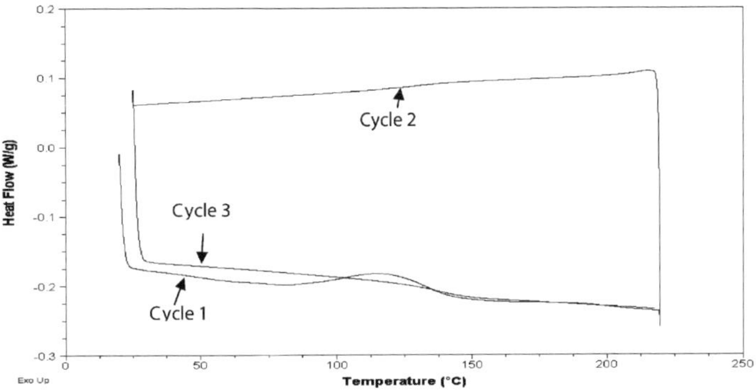

FIGURE 2.14 Comparison of DSC traces for two cycles of test on the same resin sample.

dergo curing during the test. The curing reaction will generate heat due to the formation of chemical bonds.

Figure 2.15 shows schematics of the heat flow curves obtained from two different types of DSC tests: the dynamic scanning test and the isothermal scanning test. In the dynamic DSC test, the temperature increases continuously from the start temperature up to the maximum temperature. The heat is measured as a function of time. In the isothermal DSC test, the temperature is quickly brought from the start temperature to the test temperature. The sample is then held at the test temperature for the duration of the test. Heat is measured as a function of time during the isothermal hold. Figure 2.16 shows the actual heat flow curve for an aircraft resin system.

The important information to obtain from curves such as those in Figure 2.15 is the equation modeling the rate of cure $d\alpha/dt$ where α is the degree of cure. This is essential for the heat transfer model for

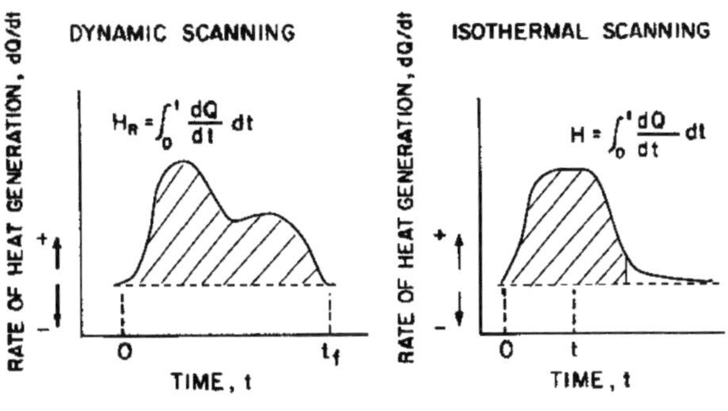

FIGURE 2.15 Schematic of heat flow curves for different types of DSC tests.

the processing of the composites as will be discussed in Chapter 4. The rate of cure is usually expressed as: $d\alpha/dt = H_R(dH/dt)$where H_R is the total amount of heat generated during the whole curing process (this is the area under the whole curve). It can be seen from Figure 2.15 that the rate of heat generation dH/dt is a function of time. This is because the degree of cure changes as time progresses. In addition the rate of heat generation is also a function of temperature. One can therefore write a general expression for the rate of cure as $d\alpha/dt = f(\alpha,T)$. By carrying out an isothermal DSC test, one can keep the temperature fixed, and the rate of cure will be only a function of the degree of cure. One can see from the isothermal curve in Figure 2.15 that the rate of heat generation (or rate of cure) is a function of time. The rate first increases with time up to a certain maximum and decreases. This shows that for low degree of cure, the rate of cure increases up to a certain value of the degree of cure. As the degree of cure gets larger, the rate of cure decreases. A significant amount of work has been done on the modeling of the kinetic behavior of the resin (equation for rate of cure as a function of the degree of cure and temperature. More details of this are presented in Chapter 4.

3.2.6. *Use of Epoxies*

One of the reasons for the widespread use of epoxy systems in advanced composites is its adaptability to manufacturing methods. The

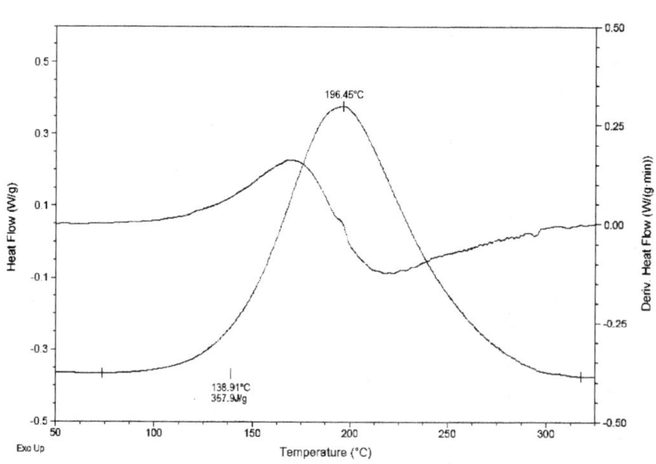

FIGURE 2.16 Heat flow and heat flow rate for Shell Epon 862 resin.

large variety of epoxy types and cure systems commercially available allow the cure rates and performance to be tailored.

High performance aircraft applications use a material form called a prepreg in which fibers are sold pre-impregnated with resin. The fibers can be unidirectional or woven. These prepregs can be made using epoxies and cure systems that are not reactive at freezer temperatures to provide long storage times for customers, yet can be cured quickly at high temperatures and provide service temperature capabilities of 121°C–177°C.

For prepregs, normally an autoclave needs to be used. Recently there is a push toward out-of-autoclave curing process in an effort to cut down the processing time. Liquid composite molding processes have been developed. For these processes, the resin needs to have low viscosity (less than 1000 cP). Special resin systems (such as RTM 6, CYCOM 890) have been developed for these applications.

Epoxy systems have also been developed for rapid cure even under cold conditions (such as gluing reflectors to road surfaces). In general, the temperature capabilities of the cured resin will be similar to the cure temperature. Room temperature cured resins will soften somewhat at temperatures just above room temperature. High temperature capability requires high temperature cures.

In the part, the high degree of crosslinking needed to achieve temperature performance reduces the toughness of the epoxy resins. However, recent advances have boosted the level of toughness of composites based on epoxies to equal or surpass the toughness of other resin systems. Techniques to improve toughness of epoxies include the incorporation of rubber particles, thermoplastic particles or nanoclays. The addition of rubber improves the toughness significantly but with reduction in glass transition temperature. The introduction of thermoplastic or nanoclay particles can provide enhanced toughness without significant impact on operating temperature. In addition, these higher toughness systems can be processed like the earlier, brittle systems. In this way, high performance composites can be fabricated with existing facilities and equipment designed for thermoset chemistry without the modification needed to handle a different curing technology.

3.3. Vinyl Ester Resins

3.3.1. Chemical Definition

Vinyl ester resins are thermosetting resins that consist of a polymer backbone R′ with a reactive termination. The general formula for the vinyl ester resin is shown in Figure 2.17.

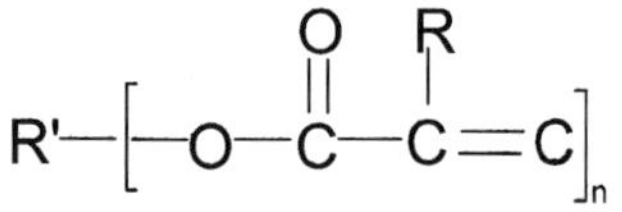

FIGURE 2.17 General formula for vinyl ester resin.

In Figure 2.17, R′ represents the rest of the molecule and R represents either the atom H (for acrylate vinyl ester) or CH_3 (for methacrylate vinyl ester). Vinyl ester resins have intermediate properties between polyester resins and epoxies. The cost of vinyl ester resins are also intermediate between the cost of polyesters and epoxies.

Figure 2.18 shows the chemical formula for a few commonly used vinyl ester resins. In this figure, the top molecule is bisphenol A epoxy based (methacrylate) vinyl ester resin. The phrase bisphenol A comes from the presence of two phenol A groups in the middle of the molecule. The word methacrylate comes from the existence of the CH_3 group at the end of the molecule. Note the star symbols (*) placed below the second carbon from the ends of the molecule. These symbols are used to denote that these are the reactive locations of the molecule. During the curing reaction (or crosslinking reaction), the double bond associated with these carbon atoms opens up and connects with the curing agent. Also in this molecule, the number n denotes the number of times the repeating unit will repeat itself. The range of $n = 0$–3 indicates the range of the size of the molecule. Other molecules in Figure 2.18 can be understood in a similar way.

Although vinyl ester resins have sometimes been classified as polyesters, they are typically diesters that (depending on the polymer backbone) contain recurring ether linkages. The backbone component of vinyl ester resins can be derived from an epoxide resin, polyester resin, and so on, but those based on epoxide resins are of particular commercial significance.

3.3.2. Vinyl Ester Resin Manufacture

Vinyl ester resins are produced by the addition of ethylenically unsaturated carboxylic acids (methacrylic or acrylic acid) (these molecules have the COOH termination) to an epoxide resin (usually of the bisphenol A-epichlorohydrin type). Molecules with the ring

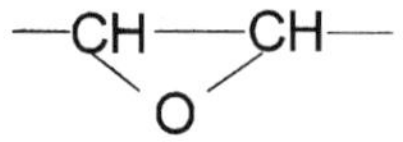

An illustration of the reaction between an epoxy resin and a carboxylic acid is shown in Figure 2.19. The reaction of acid addition to the epoxide ring (esterification) is exothermic and produces a hydroxyl group without the formation of byproducts (e.g., as in polyesterification, where water is produced.) As such the production of the vinyl ester resins is less messy than the production of polyester resins where condensation occurs. Appropriate diluents and polymerization inhibitors are added during or after esterification.

Epoxides that have been used to produce vinyl ester resins include the bisphenol A types (general purpose and heat resistance vinyl esters), and the tetrabromo bisphenol A types (fire retardant vinyl esters).

3.3.3. Curing Vinyl Ester Resins

Vinyl esters, like unsaturated polyesters, contain double bonds that react and crosslink in the presence of free radicals by chemical, thermal and radiation sources. Figure 2.20 illustrates the curing reaction. Initiators such as methyl ethyl ketone peroxide (MEKP) need to be introduced

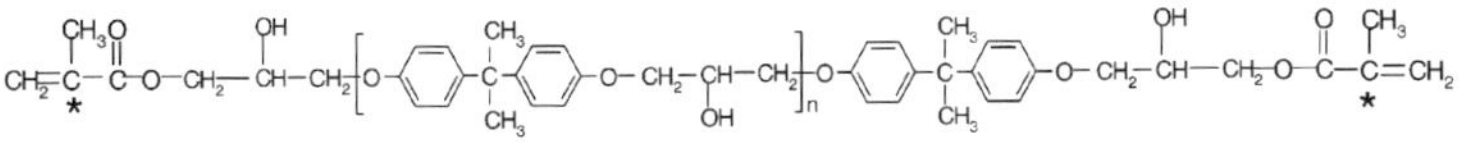

BISPHENOL A EPOXY BASED (METHACRYLATE) VINYL ESTER RESIN (*n* = 0-3)

This resin is called methacrylate due to the presence of the CH_3 group connecting to the C atom that has a radical (represented by the symbol (*) at the end of the molecule.

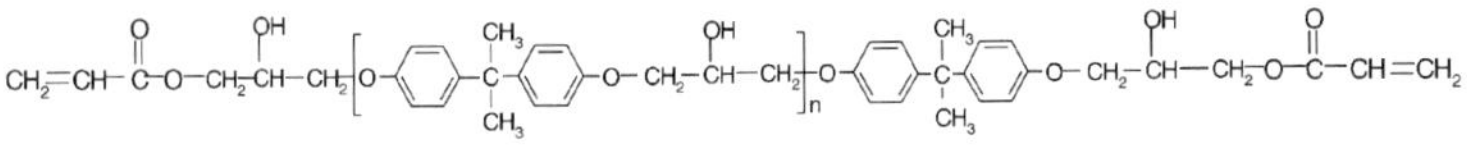

BISPHENOL A EPOXY BASED (ACRYLATE) VINYL ESTER RESIN (*n* = 0)

This resin is called acrylate due to the presence of the H atom that is connected to the C atom that has a radical (represented by the symbol (*) at the end of the molecule.

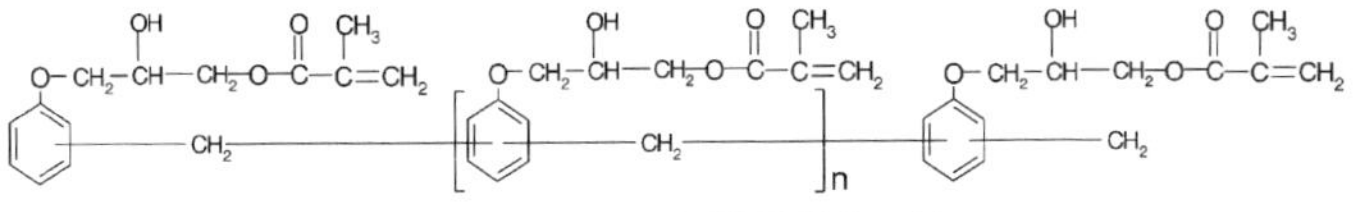

PHENOLIC-NOVOLAC EPOXY BASED VINYL ESTER RESIN (*n* = 0-1)

This resin has branches which give better mechanical and physical properties

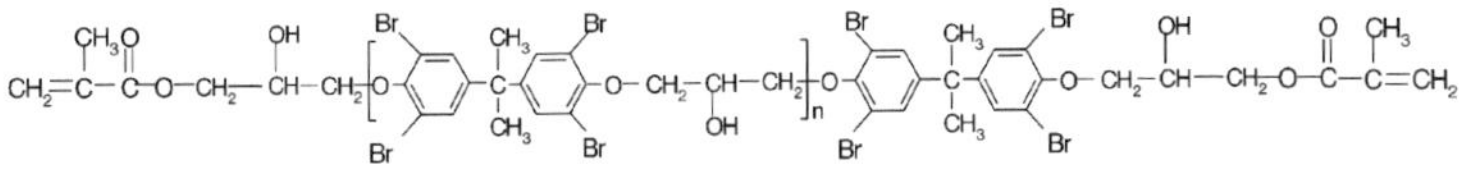

TETRABROMO BISPHENOL A EPOXY BASED VINYL ESTER RESIN (*n* = 0-3)

This resin has bromine which gives its flammability resistance. However bromine is toxic and is banned fromr use in Europe.

FIGURE 2.18 Examples of commonly used vinyl ester resins [3].

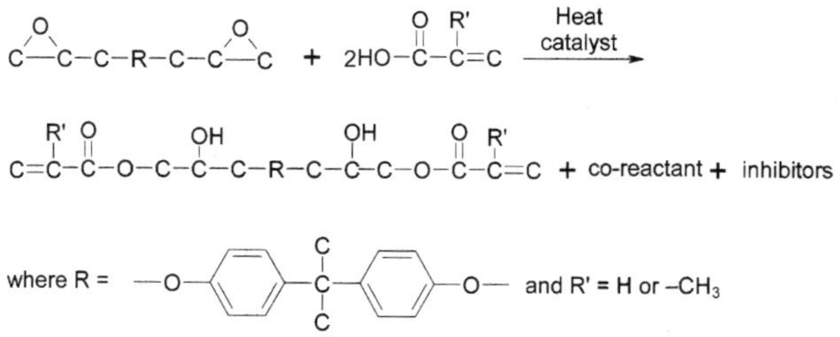

FIGURE 2.19 Reaction between an epoxy and a carboxylic acid to make vinyl ester [1].

to start the reaction. The curing proceeds by a free radical mechanism comprised of initiation (induction period), propagation, and termination. The rate-determining step is initiation, in which the catalyst swamps out the inhibitors, followed by reaction with the double bond of the vinyl esters and the curing agents (e.g., styrene).

3.4. Polyimide Resins

For applications requiring temperature stability up to 600°F (316°C), or even up to 700°F (371°C), the resin of choice is generally a polyimide rather than an epoxy, as epoxies have a normal upper use temperature of 350°F (177°C) to 400°F (204°C). Polyimides are higher in cost than the epoxies. They are often limited to aerospace applications where the higher performance can justify the increase in cost.

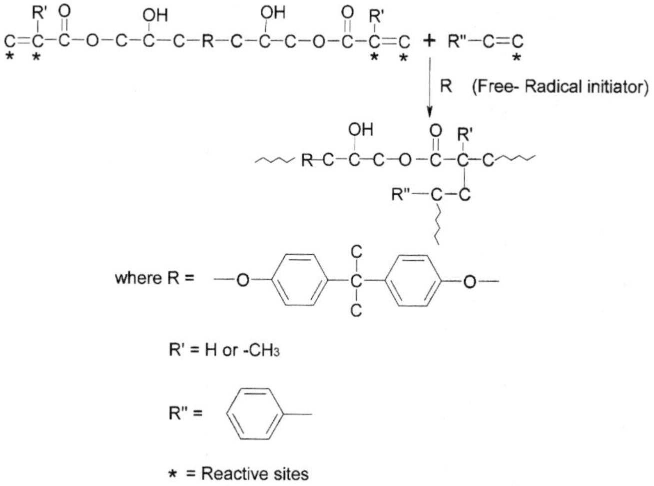

FIGURE 2.20 Curing reaction of vinyl ester resin [1].

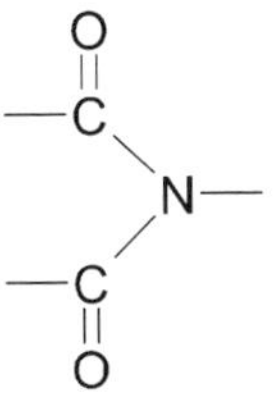

FIGURE 2.21 Chemical formula for the imide group.

Polyimides are characterized by the cyclic group containing a nitrogen and two carboxyl groups (carbons that are double-bonded to oxygens). The imide groups are structurally planar and are very rigid, which contribute to the high stiffness and stability of polymers containing these groups. Figure 2.21 shows the imide chemical group. Usually large aromatic groups are incorporated into the polymer to further enhance the temperature and stiffness properties.

3.4.1. Bismaleimide (BMI) Resins

Another type of addition polymerization is based on reactions involving bismaleimide (BMI) derivatives. The major advantage of BMI resins is that this resin can be cured at 350°F (177°C), which is the same range as is used for epoxies, therefore standard process equipment used for epoxies can be used. However, a postcure at 475°F (246°C) must be performed on the BMI resins to complete the polymerization and achieve properties that are significantly higher than epoxies. For postcure equipment, only an oven is required, without the need for vacuum and pressure as normally required for an autoclave. This is because the part already has taken shape after the preliminary curing.

3.5. Phenolic and Carbon Matrices

For high temperature applications, phenolic thermoset resins and carbon matrices can be used. Phenolic resins are manufactured by the reaction between phenol and formaldehyde. The resin is brittle and has high shrinkage. As for most applications, fillers need to be added. The vast majority of phenolic resin is filled with particular materials such as sawdust, nutshells, or carbon black. Phenolics have been used for many years as a general, unreinforced thermoset plastic in applications such as electrical switches, junction boxes, automotive molded parts, consumer appliance parts, handles, and even billiard balls. When subjected to high temperature, phenolic resins ablate (transform directly from solid to gas). The principal use for long fiber reinforced phenolic resins is for rocket nozzles, such as nose cones, where the ablative nature of the phenolic can be utilized.

Carbon matrix (carbon/carbon) composites have applications that are similar to phenolics and are used when phenolics may not provide the specific characteristics of very high temperature protection, or, occasionally, toughness that is needed. Applications of carbon matrix composites include rocket nozzles, tiles for space shuttle nose cones, aircraft, race car and truck brakes. Both the opposing parts are made of carbon matrix material reinforced with carbon fibers. This application takes advantage of the fact that carbon/carbon composites have the highest energy absorption (specific heat capacity) of any known materials.

Carbon/carbon composites are most often made from carbon fiber reinforced phenolics that may have been tailored slightly and then have been carefully charred. This process of charring is called pyrolysis. The charring process results in a porous structure because of the vaporization of certain parts of the phenolic polymer. The porous material is then reimpregnated with pitch, phenolics, or directly with carbon by vapor deposition. The resulting material is then carefully pyrolyzed again and the process is repeated as many times as necessary to fill the remaining void space with material. The entire process may take as long as six months because each charring step must be done slowly to avoid rapid vaporization and possible matrix damage.

4. THERMOPLASTIC MATRIX

As mentioned in Chapter 1, the composition of thermoplastic resins is different from that of thermoset resins. Whereas thermoset resins depend on the crosslinking of many small molecules to provide rigidity to the material, thermoplastic resins depend on the size (length) of the molecules (these molecules are non-reactive) to provide stiffness and strength for the material. Due to this inherent difference in microstructures, thermoplastic resins have advantages and disadvantages as compared to thermoset resins. The advantages of thermoplastic resins in thermoplastic composites are:

- There is no shelf life issue. The thermoplastic resin can be placed on the shelf infinitely without concern about the material becoming hard.
- The processing cycle can be short. Once the preforms have been arranged into a mold, the heating and cooling steps can be done on the order of minutes (rather than hours for thermosets).

- Thermoplastic composites have higher ductility and fracture toughness as compared to thermoset matrix composites.
- Thermoplastic resins can be melted and reused, making them recyclable.
- If there are defects in the thermoplastic composites, the material can be heated up to heal the damage.

The disadvantages of thermoplastic resin in thermoplastic composites are:

- The preform of thermoplastic composites (tapes made of thermoplastic resin and fibers) is fairly stiff and boardy. Placing sheets of these preforms requires efforts in alignment.
- The viscosity of thermoplastic resin is very high as compared to that of thermosets. This high viscosity requires high temperature and pressure for processing. It also creates the problem of occurrence of voids in the composite material.

The vast number of thermoplastic resins used in composite materials can be conveniently separated into two categories:

1. The traditional industrial thermoplastics that have been used with short fiber reinforcement for many years. Examples for these are polyethylene (PE), poly vinyl chloride (PVC), poly propylene (PP), polystyrene (PS), poly methyl methacrylate (PMMA), poly acrylonitrile butadiene styrene (ABS). The subject of traditional industrial thermoplastics has been covered extensively in the literature and it is not repeated here.
2. A new set of high performance thermoplastics that have been developed specifically for use in advanced applications. This is discussed in more detail below.

4.1. High Performance Thermoplastic Matrices

Several unique thermoplastic resin materials have been developed for use in composite materials. These resins generally have inherent thermal and mechanical capabilities beyond the conventional industrial thermoplastics. The high performance thermoplastics are usually more costly. The most important of these high performance thermoplastic matrices developed to date include the following:

- Polyetheretherketone (PEEK) and related molecules
- Thermoplastic polyimides

The current high performance resins are characterized by their high T_g's, which give good mechanical performance at temperatures much higher than the conventional thermoplastics, and in some cases, better than the polyester and epoxy thermosets.

4.1.1. Chemical Structure

The chemical structure of the most common high performance thermoplastics is given in Figure 2.22. The names of some of the molecules, such as PEEK, are taken from the type of linkage between the benzene rings. Therefore, PEEK has an ether (C–O–C), then another ether linkage, then a ketone linkage (C–C=O) and is named polyetheretherketone.

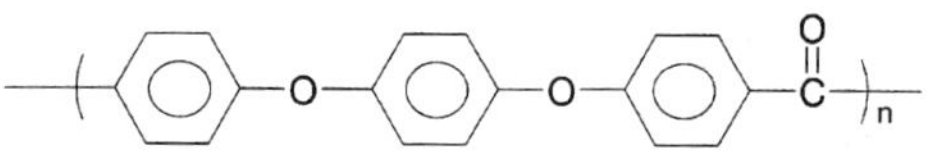

Polyetheretherketone (PEEK)

Note that the C–O–C bond is an ether bond andthe O–C=O is a ketone bond. There are 2 either bond and one ketone bond in PEEK.

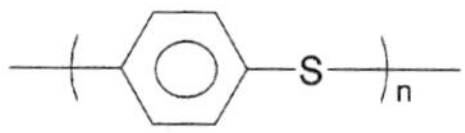

Polyphenylene sulfide

Note that C–S is a phenylene sulfide bond.

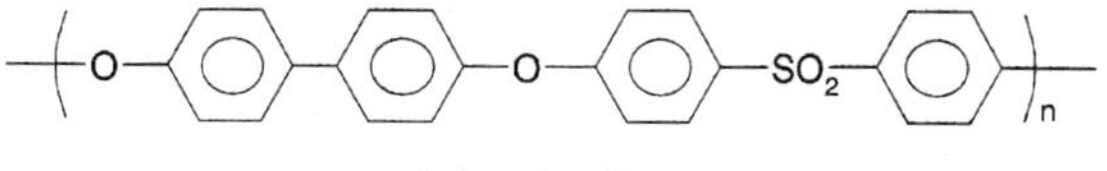

Polyaryl sulfone

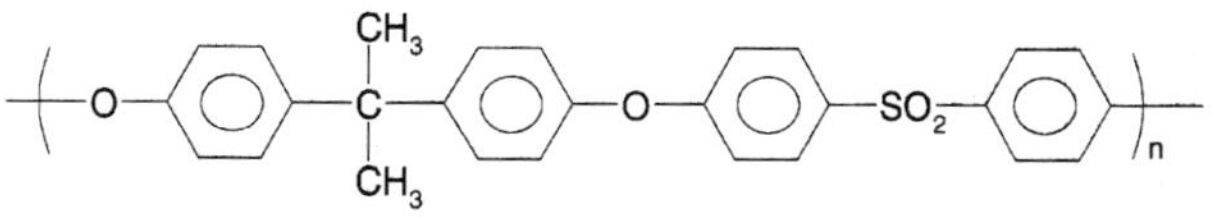

Polysulfone

Note that C-SO_2 is a sulfone bond.

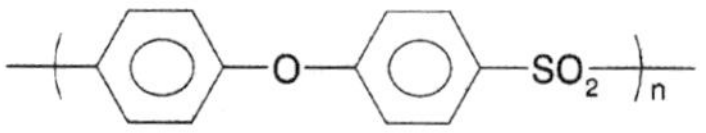

Polyethersulfone

The name of this molecule comes from the either bond C–O–C and sulfone bond C–SO_2

FIGURE 2.22 Chemical structure of a few high performance thermoplastics.

Other PEEK-related polymers are PEK with one ether linkage and PEKK with one ether linkage and two ketone linkages. Each of these polymers is characterized as highly aromatic, which increases the T_g and givse good thermal stability. The polymers generally have long repeating units that lead to a high degree of orientation in the liquid (melt) phase. Therefore, when these polymers solidify, they easily assume a crystal structure. This high crystallinity results in high modulus and low creep for thermoplastics because the tensile forces are immediately resisted by the rigid and strong crystal structure rather than by amorphous regions, as in many other thermoplastics. In addition, the crystal structure resists many solvents because the solvent molecules must also encounter these interatomic forces of the crystal structure to have any effect on the polymer. It must be remembered, however, that the polymers are not completely crystalline, so the effects of the amorphous regions can also be seen. Hence these thermoplastic polymers show much more elongation than thermosets (30%–100% versus 1%–10%) and this results in far greater toughness and impact resistance. This toughness is more pronounced at higher temperatures. In fact, at low temperatures, the thermoplastic becomes brittle.

The properties of a few commercial high performance thermoplastic resins are shown in Table 2.6. In this table, polyetherimide (PEI) has a high T_g with moderate processing temperature but low environmental resistance. Polyphenylene sulfide (PPS) has moderate processing temperature and good environmental resistance but low fracture toughness and low T_g. Polyetheretherketone has good fracture toughness, high T_g, good environmental resistance, and a large database but high processing temperature and high cost. A relatively new material, polyetherketoneketone (PEKK), has high T_g, moderate processing temperature, high toughness, and good environmental resistance, but a limited amount of information is available.

4.1.2. Fabrication of Composites

Because of the high melting point of these resins, they cannot be processed using typical plastics processing equipment unless this equipment has been upgraded to handle higher temperatures and pressures.

Many of the typical thermoset processes, such as filament winding and pultrusion, are also difficult to apply to high performance thermoplastics because of the high viscosity of their melts, 10^4–10^7 poise versus 1000 poise for noncured epoxy (molasses is about 10^5 poise).

The high viscosity in thermoplastics makes wet-out of the reinforcement very difficult. Increasing the temperature lowers the viscosity, but

TABLE 2.6 Properties of a Few Thermoplastic Resins (with permission from Cytec Engineered Materials Ltd.).

	PIE	PPS	PEEK	PEKK (DS)
Morphology	Amorphous	Semi-Crystalline	Semi-Crystalline	Semi-Crystalline
T_g (°C)	217	90	143	156
Typical Process Temperature (°C)	330	325	390	340
Comments	✓High T_g ✓Moderate Processing Temperature * *Environmental resistance*	✓Excellent environmental resistance ✓Moderate Processing Temperature * *Low* T_g * *Low Toughness* * *Poor Paint Adhesion*	✓Extensive database ✓Excellent environmental resistance ✓High toughness * *High process temperature* * *High polymer cost*	✓Excellent environmental resistance ✓High toughness ✓Lower process temperature than PEEK ✓Bonding and painting * *Limited database in composite form*

too high temperature can cause decomposition before sufficiently low viscosities are reached. High consolidation pressures are commonly employed to enhance flow of the resin.

4.1.3. Shear Thinning

One method to alleviate this wet-out problem is to take advantage of the shear sensitivity of some thermoplastics. These resins are non-Newtonian and undergo shear thinning. Therefore reinforcements can be impregnated by forcing the resin and fibers through a die at high temperature and under conditions that create high shear.

Strategies addressing the high viscosity issue for the manufacturing of thermoplastic composites will be discussed in Chapter 8.

5. FILLERS, COLORANTS AND OTHER RESIN MODIFIERS

Fillers, colorants and other resin modifiers are added into the resin for various purposes. These may produce side effects such as the reduction of mechanical properties or adhesive ability of the resin. As such they are only added into resins used for applications where load bearing is not critical. Normally these fillers are added into polyester or vinyl ester resins but not very often in epoxy resins used for aerospace applications.

5.1. Reasons for Use of Fillers

5.1.1. Cost Reduction

The most common purpose for adding fillers to resin is to lower the cost since the fillers are generally much less costly than the resin or reinforcements. Common fillers used for this purpose are powdered inorganic materials such as calcium carbonate, talc, clay in aggregate form, alumina trihydrate and glass microspheres.

5.1.2. Shrinkage Reduction

Many of the fillers also help prevent problems such as shrinkage when curing the resins. In this application, the fillers are called *low profile additives* (LPA) and are most important in polyester formulation. LPA are normally particles of thermoplastic resins.

5.1.3. Improvement of Flame Resistance

Fillers may also impart flame resistance, especially if the fillers contain chlorine, bromine, iodine, or water molecules (hydrated). Nanoclay particles can also provide resistance against flammability.

5.1.4. Alter Mechanical Properties

Fillers can also influence stiffness, electrical properties, expansion coefficients, and uncured resin viscosity. Addition of carbon nanotubes can increase the electrical conductivity and thermal conductivity of the resin. Thixotropic fillers are added to increase the viscosity of the uncured resins.

5.2. Colorants, Dyes, and Pigments

Colorants, dyes, and pigments can be readily added to most resin systems. Addition of dyes or pigments affects the cure rate. Some dye systems are also affected by the temperature changes that occur during curing.

Some pigments affect the weather resistance of the composites. Carbon black and titanium dioxide, for example, will absorb or modify ultraviolet light and prevent or diminish ultraviolet degradation of the resin.

6. CERAMIC MATRICES

Ceramic materials are attractive due to their strong resistance against high temperature and corrosion. A high degree of chemical and thermal stability is characteristic of the oxides, carbides, nitrides, and borides that form the basis for ceramic materials. However, as mentioned in Chapter 1, due to their high surface energy, it is difficult for ceramic in liquid form to adhere to fibers. As such the number of available systems is limited. One example is carbon/carbon composites where the carbon matrix is deposited on a bed of carbon fibers by chemical vapor deposition. Other forms are short fibers incorporated into a bulk ceramic for fracture toughness improvement.

7. METAL MATRIX [3]

Due to their high surface energy, it is difficult for liquid metal to bond with fibers. However a few systems of metal matrix composites have been developed. High pressure and introduction of roughness on the surface of the fibers is usually used to compensate for the lack of wetting. Metal matrix composites (MMC) have the advantage of high temperature resistance. They can be classified into two forms: short fiber metal matrix composites and long fiber metal matrix composites.

Short fiber MMC consist of high performance reinforcements (such as SiC) in a metallic matrix (aluminum, titanium, magnesium, copper). The reinforcements can be in the form of particles, whiskers, or fibers. MMCs have been used (or investigated for use) as piston ring inserts, pistons, connecting rods, impellers, brake calipers, sway bars, critical suspension components, and tennis racquets. The incorporation of a second phase (the reinforcement) into a metal significantly affects the propagation of pressure waves through the material by acting as sites for scattering and attenuation. This provides good dynamic damping capacity. The high thermal conductivity of the metal matrix, combined with the low thermal expansion coefficient of the composite, allow the design of high performance heat sinks for electronic applications. Carbon reinforced copper is of particular interest for this application.

To manufacture long fiber reinforced MMCs, continuous fibers are aligned to form a bed of fibers. Metal matrix is introduced by casting, hot molding, or plasma spraying around the fibers. Individual layers can be formed and then stacked in any ply orientation to give the desired mechanical properties.

8. REFERENCES

1. Lubin G. *Handbook of Composites,* Van Nostrand Rheinhold, 1982.
2. Meyer R.W. *Handbook of Pultrusion Technology,* Chapman and Hall, 1985.
3. Strong A.B., *Fundamentals of Composites Manufacturing,* Society of Manufacturing Engineers, 1989.
4. Hoa S.V., Ouellette P. and Ngo T.D. Monitoring cure of epoxy resins *J. of Composite Materials,* accepted for publication 2008.

9. HOMEWORK

1. Show the chemical reaction that forms a polyester made of maleic acid and propylene glycol. Also show the chemical reaction for crosslinking using styrene.

2. It is desired to make a polyester using 100 g of maleic acid and propylene glycol. A stoichiometric amount of propylene glycol is used. Crosslinking is done using styrene.

   a. After the condensate is removed, how many grams of polyester are obtained?

   b. How much heat is generated?

CHAPTER 3

Reinforcements—Fibers

1. GENERAL

Reinforcements for composite materials can be in the form of fibers, particles, or flakes. Each has its own unique application, although fibers are the most common in composites and have the most influence on properties. The reasons for this, discussed in Chapter 1, are as follows:

- The large aspect ratio (length over diameter) of the fiber, which gives rise to effective shear stress transfer between the matrix and the reinforcement
- The ability for fibers to be bent to sharp radius, allowing them to fit into sharp radius of curvature of the parts.
- Numerous techniques to manufacture fibers such as spinning, chemical vapor deposition, or oxidation.

The fibers used in composite materials appear at different scales. The manufacturing of fiber composites involves the use of fibers from the micrometer level up to the centimeter level. Figure 3.1 shows the different scales of fiber forms. At the smaller scale (level *a*) are the individual filaments with diameters of about 10 microns. These are usually bundled together to form tows consisting of thousands of individual filaments (level *b*). There are tows of 3 k (3000) filaments, 6 k, 12 k etc. These tows can then be combined either with or without the addition of resin for adhesiveness. If resin is used, the tows can be combined to form tapes. When resin is not used, the tows can be woven together to make dry woven fabrics, or the tows can be braided or knitted together to make dry fiber preforms (level *c*). The final composite part is made by placing many of

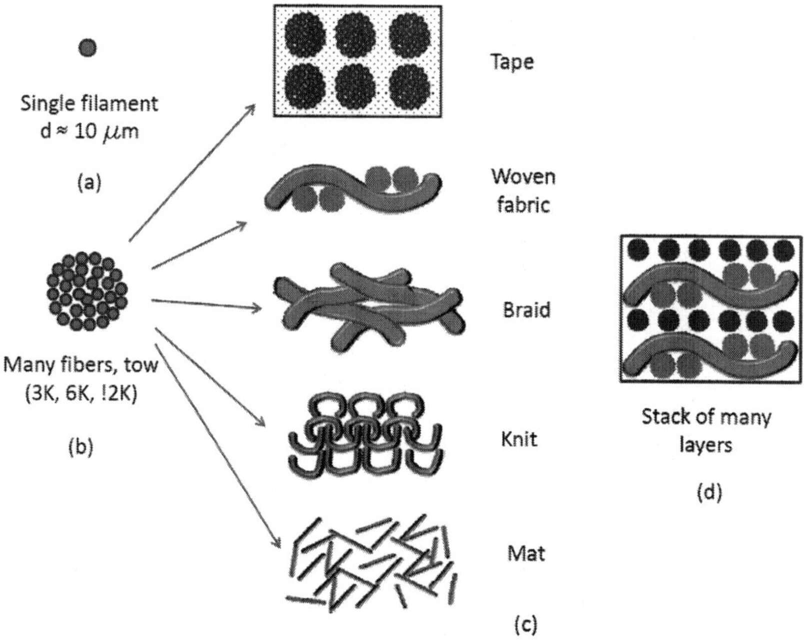

FIGURE 3.1 Fiber forms at different scales.

these layers on each other. Pressure is usually applied to press these fiber layers together (level *d*). Note that the load applied in this case is along the thickness direction (z direction) of the fiber bed. This is different from the situation where the load is applied within the plane of the fiber bed (such as along the fiber direction or transverse to the fiber direction as in normal operating conditions for load bearing composite structures). The load during the manufacturing stage is different from the loads during the operation stage and it is important to distinguish this.

The discussion on the fibers will be presented in the sequence of increasing scale level, beginning with the individual filaments.

2. INDIVIDUAL FILAMENTS

Individual filaments are normally produced by drawing from a melt of the material (such as glass) or by drawing from thermoplastic molecules and then chopping away secondary atoms from the main backbone of the fiber (such as carbon fibers). The three common fibers used for making composites are glass, carbon, Kevlar and thermoplastic fibers.

2.1. Glass Fibers

2.1.1. Glass Fiber Manufacturing Process

The raw materials for glass fibers are silica sand, boric acid, and other minor ingredients (e.g., clay, coal, and fluorospar). These are dry mixed in a high temperature refractory furnace. The temperature of this melt varies for each glass composition, but is generally about 2300°F (1260°C).

The process used for the manufacture of glass fibers from the molten glass mixture is illustrated schematically in Figure 3.2. First the mixture of silica sand and ingredients is put in a batch silo. They are then mixed and weighed. The mixture is transported to a batch charging unit where the mixture is fed into the furnace. The temperature in the furnace varies from 1540°C at the melting stage to 1425°C at the refining stage. The molten glass flows to the forehearth stage where the temperature is reduced to 1260°C. At the bottom of the forehearth station, there is a plate made of platinum which contains many tiny holes. The molten glass flows through these tiny holes and forms filaments. These filaments are then pulled mechanically to make smaller filaments. The diameter of the

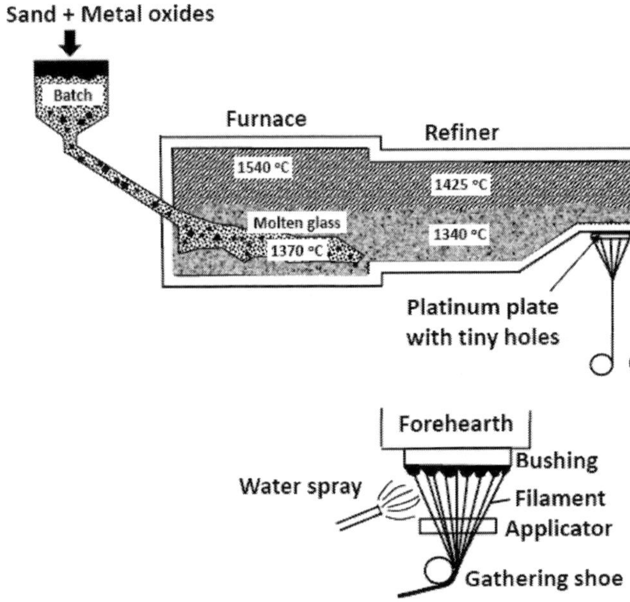

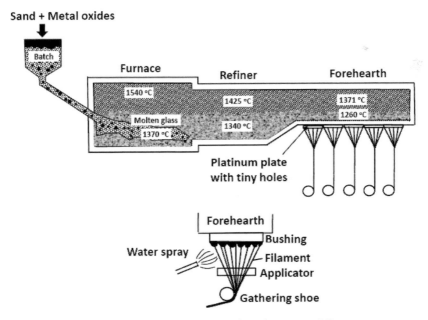

FIGURE 3.2 Fiberglass manufacturing process [4].

filaments depends on the speed of drawing. A chemical called *sizing* is applied on the surface of the fibers at this stage. The sizing is used to reduce the friction between the fibers to prevent breakage. A finish can also be applied to the fiber. Finish is a type of chemical used to protect the surface of the fibers and to provide good bonding with the matrix material when the composite is made. Many filaments are bundled together to form tows or strands. These tows or strands are then wound onto a creel for shipping purpose, or they can be cut into short fibers.

2.1.2. Types of Glass Fibers

Glass is an amorphous material that consists of a silica (SiO_2) backbone with various oxide components to give specific compositions and properties. Several types of glass fibers are manufactured but only three are used often in composites: E glass, S glass (and its variation S_2), and C glass. Table 3.1 shows the composition of the different types of glasses.

E glass (E for electrical grade) fibers have a composition of calcium aluminoborosilicate and calcium oxide, used when strength and electrical resistivity are required. The most common fiberglass used in composites, E glass is inexpensive in comparison with other types. E glass fibers are used as short fiber reinforcements for engineering thermoplastics; as fibers used with polyester or vinyl ester matrix for automotive composite components; and for fiber reinforced rods used for civil applications such as boats, seats, or trays. S glass (S for strength) is approximately

TABLE 3.1 Composition of Glasses Used in Composite Materials [3].

	E Glass Range (%)	S Glass Range (%)	C Glass Range (%)
Silicon oxide	52–56	65	64–68
Calcium oxide	16–25	—	11–15
Aluminum oxide	12–16	25	3–5
Boric oxide	5–10	—	4–6
Magnesium oxide	0–5	10	2–4
Sodium oxide and potassium oxide	0–2	—	7–10
Titanium oxide	0–15	—	—
Iron	0–1	—	—
Iron oxide	0–0.8	—	0–0.8
Barium oxide	—	—	0–1

40% higher in strength than E glass and offers better retention of properties at elevated temperatures. S glass is often used in advanced composites when strength is a premium. C glass (C for corrosion) is used in corrosive environments because of the chemical stability of its soda lime borosilicate composition.

Table 3.2 gives the physical, mechanical, thermal, electrical, and optical properties of glass fibers. Some properties, such as tensile strength and tensile modulus, are measured on the fibers directly. Other physical properties are measured on glass that has been formed into a patty or block sample and then annealed to relieve the forming stress. Note that the modulus of glass (about 70 GPa) is about the same as that of aluminum; however the strength of glass is much larger than that of aluminum.

Glass fibers normally have many defects on their surfaces. This is due to the abrasion between the fibers. The longer the fiber, the more defects there are. As such, the tensile strength of the fibers depends on their length. Figure 3.3 shows the effect of length on the tensile strengths of fibers.

Moisture has a detrimental effect on glass strength. The decrease in strength with increasing temperature is more pronounced in E glass than in S glass. However, the modulus decreases at about the same rate with increasing temperature for both E glass and S glass. The decrease is due to the rearrangement of the molecules into a less compact and hence more flexible configuration.

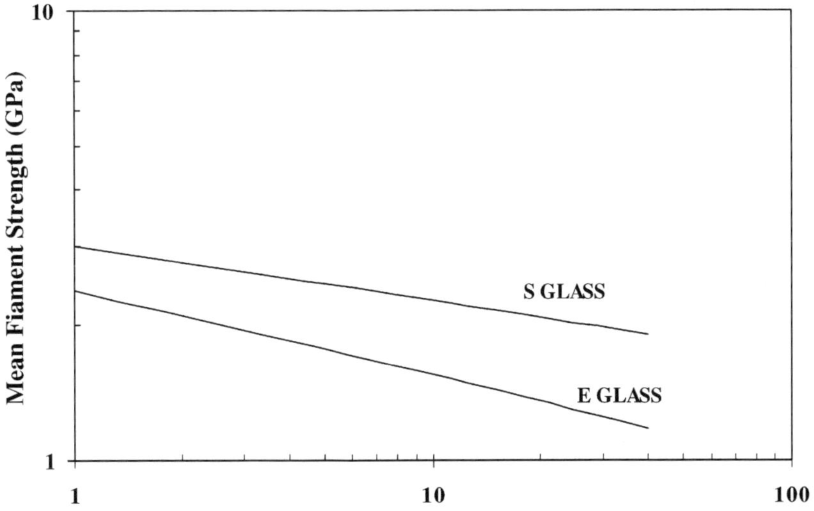

FIGURE 3.3 Effect of fiber length on tensile strength of fiber.

TABLE 3.2 Properties of Glasses [1].

Property	Type of Glass C	E	S
Density (g/cm^3)	2.49–2.50	2.54–2.62	2.48–2.50
Tensile strength (MPa)			
@22°C	3006–3280	3417	4544
@371°C	—	2597	3724–4408
@538°C	—	1708	2392
Tensile modulus (GPa)			
@22°C	68.3	71.8	84.7
@538°C	—	80.6	88.2
Elongation	0.03	0.035	0.04
Coefficient of thermal expansion (10^{-6}m/m/°C)	7.2	5.0	5.6
Heat Capacity (J/kg.C)@22°C)	800	800	800
Softening point, °C	749–750	841–846	970

2.1.3. Surface Treatment [1]

Glass fibers are extremely fragile and abrade easily during processing. The problem is especially evident in processes such as weaving, although almost any handling or moving process will cause abrasion of the glass fibers.

To guard against loss of strength, which depends strongly on surface defects that may be caused during handling, a chemical sizing (or coating) is applied to the fibers. This sizing protects the fibers during handling, and also holds the individual fibers together.

Usually sizing is temporary and after it is removed a finish is added. However in other cases the sizing also acts as the finish. The finish improves the compatibility of the fiber with the matrix. Typical finishes would be polyvinyl acetate modified with chromic chloride complex and/or organosilane coupling agents.

Coupling agents, molecules which are compatible at one end with the silane structure of the glass and at the other with the matrix, can be thought of as bridges connecting the reinforcement and the matrix. Figure 3.4 shows the bridging nature of the coupling agent between the glass fiber and the polymer matrix material. The coupling agents can combine with both the glass fiber and the polymer matrix material to form a separate phase called the *interphase*. This interphase may have different properties from either the glass or the polymer material. Figure 3.5 shows a schematic of the interphase.

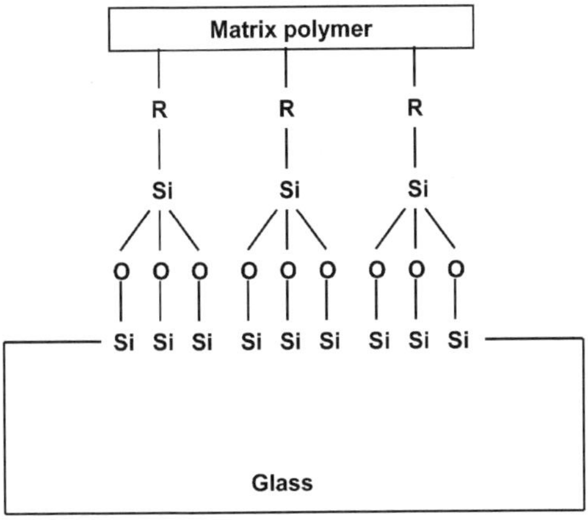

FIGURE 3.4 Idealized coupling of matrix and glass by organofunctional silane.

The use of a coupling agent can have significant effect on the mechanical properties of a composite. Changes of over 100% of the composites tensile, flexural or compressive strength with different choices of coupling agents are not uncommon for dry specimens. Because glass fibers are somewhat sensitive to moisture, the proper bonding of the glass with the matrix can also improve the mechanical properties in adverse environments. Therefore, both the application of the part and the matrix to be used should be known before specifying the fiberglass and the finish.

When glass fibers are used in pultrusion or filament winding, the strands must have high integrity, thorough wetting by the resin, and uniform processability under constant applied strain.

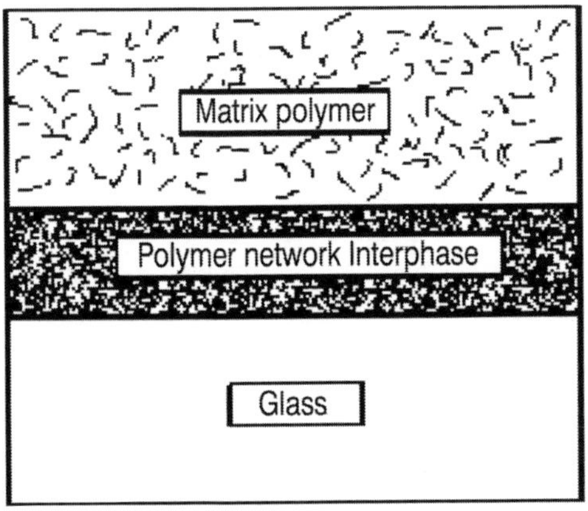

FIGURE 3.5 Silane and matrix interphase polymer network.

2.2. Carbon/Graphite Fibers [1]

Carbon/graphite fibers are used extensively in making composites for aerospace applications. Carbon and graphite are both based on layered structures of hexagonal rings of carbon. Structures of this type are called grapheme and are related to true graphite, although some differences exist in the structure.

While the terms are often used interchangeably, carbon and graphite fibers are different, at least in theory. Graphite fibers are those carbon fibers that have been subjected to heat treatment in excess of 3000°F (1650°C), possess 3-D ordering of the atoms, have carbon content in excess of 99% (although the graphite structure is still less than 75%) and have tensile modulus on the order or 344 GPa (50 Msi).

2.2.1. Fabrication Process for Carbon/Graphite Fibers

Carbon fibers are made using a raw material called the precursor. Theoretically there are three types of precursors. These are polyacrylonitrile (PAN), pitch and cellulose. However the PAN based and Pitch based precursors are more common and are discussed below.

2.2.1.1. PAN Based Precursor [1]

The principle of making carbon fiber made from the PAN precursor is outlined in Figure 3.6. In this process, one begins with the polyacrylonitrile (PAN) molecules. PAN molecules are thermoplastic polymers made by addition polymerization (first row, left, in Figure 3.6). With the application of heat, the triple bond between the carbon and the nitrogen atoms is broken. This is replaced by the double bond between the carbon and nitrogen, and a single bond between the nitrogen atom and another carbon atom, forming a ring structure (first row, right). Upon heating to between 400°C–600°C, the process of dehydrogenation (removal of hydrogen atoms) takes place in which many of the hydrogen atoms are removed. The structure becomes two dimensional with many hexagons formed (second row in Figure 3.6). Upon further heating to between 600°C–1300°C, the process of denitrogenation takes place, wherein the nitrogen atoms are removed, leaving a structure consisting mainly of carbon atoms (third and fourth row in Figure 3.6). Sheets of many hexagons are stacked against each other.

A microscopic cross section of a carbon fiber is shown in Figure 3.7. This shows about one-quarter of a carbon fiber, where the vertical direction coincides with the axis of the fiber. Note that the fiber consists of many sheets containing the hexagonal arrangement. The considerable strength and stiffness within the plane of this sheet are based on the strength and stiffness of the carbon-carbon bond, which is the same as the bond in diamond, except that diamond has bonds in three directions and the fiber only in two. The strength transverse to the sheet (normal to the axis of the fiber) is low. This is similar to onion layers, where one can peel them from each other rather easily. Carbon (or graphite) fibers, in which the strength and stiffness along the third direction is much weaker than

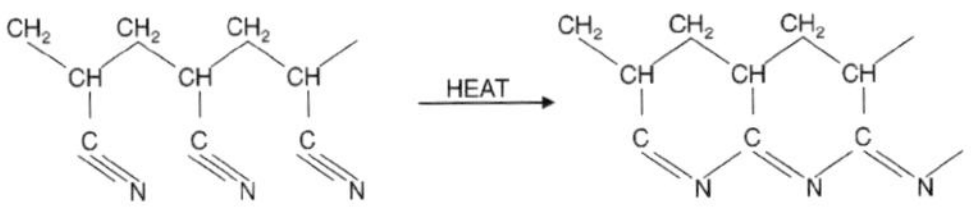

In the above, polyacrolynitrile molecule, which is a thermoplastic, is subjected to heat. This breaks the triple bond between C and N atoms and forms double bond between C and N atoms. In addition, rings are formed.

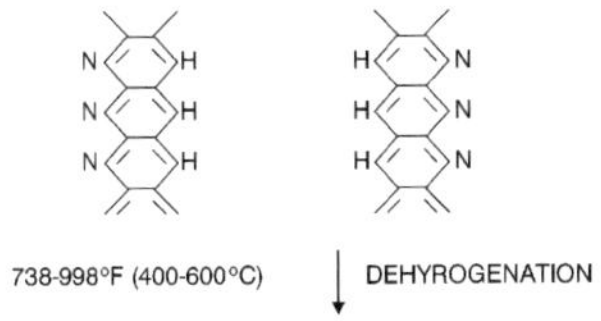

In the above, application of higher temperature removes the hydrogen atoms (dehydrogenation process)

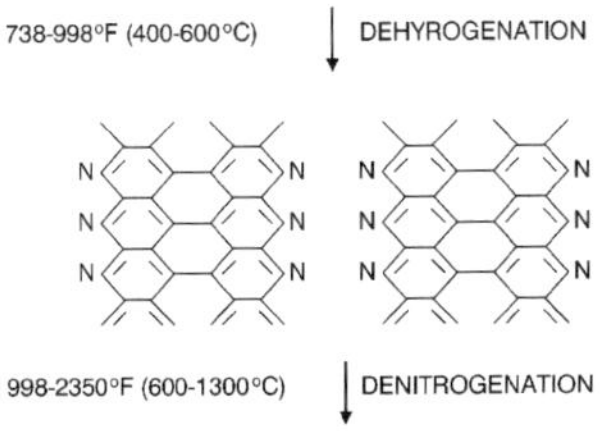

In the above, increasing the temperature to more than 600°C removes the nitrogen (denitrogenation) and links between hexagons of carbon atoms are formed.

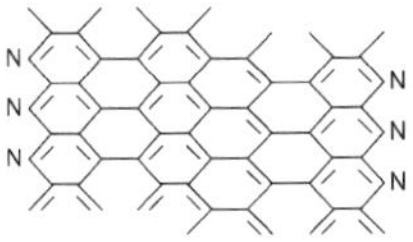

In the above, most nitrogen atoms are removed. What remains are sheets of hexagons of carbon atoms.

FIGURE 3.6 Formation of the carbon fibers from PAN [1].

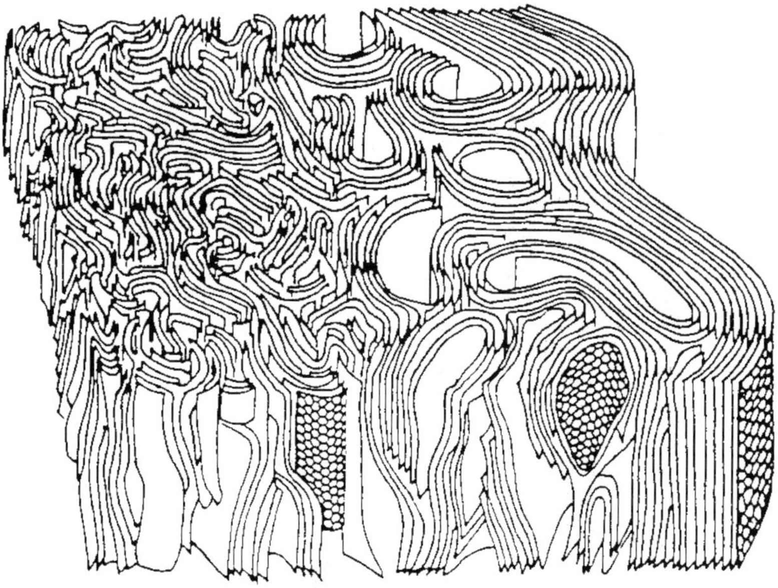

FIGURE 3.7 Configuration of a microscopic cross section of a carbon fiber (Reproduced from S.C. Bennet, D.L. Johnson, and W. Johnson, *Journal of Materials Science,* 18, 1983, p. 3337, with kind permission from Springer).

those in the in-plane directions, are therefore anisotropic, their properties depending upon direction.

The different steps in the manufacturing of carbon fibers from PAN is shown in Figure 3.8. The PAN-based precursor is first stabilized by thermosetting (crosslinking) so that the polymers do not melt in subsequent processing steps. This thermosetting step, which requires moderate heat, must be accompanied by a stretching of the fibers (or, more appropriately, a holding of the fibers at constant length against their inherent shrinkage as they become stabilized).

The fibers are then carbonized or, in other words, pyrolyzed, until they are essentially transformed into all-carbon fibers. It is during this stage that the high mechanical property levels are developed. The rapid evolution of gases up to 1800°F (982°C) requires that the heatup rate be quite slow to avoided forming voids and other defects. At about 1800°F (982°C), PAN-based fibers are approximately 94% carbon and 6% nitrogen, with further reductions in the nitrogen content until approximately 2300°F (1260°C), when the carbon content is over 99.7%.

Graphitization is carried out at temperatures in excess of 3200°F (1760°C) to improve the tensile modulus of the fiber by improving the crystalline structure and the 3-D nature of the structure. After graphitization, the fibers are surface treated, a sizing or finish is applied, and the fibers are then wound for shipment.

2.2.1.2. Pitch Precursor [1]

Pitch is the lowest grade obtained from the distillation of petroleum products. It is also the least expensive as compared to other grades. At the beginning of the fiber-making process, pitch must be converted into a suitable fiber. In the conversion process, pitch goes through a mesophase, called liquid crystal phase, in which the polymer chains are somewhat oriented even though the material is liquid. This orientation is responsible for the ease of consolidation of the pitch-based product into a carbon/graphite form. Subsequently the process follows a similar sequence as the PAN precursor, as shown in Figure 3.8.

2.2.2. Surface Treatment of Carbon Fibers [1]

Surface treatment of the carbon fibers improves the bonding between the fibers and matrix and thus improves the interlaminar shear strength (ILSS). Surface treatment can be divided into two types: oxidative and non-oxidative.

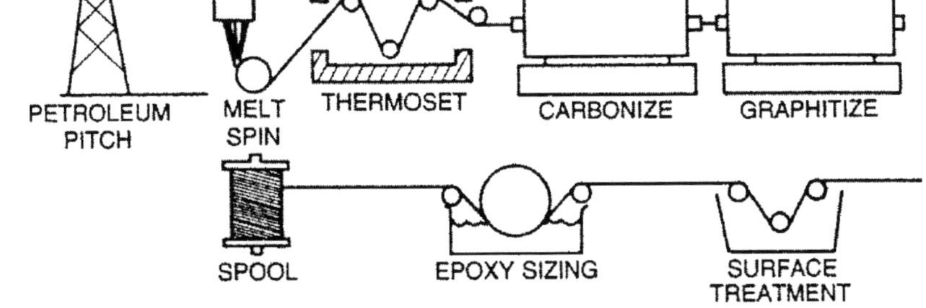

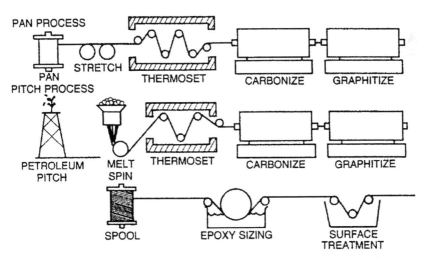

FIGURE 3.8 Process of making carbon fibers (Reproduced from Carbon/Graphite Fibers by Russell J. Diefendorf in *Engineered Materials Handbook, Vol. 1, Composites*, 1987, with permission from ASM International).

Liquid phase oxidative treatment can be done by simply drawing the fibers through a bath of some convenient oxidative agent (such as nitric acid, potassium permanganate, or sodium hypochlorite), or drawing the fiber through an acidic or alkaline electrolytic bath. The electrolytic bath is preferred since the process can be done continuously. The net result of this method is to clean the carbon fiber surface and then to attach chemical groups, such as hydroxides, which can bond with the matrix or a surface coating (sizing). Excessive oxidation, however, can result in a decrease in fiber tensile strength, presumably due to pitting of the fiber surface.

Non-oxidative treatments are of three types: whiskerization, pyrolytic surface coatings, and polymer grafting.In whiskerization, single crystals of ceramic materials (SiC, TiO_2, or Si_3N_4) are grown on the surface of the fibers. These whiskers are perpendicular to the fiber axis. The resulting improvement in interlaminar shear strength (ILSS) is excellent but the process is very expensive and is not commercially available at present. Pyrolytic coating consists of a vapor phase deposition of pyrolytic carbon on the fiber surface. This method shows good increase in ILSS but is expensive and unavailable commercially. The attachment of a polymer graft involves the preliminary attachment of some group that can be used as an anchor for the polymer. These reactive groups are generally attached by one of the oxidative methods, so the polymer graft method is really a subgroup of the oxidative method in which a polymer is introduced to act as a bridge between the fiber and the size of matrix.

The effectiveness of the fiber is often measured by the wettability of the fiber. The extent of wetting of the fiber is directly proportional to the number of chemical groups attached, and therefore the effectiveness of the surface treatment. This wettability is most often measured by looking at the contact angle of the fiber/water interface.

2.2.3. Surface Coatings (Sizing) of Carbon Fibers

In some cases organic coatings (sizings) are also added to the fibers to further improve the fiber/matrix bonding and to protect the fibers from damage during subsequent processing. These sizings are supplied to both treated and untreated fibers by passing the fibers through a heated bath of the sizing agent. The most common sizing is an epoxy, although polyvinyl alcohol and polyimides have also been used.

The value of the sizing or finish on carbon fibers is not as dramatic as it is in glass fibers in improving the mechanical properties in adverse environmental conditions. The reason for this is the inherently better moisture resistance of the carbon fibers compared to glass fibers.

2.2.4. *Properties of Carbon Fibers [1]*

Table 3.3 shows the properties of different types of fibers. The elastic modulus of carbon fibers can be in two categories: low modulus and high modulus. The modulus can be as low as 157 GPa for pitch based carbon fibers to as high as 383 GPa for PAN-based fibers. Tensile strength also varies from 1367 MPa for pitch based fibers to as high as 3280 MPa for PAN-based fibers. The elongation of carbon fibers is low (about 1%). The higher modulus fibers have lower elongation than the lower modulus fibers. Carbon fiber composites are therefore brittle due to the limited elongation of the fibers.

Carbon fibers are moderately good conductors of electricity. The conductivity of carbon fiber is in the order of 10^4 S/cm. This conductivity is assumed to arise from the natural conductivity of graphite and the orientation of the graphene ring, which is parallel to the axis of the fiber. Pitch fibers have better electrical conductivity than PAN-based fibers. The coefficient or thermal expansion (longitudinal) of carbon fibers is slightly negative and becomes more negative with increasing modulus. This contraction can be combined with the positive coefficient of thermal expansion of the matrix to yield a near zero coefficient of thermal expansion.

2.3. Organic Fibers

The most common organic fibers used for reinforcements are the aramids, with Kevlar, a Dupont fiber, currently being the major brand. A

TABLE 3.3 Different Types of Carbon Fibers and Their Properties [1].

PAN-Based Fibers	Low Modulus	High Modulus
Tensile modulus (GPa)	226	383
Tensile strength (MPa)	3280	2392
Elongation (%)	1.4	0.6
Density (g/cc)	1.8	1.9
Carbon assay (%)	92–97	100
Pitch-Based Fibers		
Tensile modulus (GPa)	157	376
Tensile strength (MPa)	1367	1708
Elongation (%)	0.9	0.4
Density (g/cc)	1.9	2.0
Carbon assay (%)	97	97

new ultra-high-orientation polyethylene fiber (Spectra) with some interesting and useful properties has been introduced into the market.

2.3.1. Aramid Fibers [1]

Aramid fibers are made by mixing paraphenylene diamine and terephthaloyl chloride in an organic solvent to form poly-paraphenylene-terphthalamide (aramid) (Figure 3.9). After polymerization, the polymer is washed and then dissolved in sulfuric acid. At this point, the polymer is a partially oriented, liquid crystal form. Then the polymer solution is extruded through small die holes (spinnerettes) in a process known as solution spinning. The fibers are then washed, dried and wound up for shipment.

Although the aramid molecule is strong due to the presence of the aromatic rings, the molecules are linked to one another only by weak hydrogen bonds, with the result that aramid fibers are strong in tension but weak in compression. When aramid fibers fail, they break into small fibrils which are like fibers within the fiber. These fibrils arise from the rod-like structure of the liquid crystals as they are being spun into fibers and from the weak intermolecular bonds. Tensile strength failure is believed to initiate at the ends of the fibrils and is propagated through the fiber by shear. This unique failure mechanism is responsible for the very high strength of aramid fibers versus conventional fibers, but is also the reason aramid composites have compression strength one-half of carbon composites. The positive point from this type of behavior is

Poly Para-Phenyleneterephthalamide (Aramid)

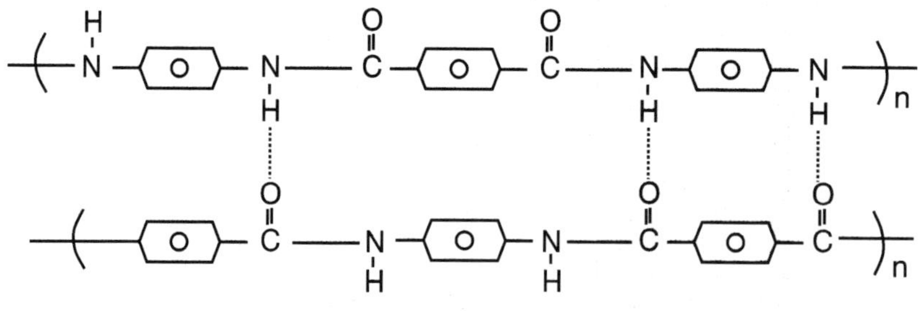

Intermolecular bonding in aramids

FIGURE 3.9 Formulation of aramid fibers [1].

that aramid composites can absorb a large amount of energy upon failure by impact.

2.3.1.1. Properties of Aramid Fibers [1]

Kevlar fibers are made in three different types: Kevlar 29 (high toughness), Kevlar 49 (high modulus) and Kevlar 149 (ultra high modulus). Table 3.4 gives the properties of Kevlar fibers.The modulus increases from Kevlar 29 to Kevlar 49 to Kevlar 149 and is due to changes in process conditions that promote additional crystallinity in the high modulus and ultra high modulus products. The tensile strength and modulus of aramids are substantially higher than for other organic (textile) fibers, although not as high as carbon fibers.

Aramid fibers are less brittle than carbon fibers or glass fibers. The elongation of aramid fibers is about 2% to 4%, and this combination of good strength, low weight, and toughness has led to the key applications of aramid composites. Aramid fibers have negative coefficient of thermal expansion in the fiber direction, and this has led to some applications where limited thermal expansion is a key desired property.

2.3.1.2. Bonding to Matrix [1]

Aramids do not bond as well to matrices as do glass or carbon fibers. Therefore, the ILSS is low for aramids. These low values may also arise, in part, from the relatively low shear resistance of the fiber itself. This poor bonding may give rise to good absorption of energy upon impact.

2.3.1.3. Use of Aramid Fibers [1]

Aramid fibers have been used principally as reinforcements for tires, belts, and other rubber related goods as well as for bullet proof vests, high strength cloth such as sails for racing boats, and reinforcements

TABLE 3.4 Properties of Aramid Fibers [1].

Type of Kevlar Fiber	29	49	149
Tensile modulus (GPa)	83	131	186
Tensile strength (MPa)	3.6	3.6	3.4
Elongation (%)	4	2.8	2
Density (g/cc)	1.44	1.44	1.47

for composite materials. For high load applications, they have been used in making pressure vessels where tensile strength is more important than compressive strength. Kevlar 49 with epoxy has hoop stress value of 450 ksi (307.5 MPa) which is higher than other high modulus fibers.

2.3.2. Ultra-High Orientation Polyethylene Fibers

Polyethylene is a low strength, low modulus thermoplastic polymer. It is used to make low cost items such as toys. Its microstructure is as shown in the left of Figure 3.10. When a piece of material made of this polymer is subjected to a tensile load, the molecules slide across each other, the low resistance coming from low intermolecular attraction. If the molecules can be straightened so that they are mostly aligned, as shown in the right side of Figure 3.10, the interatomic forces in the carbon-carbon backbone becomes much more resistant when a load is ap-

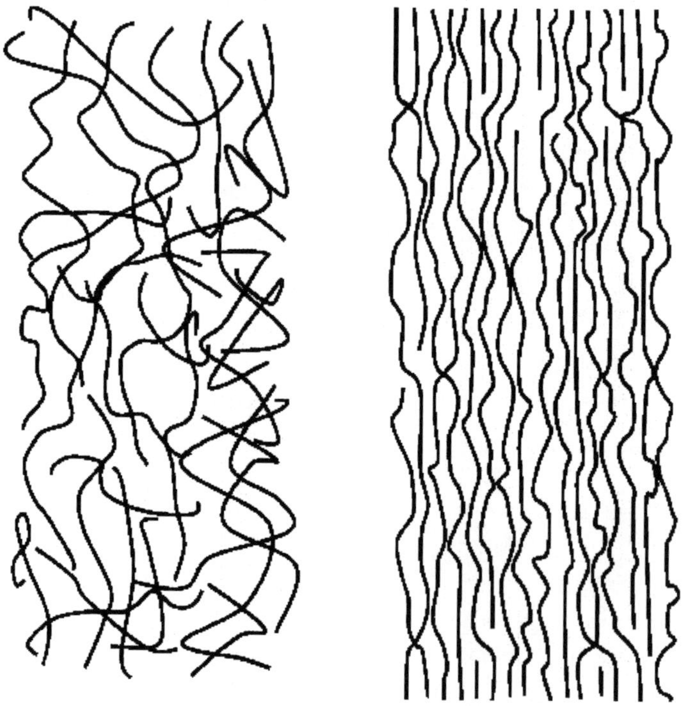

FIGURE 3.10 Formation of polyethylene fibers.

plied. A polyethylene fiber (Spectra) was developed based on this concept by Allied Corporation. It has excellent strength to weight and modulus properties.

The strength and modulus of this fiber are approximately the same as aramid but because of the lower density, its specific strength and modulus are higher than aramid and nearly as high as high modulus carbon fibers. The solvent resistance is superior to aramids, but temperature performance is inferior to aramids.

Major uses include: ballistic penetration prevention, radomes, water skis, bicycles, kayaks, and uses where the low dielectric constant is of value.

2.4. Boron and Silicon Carbide Fibers [1]

Boron and silicon carbide (SiC) are the most common of the very high modulus reinforcements. Boron is used as a reinforcement for epoxy in high performance uses. SiC is used as a reinforcement for ceramic and metal matrices. These materials are much more expensive than the other conventional reinforcement materials.

These fibers are made by chemical vapor deposition (CVD), in which a substrate filament, normally tungsten for boron and carbon for SiC, is pulled through a cleaning section and then a deposition chamber. The substrate element is heated electrically to approximately 2500°F (1370°C). In the deposition chamber, the chemicals that will give the coating are introduced and the reaction occurs on the surface of the heated filament. For boron, the reactants are boron trichloride and hydrogen, while for SiC the reactants are alkyl silanes having the Si–C–Si structure (with various side groups) and hydrogen. The tungsten filament is typically 0.5 mils (0.013 mm) in diameter and the resulting boron fibers are 4–8 mils (0.1 mm–0.2 mm) in diameter. The carbon filament is typically 1.3 mils (0.033 mm) in diameter and the resulting SiC fibers are 5–6 mils (0.13 mm–0.15 mm).

3. TOWS

Refer to Figure 3.1. The previous presentation was on individual filaments. At the end of the fiber manufacturing process, the individual filaments are normally grouped together in the form of tows consisting of thousands of filaments. The commercially available tows are in 3 k (3000 filaments), 6 k, or 12 k. The larger the number of filaments in a tow, the more efficient the deposition of the fiber to make a part (it takes less time to deposit a certain amount of material). Larger tows, however, mean more filaments are bundled together and this may make it difficult to get the liquid matrix to flow to the surface of individual filaments for wetting. The balance between the size of the tows and material deposition rate depends upon the application.

4. FABRICS AND OTHER REINFORCEMENT FORMS

Fibers are sold and used in many different forms. Rarely, if ever, are fibers used in single filament form except for laboratory testing and experimentation. Discussion of reinforcement forms will examine how fibers can be assembled together. The method of assembling the fibers (i.e., the arrangement of the fibers when they are used) often has a major effect on the manufacturing process and mechanical properties of the composite. The method to be used in the fabrication of the composite often determines the shape of the reinforcement. Following is a summary of the different forms of the reinforcement.

- *Fibers:* A general term for a material that has a long axis many times greater than its radius.
- *Filament:* A single fiber. This is the unit formed by a single hole in the spinning process.
- *Strand:* A general and somewhat imprecise term. Usually refers to a bundle or group of untwisted filaments but has also been used interchangeably with fiber and filament.
- *Tow:* An untwisted bundle of continuous filaments, usually with a specific count [such as 3k tow (3000 filaments per tow) or 6k tow (6000 filaments per tow].
- *Yarn:* A twisted bundle of continuous fibers, hence a twisted tow. Often used for weaving.
- *Roving:* A number of yarns or tows collected into a parallel bundle without twisting.
- *Tape:* A collection of parallel filaments (often made from tow) held together by a binder (usually the composite matrix). The length of the tape, in the direction of the fibers, is much greater than the width and the width is much greater than the thickness. Figure 3.11 shows a photograph of unidirectional tape.
- *Woven fabric:* A planar material made by interlacing yarns or tows in various specific patterns. Figure 3.12 shows a few common fabric forms.

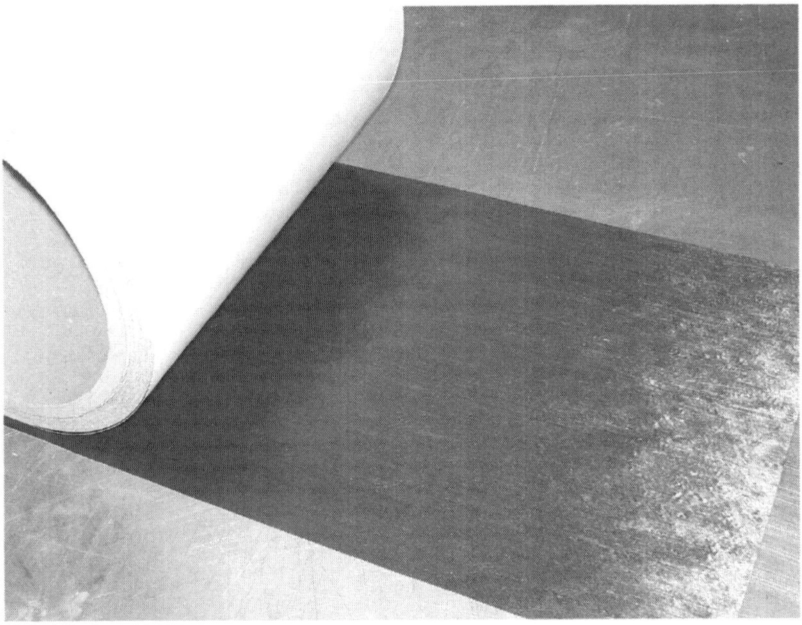

FIGURE 3.11 Photograph of a roll of unidirectional tape.

- *Braid:* The interlacing of yarns or tows into a tubular shape instead of a flat fabric.
- *Mat:* A sheet like material (planar form) consisting of randomly oriented chopped fibers or swirled continuous fibers held together loosely by a binder.
-

4.1.Tows and Rovings

The simplest form of reinforcement commonly used is a tow. Tows can be laid down as parallel fibers to form a tape. Tows can also be wound around a shape, as in filament winding, or they can be chopped into short fiber segments. A tow can also be twisted into a yarn, or several tows combined into a roving. Tows are sold on spools with a particular filament count for each tow end.

4.2. Weave Types

Figure 3.12 shows the most common weave types. In the weave there are two directions: the fabric, or long direction, is called the warp; and the cross, or width direction, is called the fill, weft, or woof.

The plain weave, the simplest weave form, is made by interlacing yarns in an alternating over-and-under pattern. There is one warp fiber for one fill fiber without skipping. The maximum fabric stability and firmness with minimum yarn slippage results from this weave. The pattern gives uniform strength in two directions when yarn size and count are similar in warp and fill. This weave type is the most resistant in shear and is therefore considered to be a rather stiff weave. Because the weave is stable, it is usually left moderately open so resin penetration and air removal are fair to good. Plain weave fabrics are used for flat laminates,

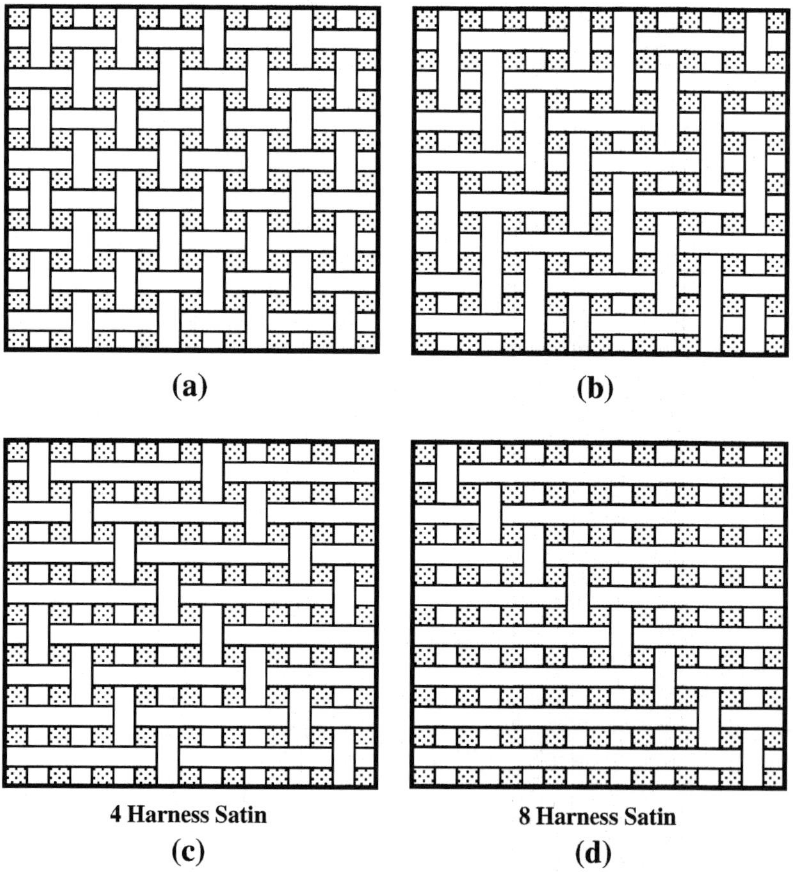

FIGURE 3.12 A few common fabric forms: (a) plain weave, (b) basket weave,(c) 4-harness-satin, (d) 8-harness-satin.

printed circuit boards, narrow fabrics, tooling, and for covering wood structures such as boats.

The basket weave is similar to the plain weave except that two warp yarns are woven as one over and under two fill yarns. This weave is less stable than plain weave so it is more pliable. This fabric is flatter and stronger than an equivalent weight and count of plain weave. The uses are similar to those for plain weave but with better drape on mild contours.

The long shaft satin weave or harness weave has one warp yarn weaving over four or more fill yarns. For example, in the 4-harness-satin weave, one warp yarn goes over three fill yarns. In the 8-harness-satin weave, one warp yarn goes over seven fill yarns, etc. This weave has high degree of drape and stretch in all directions. A high yarn density is possible. The weave is less stable than plain weave (this means that if the weave is held by hands, yarns may not keep their positions and may move). Also the weave is less open than most others so wetting and air removal can be a problem unless vacuum is used. This weave is used extensively in the aircraft industry where complex shapes are common. It is also used for housings, radomes, ducts, and other contoured surfaces.

4.3. Mats

Mat is normally made of relatively short fibers and is used in noncritical applications. Most applications for composites are noncritical, so in total consumption, mat materials are widely used. Laminates made with mats are only 33–50% as strong as fabric laminates of comparable thickness because of the lower fiber content and because the mat relies on the matrix for much of its structural integrity. Mat costs much less than woven fabrics. In general, mat requires more resin to fill interstices and more positive methods for removing the trapped air. Mat might typically be used for the inner layers of boats where thickness buildup is the critical function.

Mats can be made from continuous strands to achieve increased strength (though not as high as fabrics). This continuous mat is produced by swirling strands of continuous fibers onto a belt, spraying a binder over them, and drying the binder. This is called *swirl mat* and Figure 3.13 shows its configuration.

Some very fine mats (called surfacing mat or veil) made from glass and carbon fibers are used for the top layer in a composite and provide a smooth texture for the surface. These fine mats are also used as carriers

FIGURE 3.13 Configuration of swirl mat.

for film adhesives. Fine mats are available with almost any fiber used as a reinforcement in composites. Surface mats also prevent fiber blooming (which is the breaking of the resin surface by a fiber, usually due to environment exposure).

4.4. Braids

Interest in braids has increased because of the effort to maximize properties in all directions of a part. Braid reinforcement is intended to improve torsional load capability to contain an inner core, to provide impact resistance or damage tolerance, or to enhance product appearance. Figure 3.14 shows the configuration of a braid.

Braids can either be deposited directly onto a mandrel or wound on a spool, removed, and slipped onto a mandrel when needed. Because the speed of slipping onto a mandrel is much faster than winding directly onto a mandrel, the spool winding method is generally preferred. Mandrels can assume many shapes other than round. Generally, braided reinforcements are stronger than other types of reinforcement patterns.

4.5. 3-D Weaves

Many types of 3-D weaves are available and are used to improve the performance of shaped composite parts that can be made from preformed reinforcements. Several weaving machines are available for the fabrication of these preforms. Therefore, whenever strength in the third direction is desired and can best be given by fibers rather than by a structural member (such as I beam), 3-D woven material should be considered.

4.6. Hybrids

Reinforcement schemes that combine two or more types of reinforcements are called hybrids. Typical examples would be carbon/aramid, glass/carbon, and glass/aramid. The advantages of hybrids is that in some weaves and braids, the best properties of each of the types of reinforcements can be utilized. Figure 3.15 shows a hybrid fabric of carbon/glass.

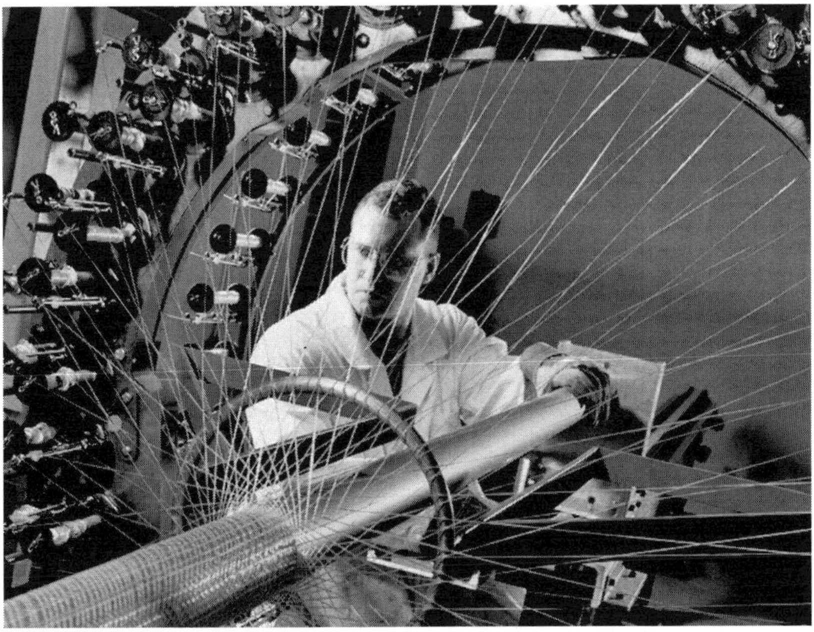

FIGURE 3.14 A braiding machine.

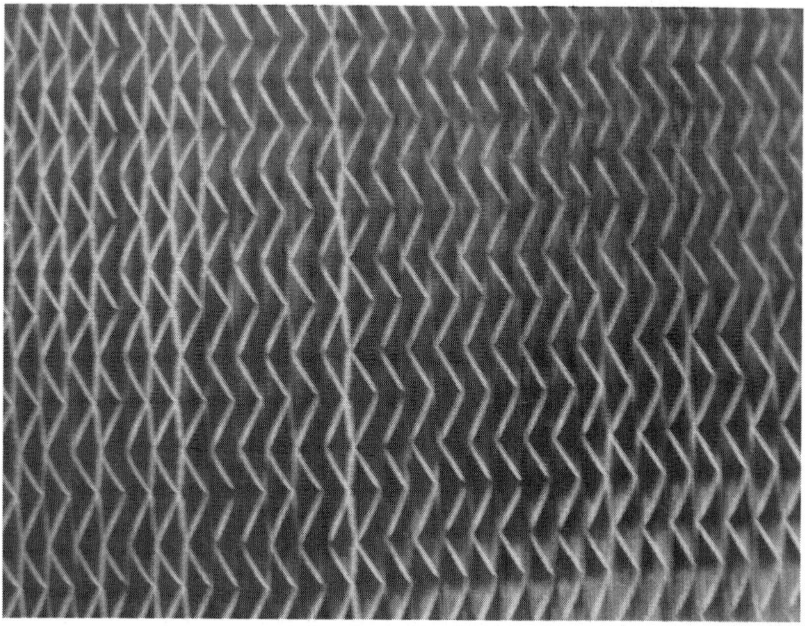

FIGURE 3.15 Hybrid fabric of carbon/glass.

5. DEFORMATION OF A BED OF FIBERS [2]

5.1. Introduction

Advanced composite materials use fibers in very high fiber volume fractions, usually in the range $0.5 \le V_f \le 0.7$. Two important consequences of these high fiber volume fractions are: (1) the fibers must be aligned, which leads to anisotropic behavior; and (2) the interfiber spacing is very small, which leads to significant fiber to fiber contact. The mean interfiber spacing δ_f can be calculated as:

$$\frac{\delta_f}{d} = \left(\sqrt{\frac{V_a}{V_f}} - 1 \right) \tag{3.1}$$

where V_a is the maximum allowable fiber volume fraction and d is the fiber diameter. Maximum allowable fiber volume fraction means the volume fraction when fibers touch each other.

Equation (3.1) can be derived as follows:

Assuming square packing of fibers as shown in Figure 3.16. The fiber volume fraction is calculated to be:
Let a denote the side of the square, d the fiber diameter and δ_f the interfiber spacing. The volume fraction is calculated to be:

$$V_f = \frac{n^2 \pi d^2 / 4}{[n(d+\delta_f)]^2} = \frac{\pi}{4} \frac{d^2}{(d+\delta_f)^2}$$

which can be rearranged to be:

$$\frac{\delta_f}{d} = \left(\sqrt{\frac{\pi / 4}{V_f}} - 1 \right)$$

When the fibers are touching each other, $\delta_f = d$ and $V_f = \pi/4$. $\pi/4$ can be considered as the maximum fiber volume fraction for the square packed array. For this type of array or other type of array (such as the hexagonal array), the maximum fiber volume fraction can be denoted as V_a, and Equation (3.1) can be obtained.

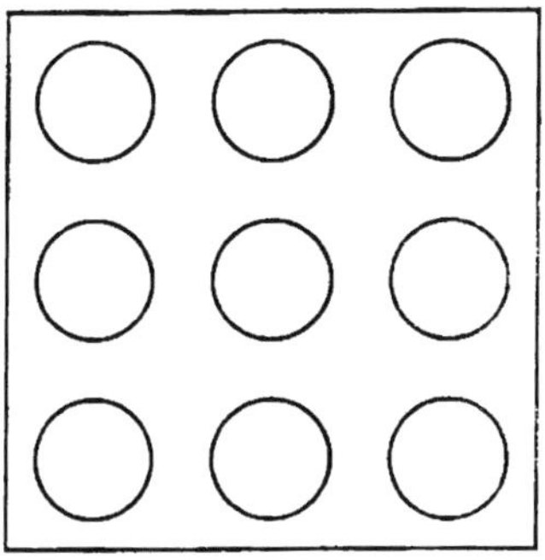

FIGURE 3.16 Square array of fibers.

For a square packing, $V_a = \pi/4$. Now if $V_f = 0.65$, then $\delta_f/d \sim 0.10$. For typical carbon fibers, $d \sim 10^{-5}$ m, which leads to a very small interfiber spacing of $\delta_f \sim 10^{-6}$ m. Given the fact that the fibers are neither perfectly aligned nor perfectly straight, these small average interfiber distances suggest that there are multiple fiber-fiber contacts. This, in fact, is the case. It has been shown that the fiber bundle can carry finite transverse loads in the volume fraction range for advanced composites. Furthermore, there are processing conditions in which the load carrying capacity of the fiber will dominate the processing behavior. Some examples of processing geometries are shown in Figure 3.17.

These suggest that the fiber bundles may be subjected to complex, 3-D state of stress during the process. (Note that the stresses considered here

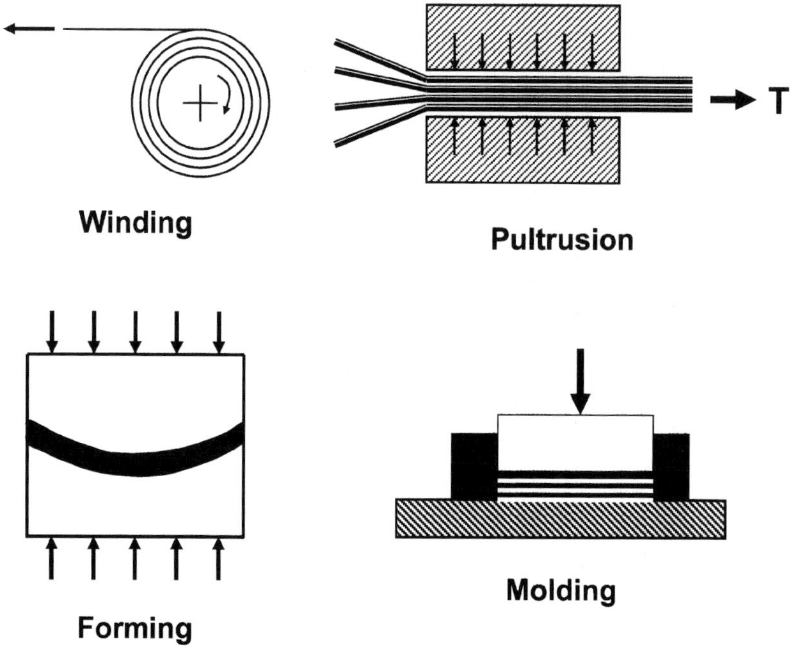

FIGURE 3.17 Examples of different loading patterns on fiber bundles [2].

are the stresses required for manufacturing, and not the stresses created due to the loading of fibers during load bearing operation).

In the case of filament winding, as the tows are being wound onto the mandrel, there is a slight tension applied on the tows. This tension serves to keep the tows tight and to compact the many layers together on the mandrel. The axial tension in the fiber bundle leads to a contracting of the fiber bundle and a corresponding increase in fiber volume fraction. This phenomenon has a large effect on the wetting behavior for the fiber bundle. In addition, the axial tension shown in winding results in a radial compression, particularly for the innermost fiber bundles. This, in turn, can reduce the axial tension in the bundles close to the mandrel and even result in axial compression.

For the case of pultrusion, the fiber bundles are pulled from the exit end. This serves to compact the many tows coming in from the creels. There can be a significant amount of shrinkage in the resin during the curing process in the die. In order to compensate for this, the amount of lateral pressure is necessary in order to avoid the formation of voids. The relation between axial tension and lateral compression needs to be understood to ensure good quality.

The molding example implies that the bundles will be subjected to a transverse compression and hence the load carrying capacity of the fiber bundle will be of utmost importance. In forming, transverse compression as well as viscous shearing will be dominant effects.

To a first approximation, the bundle of lubricated aligned fibers can respond to these applied loads either elastically or viscously. Basically the viscous loadings correspond to shear loading, while the elastic modes correspond to normal loading. A typical element of such a fiber bundle as well as the material coordinate system is shown in Figure 3.18. In general, one is interested in the average behavior of the fiber bundle and therefore will treat it as a continuum, as has been suggested by various researchers in the textile industry.

5.2. The Elastic Deformation of Fiber Bundle

5.2.1. Model Development [2]

In this section, following Gutowski [2], a general elastic deformation model for an aligned fiber bundle subjected to stresses σ_1 and σ_b was developed. It is assumed that the fibers are lubricated, but in a drained state ($p_r = 0$) and make multiple contacts with their neighbors. From earlier measurements, it was observed that the fibers have a slight waviness of a sinusoidal character. Later we will relate this waviness to the typical

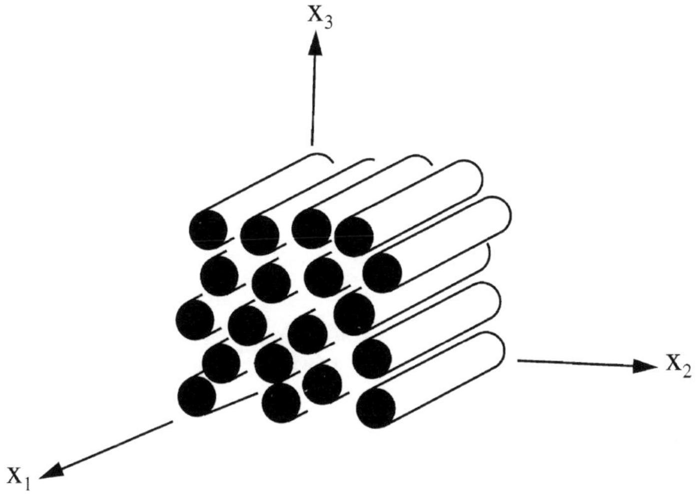

FIGURE 3.18 Representative element of an aligned fiber bundle.

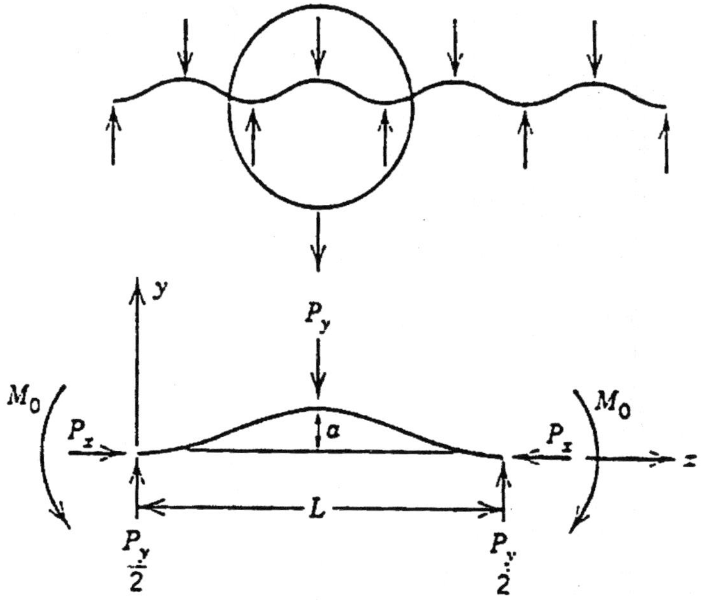

FIGURE 3.19 Slightly curved fiber with $L/a \geq 100$ and representative element for the fiber bundle [2].

length between contacts and construct a cell around a fiber, using these dimensions. First, consider the deformation behavior of a segment of a slightly curved fiber as shown in Figure 3.19.

The arch length over height L/a is on the order of 300. The fiber shown is just one small segment of the continuous fiber so the slopes at the ends are zero. When needed, it is assumed that the fiber has a shape as given by:

$$y = \frac{a}{2}\left[1 - \cos\left(\frac{2\pi x}{L}\right)\right] \tag{3.2}$$

Note that other reasonable assumed shapes for the fiber can be used; these will give similar results, with similar scaling but slightly different constants. Since these constants are adjusted experimentally for the final model, the exact shape is not critical.

The fiber will be deformed by the application of two loads, one along the axis P_x and the other transverse to axis x, in the plane of the arch P_y. The deflection of this fiber in both x and y directions (Δx and Δy) can be obtained by direct calculation as:

$$\begin{bmatrix} \Delta x \\ \Delta y \end{bmatrix} = \begin{bmatrix} \dfrac{a^2 L}{8EI} + \dfrac{L}{EA} & -\dfrac{aL^2}{4\pi^2 EI} \\ -\dfrac{aL^2}{4\pi^2 EI} & \dfrac{L^3}{192EI} \end{bmatrix} \begin{bmatrix} P_x \\ P_y \end{bmatrix} \tag{3.3}$$

Here, E and I correspond to the bending modulus and the moment of inertia of the fiber, respectively.

This result can be related to the deformation behavior of a fiber bundle if one constructs a suitable unit cell around the fiber and then modify Equation (3.3) to represent the deformation of this unit cell. For example, a square cell is shown in Figure 3.20. The dimensions L and h will change during deformation. Because the fiber is only slightly arched, the actual length of the fiber and the cell length L will be almost identical. In the transverse direction, one can make the height and the width of the box equal. Hence, even though fibers may be thought of as simple arches, this assumption will give the bundle transverse isotropy in the 2-3 plane. This assumption is rooted in the earlier assumption that the bundle has no shear strength in the 2-3 plane. Having established these dimensions, one can now define the cell stresses, which are the same as the bundle stresses:

$$\sigma_1 = \frac{P_x}{h^2} \tag{3.4}$$

$$\sigma_b = \frac{P_y}{hL} \tag{3.5}$$

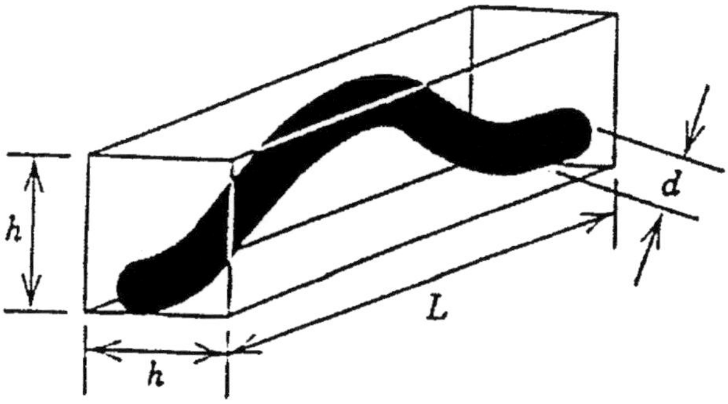

FIGURE 3.20 Representative fiber cell [2].

and two measures of deformation of the bundle:

$$e_1 = \frac{\Delta_x}{L} \tag{3.6}$$

$$e_b = \frac{\Delta_y}{h} \tag{3.7}$$

Note that these quantities are normalized by the current values of the cell dimensions. An expression for e_b may be directly related to the cross-sectional area, and the fiber volume fraction of the fiber bundle as:

$$e_b = \frac{h - h_o}{h} = 1 - \frac{h_o}{h}$$

Assuming constant volume, one has:

$$e_b = 1 - \sqrt{\frac{A_o}{A}} = 1 - \sqrt{\frac{V_f}{V_o}} \tag{3.8}$$

where the subscript o represents an initial state, and the subscript f represents the instantaneous state.

In order to relate the forces and the deflections for a single fiber to the deformation of a cell and hence the fiber bundle, two other relationships are needed.

The first relation is derived as follows:

$$\frac{a}{d} = \frac{h - d}{d} = \frac{h}{d} - 1$$

In order to obtain the expression h/d, one can consider the case of square packing. It can be shown that for this fiber arrangement, the fiber volume fraction can be expressed as:

$$V_f = \frac{\pi d^2 / 4}{h^2} = \frac{\pi}{4}\left(\frac{d}{h}\right)^2$$

from which

$$\frac{h}{d} = \sqrt{\frac{\pi / 4}{V_f}}$$

In the extreme case where $h = d$, we have $V_f = \pi/4$. This maximum fiber volume fraction can be denoted with a more general name as V_a. If one assumes some other form of array (such as the hexagonal array), V_a can have a different value. For the general case of maximum allowable fiber volume fraction V_a, one has:

$$\frac{h}{d} = \sqrt{\frac{V_a}{V_f}}$$

This gives:

$$\frac{a}{d} = \sqrt{\frac{V_a}{V_f}} - 1 \tag{3.9}$$

The second relationship is:

$$\frac{L}{d} = \frac{L}{a}\frac{a}{d} = \beta\left(\sqrt{\frac{V_a}{V_f}} - 1\right) \tag{3.10}$$

Where the coefficient β is defined as:

$$\beta = \frac{L}{a} \tag{3.11}$$

This quantity, assumed to be constant, can be termed the *waviness ratio* and it is determined empirically. Equation (3.10) relates the contact frequency, or length between contacts to transverse compression.

Now applying Equations (3.4)–(3.8) to Equation (3.3), one may obtain the functional relationship between the applied stresses and the deformations for a fiber bundle as:

$$\begin{bmatrix} Le_1 \\ he_b \end{bmatrix} = \begin{bmatrix} \dfrac{a^2L}{8EI} + \dfrac{L}{EA} & -\dfrac{aL^2}{4\pi^2 EI} \\ -\dfrac{aL^2}{4\pi^2 EI} & \dfrac{L^3}{192EI} \end{bmatrix} \begin{bmatrix} \sigma_1 h^2 \\ \sigma_b hL \end{bmatrix}$$

or

$$\begin{bmatrix} e_1 \\ e_b \end{bmatrix} = \begin{bmatrix} \dfrac{a^2h^2}{8EI} + \dfrac{h^2}{EA} & -\dfrac{aL^2h}{4\pi^2 EI} \\ -\dfrac{aL^2h}{4\pi^2 EI} & \dfrac{L^4}{192EI} \end{bmatrix} \begin{bmatrix} \sigma_1 \\ \sigma_b \end{bmatrix} \tag{3.12}$$

One now defines a new variable, which is related to the maximum allowable fiber volume fraction and the instantaneous fiber volume fraction (called the fiber volume parameter), as:

$$\varsigma = \sqrt{\frac{V_a}{V_f}}$$

Equation (3.12) can be shown to be,

$$\begin{bmatrix} e_1 \\ e_b \end{bmatrix} = \begin{bmatrix} F_{11} & F_{1b} \\ F_{b1} & F_{bb} \end{bmatrix} \begin{bmatrix} \sigma_1 \\ \sigma_b \end{bmatrix} \tag{3.13}$$

where

$$F_{11} = \frac{4}{\pi}\frac{1}{E}\varsigma^2[1 + 2(\varsigma - 1)^2] \tag{3.14a}$$

$$F_{1b} = F_{b1} = -\frac{16}{\pi^3}\frac{\beta^2}{E}\varsigma(\varsigma - 1)^3 \tag{3.14b}$$

and

$$F_{bb} = \frac{\beta^4}{3\pi E}(\varsigma - 1)^4 \tag{3.14c}$$

These equations give the 3-D elastic behavior of a fiber bundle. Since both the F_{ij} values and e_b are functions of V_f, Equation (3.13) represents two equations and two unknowns: e_1 and V_f. From this expression, the deformation behavior of a fiber bundle can be obtained for any applied stresses σ_{01} and σb.

The solution procedure is to first write:

$$e_b = F_{b1}\sigma_1 + F_{bb}\sigma_b \tag{3.15}$$

and then use Equation (3.13) to solve for V_f. The solution, say V_f^* is then substituted into Equation (3.13) to obtain the axial strain:

$$e_1 = F_{11}(V_f^*)\sigma_1 + F_{1b}(V_f^*)\sigma_b \tag{3.16}$$

Note that the subscript o used in Equation (3.8) for the definition of e_b refers to some original value. This means a specific reference value corresponding to the condition when the unstressed fibers can just begin carrying transverse loads. More specifically, $V_f = V_o$ would indicate the maximum fiber volume fraction for which $\sigma_1 = 0$ and $\sigma_b = 0$. The cell dimension h_o and cell or bundle area A_o then correspond to V_o. The term V_o is experimentally determined.

The relationships given in Equation (3.13) represent the elastic deformation of fiber bundles. As $V_f \rightarrow V_a$, the "bundle" nature of the fibers is lost. For example, when $V_f = V_a = \pi/4$, one obtains:

$$F_{11} = \frac{1}{V_f E_f} \tag{3.17}$$

which is a familiar result corresponding to a composite with zero matrix stiffness. For the same conditions:

$$F_{1b} = F_{b1} = F_{bb} = 0 \tag{3.18}$$

which implies that there is no functional relationship between the stresses and the deformations in this region. Hence in this region the deformation of the fibers does not depend on their "bundle" nature. Probably a better model for the fibers in this region would be based on their bulk properties. Of course, fibers are not used in this region in real composites. From Equation (3.13) several useful results with practical implications can be derived.

5.2.2. Axial Extension of a Fiber Bundle

The axial extension behavior of a fiber bundle is of interest during such operations as prepregging, pultrusion and filament winding. If one applies only axial tension to the fiber bundle ($\sigma_b = 0$), then Equation (3.13) reveals a slightly nonlinear relationship between the applied force P_1, and the axial extension e_1 due to the straightening of the fibers. Of more interest, however, is the coupling behavior, or the induced fiber volume fraction increase due to this tension. This relationship can be derived by setting $\sigma_b = 0$ and solving for a relationship between P_1 and V_f, where the force P_1 is defined as:

$$P_1 = A\sigma_1 = \frac{A_o V_o}{V_f} \frac{e_b}{f_{b1}} \tag{3.19}$$

This leads to:

$$P_1 = -\frac{A_o V_o}{V_f} \frac{1-\sqrt{\frac{V_f}{V_o}}}{\frac{16}{\pi^3}\frac{\beta^2}{E}\sqrt{\frac{V_a}{V_f}}\left(\sqrt{\frac{V_a}{V_f}}-1\right)^2} \tag{3.20}$$

This equation shows that tensioning of the bundle has a strong effect on increasing the fiber volume fraction.

5.2.3. Bulk Compressive Stresses

During many advanced composite processes (e.g., autoclave molding, compression molding), bulk compressive stresses are applied to the fiber bundle. One particular case of interest is the bulk compressive stresses that are applied in the absence of any axial stress ($\sigma_1 = 0$). This may be similar to conditions during autoclave processing. In this case, an expression can be obtained that relates the bulk compressive stress with the fiber volume fraction

$$\sigma_b = \frac{e_b}{F_{bb}} = \frac{3\pi E}{\beta^4} \frac{1-\sqrt{\frac{V_f}{V_o}}}{\left(\sqrt{\frac{V_a}{V_f}}-1\right)^4} \tag{3.21}$$

This equation, or slightly modified versions of it, has been used extensively for modeling the transverse compression of graphite and glass fiber bundles in thermoplastics, thermosets and oils. It reveals a rapid stiffening effect that can lead to load transfer from the resin to the fibers during processing. The importance of this effect is that the reduced resin pressure can, in turn, lead to voids in the composite.

A variety of other important loading conditions can also be modeled using equation (3.13). One worth noting is the axial extension ($e_1 > 0$) induced by transverse compression. If this axial extension is constrained, then the applied compression bulk stress would lead to an induced axial compression. This, in turn, could lead to fiber buckling and fiber waviness.

Example 3.1

Consider the case of autoclave processing. A bed of fibers is being compressed. The initial fiber volume fraction is $V_o = 0.50$. The final fiber volume fraction is $V_f = 0.68$. Assume that the maximum allowable fiber volume fraction is $V_a = 0.785$. Assume also that $E = 234$ GPa and $L/a = 200$. What would be the compressive stress to reach this final fiber volume fraction?

Solution

Using Equation (3.21), we have:

$$\beta = L / a = 200$$

$$\frac{3\pi E}{\beta^4} = (3)(\pi)(234 \text{ Gpa}) / 200^4 = 1378 \text{ Pa}$$

$$1 - \left(\frac{V_f}{V_o}\right)^{0.5} = 1 - \left(\frac{0.68}{0.50}\right)^{0.5} = -1.66$$

$$\left[\left(\frac{V_a}{V_f}\right)^{0.5} - 1\right]^4 = \left[\left(\frac{0.785}{0.68}\right)^{0.5} - 1\right]^4 = 0.000031$$

The required compressive stress is:

$$\sigma_b = \frac{(1378 \text{ Pa})(-0.166)}{0.000031} = 7.38 \text{ MPa} = 1071 \text{ psi} \qquad \text{(a)}$$

This stress is fairly high compared to normal autoclave pressure.

It can be seen from Equation (3.21) that apart from the volume fractions, the two parameters affecting the value of the compressive stress are the modulus E and the waviness ratio $\beta = L/a$.

If one were to use the value of $L/a = 350$, and the modulus of $E = 181$ GPa, the result would be:

$$\frac{3\pi E}{\beta^4} = (3)(\pi)(181 \text{ GPa}) / 350^4 = 113.6 \text{ Pa}$$

The compressive stress is now:

$$\sigma_b = \frac{(113.6 \text{ Pa})(-0.166)}{0.000031} = 0.608 \text{ MPa} = 88.3 \text{ psi} \qquad \text{(b)}$$

If $L/a = 400$ and the modulus is $E = 181$ GPa, one has:

$$\frac{3\pi E}{\beta^4} = (3)(\pi)(181 \text{ GPa}) / 400^4 = 66.6 \text{ Pa}$$

and the compressive stress is:

$$\sigma_b = \frac{(66.6 \text{ Pa})(-0.166)}{0.000031} = 0.357 \text{ MPa} = 51.8 \text{ psi} \qquad \text{(c)}$$

5.2. Experimental Values

A number of experiments were performed by a number of researchers on different types of fabrics. The results are shown in Figure 3.21. This figure shows the variation of the compressive stress with the fiber volume fraction. It can be seen that even though the experimental results follow curves of similar shape, there is a significant degree of variation between the data. Equation (3.21) shows that the compression stress depends on the modulus of the fiber E, the aspect ratio β ($= L/a$) and the three fiber volume fractions V_o, V_f and V_a. Among the three fiber volume fractions, V_f is the fiber volume fraction of the final laminate to be manufactured. V_o represents the initial fiber volume fraction, and may depend on how tight or how loose the fiber bundles are in their initial state. The maximum fiber volume fraction, V_a, is the value at which fluid cannot flow across the fiber bundle. This may have relation with the theoretical maximum value of fiber volume fraction.

By fitting Equation (3.21) to the individual data sets shown in Figure 3.21, Gutowski [2] showed that there are good fits shown in the figure. Also by shifting all of these curves along the V_f axis, a "universal" curve can be obtained to describe the bulk compression of carbon fibers. This is shown in Figure 3.22.

In the above derivations, the concept of maximum fiber volume fraction V_a is used. This maximum fiber volume fraction depends on the following factors:

- *The state of lubrication of the fiber beds.* Dry fiber beds may have lower V_a than lubricated fiber beds.

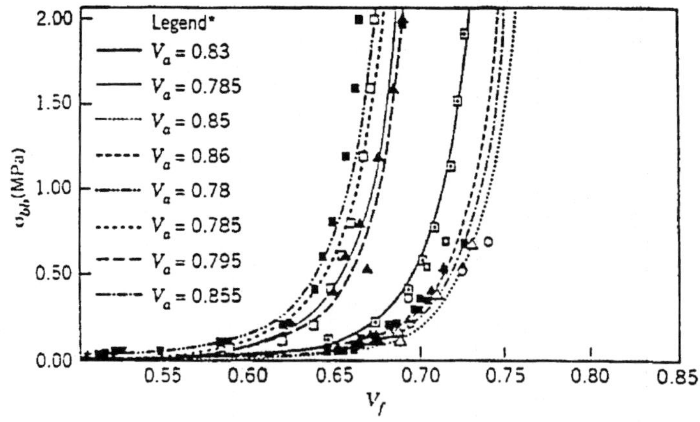

FIGURE 3.21 Fitting the experimental points with the equations [2].

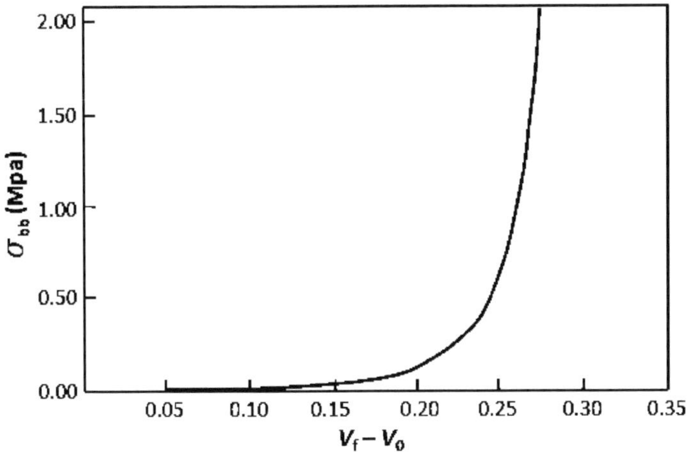

FIGURE 3.22 Master curve between compression stress and adjusted fiber volume fraction.

- *The rate of loading.* Faster rate of loading will produce lower V_a.
- *Repeat loading.* The maximum fiber volume fraction V_a can increase after a first cycle of loading. This increase will taper off after three or four cycles of loading.
- *Type of fabrics to be compressed.* Mats will show different values of V_a as compared to fabrics. Fabrics of different weave types will also show different V_a values.
- *The stacking sequence.* A fiber bed consisting of different combinations of fiber type or fabric type will show different values for V_a when the stacking sequence is different.
- *The fiber bed will also exhibit creep behavior* (deformation under constant load after some time).

Equations (3.13), (3.20) and (3.21) should be validated with a number of experiments on the specific type of fiber beds before use.

5.4. Summary and Conclusions

A number of experiments were performed by various researchers on different types of fabrics. They found that the compressive stress versus fiber volume fraction of the fiber bundles follows exponential curves after a certain critical fiber volume fraction. Different types of fabric and wetting systems follow different curves. Gutowski [2] showed that by plotting the compressive stress versus the difference between the fiber volume fraction and the initial fiber volume fraction, the curves from the different material systems fall along one single line.

6. REFERENCES

1. Strong B., *Principles of Composites Manufacturing,* Society of Manufacturing engineers, 1989.
2. Gutsowksi T.G., ed. *Advanced Composites Manufacturing,* John Wiley and Sons, 1997.
3. Lubin, G., *Handbook of Composites,* Van Nostrand Reinhold, 1982.
4. Watson, J. C. and N Raguhathi, in *Engineered Materials Handbook Composites,* ASM International, 1987.

7. HOMEWORK

1. A bed of fibers is compressed. The dimensions of the bed of fibers are 0.5 m × 0.4 m. The initial fiber volume fraction is 0.50. The maximum allowable fiber volume fraction is 0.80. Determine the compressive load required to compress the bed of fibers so that the fiber volume fraction reaches the following values:

    a. 0.60
    b. 0.65
    c. 0.70
    d. 0.75

PART 2

Techniques for Composites Manufacturing

Theoretically, manufacturing of composites can be broken down into the following items:

- Aligning of fibers
- Single filaments
- Tows
- Fabrics (mats, weaves, braids, knits)
- Bed consisting of many layers of fabrics
- Filling the interstices between filaments with liquid matrix
- Wetting the fibers
- Curing the resin

Filling interstices between the filaments with liquid resin can be done at the levels of single filaments, tows, fabrics, or a bed consisting of many layers of fabrics (refer to the list below). The arrangement of the fibers in different configurations was shown schematically in Figure 3.1. It is of no practical use to fill and wet single filaments with liquid resin so it is rarely done.

a. If resin filling happens at the level of many tows, preimpregnated tapes (for thermosets) or preimpregnated tow preg (for thermoplastics) is made.
b. If resin filling happens at the level of fabrics, preimpregnated fabrics are made. In order to make the final composite part, these preimpregnated tapes or fabrics need to be laid up on top of each other to form the thickness of the laminate.

c. If resin filling occurs at the bed of many layers of fabrics, then one can make the final composite part at the same manufacturing session with the liquid filling.

No matter which fiber level the liquid resin is introduced, it is essential that it wets the filaments to ensure good bonding between matrix and fibers. In the manufacturing process involving activities (*a*) and (*b*), the liquid resin has an opportunity to wet the fiber two times. First, when the preimpregnated tapes or fabrics are made, and second, when these layers are placed on top of each other and processed to make the final part. For processes involving activity (*c*), the liquid has only one opportunity to wet the fibers. Therefore it is necessary that sufficient time is allowed for the resin to properly wet the fibers.

The curing of resin is normally done after the resin has already filled all interstices and wet all filaments. However, due to the exothermic reaction, in order to avoid a peak in temperature rise, sometimes curing is done in stages (i.e. a thin layer may be cured first before additional layers are placed to add more thickness).

Above are the theoretical activities. For making a composite part, a manufacturer can combine or alternate these steps, depending on the requirements for quality and cost, as follows:

- For hand-lay-up in open mold for fiber glass/polyester, dry tows or dry fabrics are laid on a mold, liquid resin is then poured and spread onto the fiber beds. A few layers are wetted and left to cure in open air. After these layers are cured, more layers are added.
- For autoclave curing, pre-impregnated layers made from tows and fabrics are first manufactured by one group of people. The second group of people obtain these prepregs and lay them up to fit the the thickness and configuration of the part. These are placed inside an autoclave for curing.
- For filament winding, dry tows of fibers are run through a bath of resin to be wetted. These are then deposited onto the surface of a rotating mandrel. Many of these layers are deposited until the desired thickness is obtained. These layers are then left to cure either in room temperature or under some form of heat lamp.
- For pultrusion, the dry tows are run through a bath of resin to be wetted. These are then fed into a heated die. The fibers and resin are subjected to compaction and heating. When the assembly of fibers and resin exit the die, they are compacted and cured.

- For liquid composite molding, layers of dry fibers are stacked on top of each other in a mold. Liquid resin is then injected into the mold to wet the fibers. Heat is applied to cure.
- For thermoplastic composite processing, two different approaches are taken. In the first approach, the fiber tows are first impregnated with the liquid thermoplastic resin. It is then allowed to cool so that tow pregs of themoplastic resin and fiber are obtained. These tow pregs are then placed on top of each other to make up the thickness. The stack of tow pregs is compression molded to make the part. In the second approach, the thermoplastic resin is brought to the vicinity of the fibers. This can be done by drawing the thermoplastic resin into small filaments, calendering the liquid thermoplastic resin into thin sheets, or grinding the thermoplastic resin into powders. These forms of thermoplastic resin are placed close to the dry fibers. The whole assembly of dry fiber and thermoplastic resin is placed inside a mold. Compression molding is used to consolidate the final part.

The rule of thumb is that for good quality, the whole process should be broken down into small steps and each step done at a different time. For low cost manufacturing, many steps may be combined so the process can be done at one time or a lesser number of times. The selection of the process depends on weighing these two parameters: quality and cost. The following chapters present current techniques for manufacturing. By understanding the principles used for manufacturing, new techniques can also be developed.

CHAPTER 4

Hand Laminating (or Wet Lay-up) and the Autoclave Processing of Composites

1. HAND LAMINATING (OR WET LAY-UP)

The hand laminating or wet lay-up processing technique is presented in the same chapter with autoclave processing because these two processes have one thing in common: the depositing of the different layers is done by hand. (Initially the lay-up for autoclave has been done by hand; recently many large companies have adopted the automated fiber placement machine for laying laminates for autoclave curing.) Apart from that, the materials used, the components to be made and the quality of the parts are very different. The hand laminating or wet lay-up process is used to make low cost components such as boats and corrosion-resistant components in the chemical process industry, while the autoclave process is used to make components for the aircraft industry. Hand laminating or wet lay-up works with fiber glass and polyester (or vinyl ester) while the autoclave process mostly works with epoxy resins.

Hand laminating is a primitive but effective method that is still widely used for prototyping and small batch production. The most common materials are E glass fiber and polyester resin, although higher performance materials can also be used. The single sided mold is invariably operated at room temperature using an ambient curing resin. The reinforcement may be in the form of chopped strand mat or an aligned fabric such as woven rovings.

The usual feature of hand laminating is a single sided female mold, which is often itself made of glass fiber reinforced plastics (GRP), by taking a reversal from a male pattern. The GRP shell is often stiffened with local reinforcement, a wooden frame or light steel work to make it sufficiently stiff to withstand handling loads. The mold surface needs to be smooth enough to give an acceptable surface finish and release properties and this is provided by a tooling gel coat that is subsequently coated with a release agent. The latter prevents the matrix resin from bonding to the mold surface and facilitates the de-molding operation. It is common practice to use a surface tissue immediately after the gel coat to mask any reinforcement print-through on the outer surface.

Once the gel coat has hardened sufficiently, the reinforcement is laid in, one layer at a time. Catalyzed resin is then worked into the reinforcement using a brush or roller. This process is repeated for each layer of reinforcement until the required thickness is built up. For thick laminates, pauses need to be taken after a certain number of layers have been deposited to allow the exothermic heat to dissipate before additional layers are deposited. Local reinforcements can be used to provide stiffness in specific areas and lightweight formers such as foams or hollow sections can be laminated in for the same purpose. Figure 4.1 shows a schematic of the hand laminating process.

The major limitation of hand laminating is that the mold has only one smooth surface. The absence of direct control over part thickness, fiber content, void fraction and surface quality on the other surface means that the moldings are used in very low stress applications and in areas where dimension accuracy is noncritical. Although capital costs are low, production is labor intensive and quality control is relatively difficult. The quality of the final part is highly dependent on the skill of the operator.

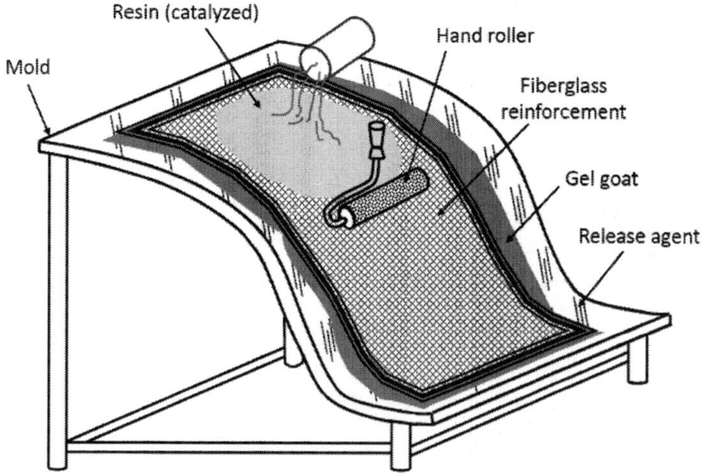

FIGURE 4.1 Schematic of the hand laminating process.

The process remains an important one for low volume manufacture, although increasingly stringent emission regulations are forcing several manufacturers to explore the use of closed mold alternatives. Hand laminating using open molds has traditionally been used for making structures out of fiberglass and polyester, but there are environmental concerns about evaporation of styrene into the atmosphere. There are new techniques of liquid composite molding that may produce similar parts with the closed mold, thus avoiding the environmental issue.

2. AUTOCLAVE PROCESSING

2.1. Introduction

Autoclave processing is commonly used for manufacturing composite components for the aerospace industry. The process produces composite components of high quality, but it requires a considerable amount of time. The main steps of the autoclave processing of composites are:

- Prepregs
- Tool preparation
- Laying up prepregs on the tool to make the part
- Curing of the part
- Removal of the part from the tool
- Inspection
- Finishing steps

In the Chapter 1, Figure 1.7(a) shows the different stages for the manufacturing of composites. For the manufacturing using autoclave, the stages involved are *a*, *b* and *c* or *a*, *b* and *d*. As one moves from stage *a* to *b*, the dry fibers and the liquid resin are combined to make a semi-finished form. This semi-finished form consists of both fibers that are uniformly spread out over a rectangular cross section area and partially cured matrix resin that holds the fibers in place. This semi-finished form is called *prepreg* (pre-impregnated). As one moves from stage *b* to stage *c* or from *b* to *d*, many layers of the prepregs are stacked on each other to form a sheet of sufficient thickness for practical use. Note that one does not move from stage *c* to stage *d*. In the case where the process involves stages *a*, *b* and *c*, flat plate laminates (as in stage *c*) are made. Flat plate laminates are usually made for the purpose of studying and characterizing the properties of the laminate. For real practical applications, normally laminates may have curvatures and more complex configuration such as those shown in stage *d*. For this latter case, the process goes di-

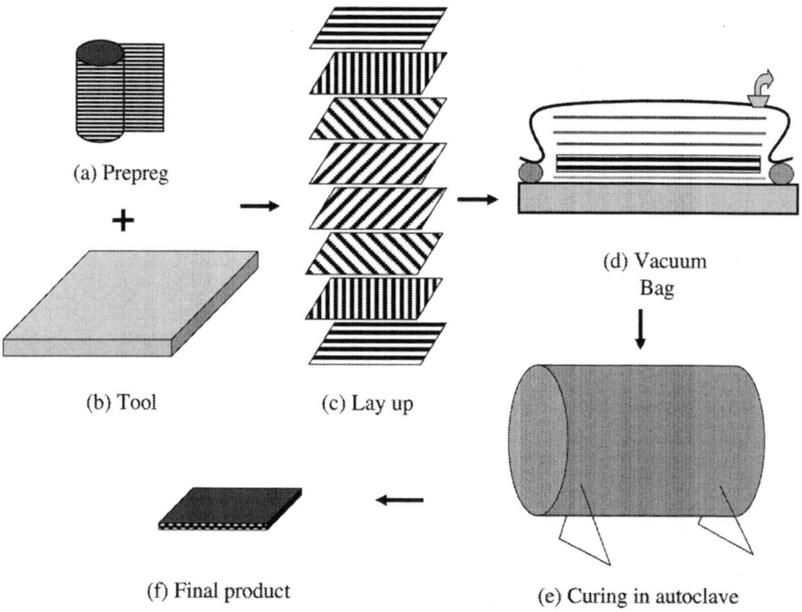

FIGURE 4.2 Main steps in the autoclave manufacturing process.

rectly from stage *b* to stage *d*. The properties of flat laminates can usually be applied to curved laminates with appropriate consideration for the change of the coordinate system. Figure 4.2 shows the main steps for autoclave manufacturing. These are described below:

2.2. Prepreg

2.2.1. Prepreg Manufacturing and Handling

Prepregging involves the incorporation of the partially cured resin with the fibers. Figure 4.3 shows a schematic of the prepregging machine. In the prepregging process,dry fibers are fed from creels through stations of combs where the fibers are spread out. The fibers then enter into a bath of wet resin where they are wetted. Subsequently the fiber/resin combination is heated to change the liquid resin into a partially cured state. The partially cured resin is viscous enough to help keep the fibers in the configuration of flat sheets. This fiber/viscous resin combination is called prepreg. Normally sheets of backing paper are placed on both sides of the prepreg for handing purposes. Then the prepregs are rolled up for storing and shipping.

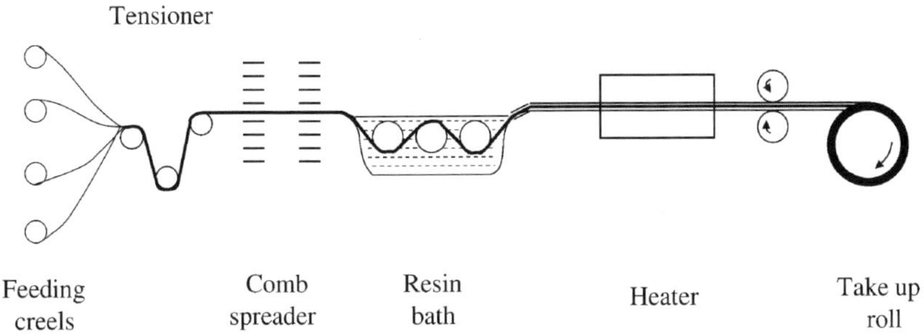

FIGURE 4.3 Schematic of a prepregging machine.

Resins are usually thermosets such as epoxies, although recently prepregs made of vinyl ester resins have also been available. The partially cured resin has about 30% of the crosslinks already formed. With the incorporation of fibers (such as carbon, glass or Kevlar at about 60% by volume), prepregs are flexible sheets of fibers about 150 mm thick. This is similar to a sheet of wallpaper except that it is sticky on both sides. Figure 4.4 shows a roll and a sheet of the carbon/epoxy prepreg.

FIGURE 4.4 A roll and a sheet of carbon/epoxy prepreg.

The thermoset resins (such as epoxy) inside the prepregs contain all of the curing agents necessary for the complete cure of the resin. Only a portion of the crosslinking, however. has taken place, due to the addition of certain amounts of inhibitors into the resin, and/or by allowing the reaction to take place at low temperature and within a limited amount of time. The time from the date of manufacturing of the prepregs to the time when the actual part is done may be several months. If the prepregs are left at room temperature on the shelf, the reaction in the resin may continue and the resin becomes hard (more crosslinking has already taken place). Once the resin is hard, it is no longer sticky and one cannot bond the prepregs together to make the composite component. Therefore it is essential to slow down the reaction of the resin until the prepregs are laid up to make the composite component. Slowing down is usually done by storing the prepregs at low temperature. Normally during shipping the prepregs are contained inside refrigerated bags. Once received, the prepregs need to be stored inside a freezer at about –5°C. Usually the supplier of the prepregs provides specifications as to the storage temperature and the maximum amount of time of storage (shelf life) at that temperature. This means that even at that low temperature, the resin in the prepregs may keep on with the process of crosslinking and become mostly cured at the end of the period.

Prepregs are available in the form of tapes (unidirectional fibers) or fabrics (woven). Widths of prepreg rolls vary from 25.4 mm (1 inch) to 305 mm (12 inches).

2.2.2. Prepreg Cutting

When one is ready to make the composite component, the prepregs are taken out of the freezer and left at room conditions for a few hours. This allows the prepregs to rise to room temperature, reducing viscosity of the resin. Since normal room conditions contain a certain amount of humidity, equilibrating between the prepregs and room conditions means that the prepreg may acquire a similar amount of moisture. This has implications for the formation of voids, which will be discussed later in this chapter.

Once the prepregs have become pliable (viscosity of the resin is low enough for the prepregs to be shaped to conform to the contour of the mold), they are cut into the desirable configuration. Cutting the fabrics of the prepregs is similar to cutting cloth fabrics to make a suit, except that here carbon fibers are involved, rather than textile fibers. This means that the contour of the cut depends on the configuration of the part to be made. For example, to make the nose cone of an airplane, the geometric configuration of the surface of the cone when it is opened up onto a flat surface needs to be worked out, in order to learn how the flat piece(s) of fabric will cover that surface. On surfaces with double curvatures, where several flat pieces of fabric are required, care should be taken to assure continuity of fiber orientation between adjacent pieces.

Cutting used to be done by hand. Recently, however, this work has been mechanized and computerized, not only to speed up the process but also to reduce waste, which ordinarily increases very significantly the cost of manufacturing a composite part. Waste occurs due to the requirement of cutting prepregs at an angle with the axis of the prepreg roll. By computerization, one may nest the different parts to minimize waste.

With the use of the prepregs, the availability issue mentioned in chapter 1 is not a concern. For the autoclave process this issue is already taken care of during manufacturing.

2.3. Tool Preparation

Manufacturing using autoclave is a molding process. As such, molds (also called tools) are required. The mold provides the shape and surface finish for the part. As such the size of the mold depends on the size of the part. Large parts require large molds and these can be very expensive. Advanced composites must be cured at about 180°C and at pressures of about 600 kPa; molds would be required to sustain these conditions for periods of several hours. In addition, there are many other considerations when designing and building tools. These include tool cost, life, accuracy, weight, machinability, strength, thermal expansion, dimensional stability, surface finish, and thermal mass and thermal conductivity.

Over a wide range of material systems and processing scenarios used for composites, there are many materials suitable for tooling. In general, the choices fall into three categories:

1. Reinforced polymers, for low to intermediate temperature ranges
2. Metals, for low to high temperatures
3. Ceramics and bulk graphite, for very high temperatures

In addition, one may use cast plaster and other inexpensive and relatively easy-to-process materials for small scale part runs like those needed for prototype verification.

For production tooling for advanced composites, the choice is usually made between metals, including aluminum, steel, nickel alloys (Invar), electroformed nickel, and graphite/epoxy tooling. Elastomeric tooling is often used as a pressure intensifier and to distribute the applied pressure over a part.

For high temperature applications, such as thermoplastic composites, much consideration has been given to bulk graphite and various ceramic systems, including a new material called "geopolymere."

2.3.1. Aluminum, Steel, and Invar Tool [1]

These materials are highly desirable for production tooling because of their good surface finish and ability to stand up to repeated production runs. In this respect, Invar is quite desirable because its hardness is similar to that of steel and greater than that of aluminum and its coefficient of thermal expansion (CTE) is below both aluminum and steel (on the order of 0.28×10^{-6} m/m °C). Conversely, aluminum is the least desirable because of its relative softness and high CTE (on the order of 26.2×10^{-6} m/m °C). On the other hand, aluminum tools can be significantly lighter than other tooling materials and therefore easier to move about on the factory floor. Table 4.1 gives the CTEs of various tooling materials relative to a $[45/0/-45/90]_s$ lay-up of the thermoplastic composite APC2/AS4. Aluminum has large thermal expansion coefficient which can result in large mismatch with that of the composite. In terms of ease of fabrication, however, aluminum is much easier to machine than is steel or Invar.

2.3.2. Electroformed Nickel [1]

Electroformed nickel tooling is produced by an electroplating process

TABLE 4.1 Coefficient of Thermal Expansion of Various Mold Materials (1) Lay-up $[45/0/-45/90]_{3s}$, (2) 0.2–0.5% Carbon Steel, (3) 18 Cr+9% Ni Steel.

Material	Coefficient of Thermal Expansion (10^{-6}/°C)
Composite APC2/AS4	3.8
Bulk graphite	3.0
Ceramic	0.7
Metal-ceramic	7.2
Polyimide	4.7
Aluminum	26.2
0.2–0.5% carbon steel	13.2
18Cr+9% Ni steel	17.8
Cast iron	11.1

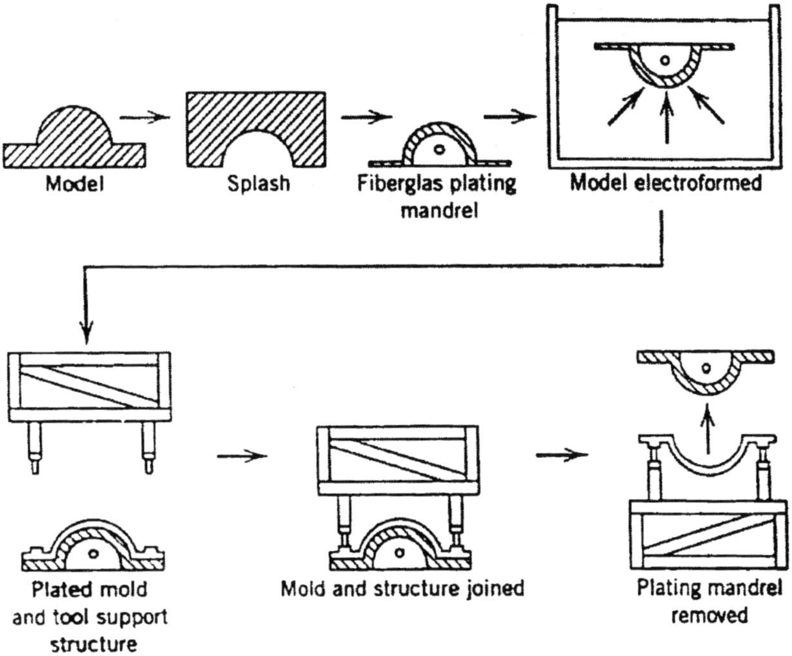

FIGURE 4.5 Stages in the production of electroformed nickel tooling (reproduced from *Advanced Composites Manufacturing* by T.G. Gutowski, 1997, with permission from John Wiley and Sons).

that deposits the material onto a master mold (made usually of plaster) which matches the dimensions of the final part (with perhaps some allowance for expected thermal distortions). The key requirement of the master mold is that it must remain dimensionally stable during the plating process. The basic stages of the process are shown in Figure 4.5.

Since the production of the master mold is the most time-consuming part of the process, one of major advantages of the technique is that duplication can be carried out at a relatively low cost and with great repeatability. Once the tool surface is produced, high polishes may be achieved using standard metal finishing techniques. The major drawback associated with electroformed nickel tooling is the high cost associated with the initial tool production.

Electroformed nickel tools are durable, scratch resistant, relatively easily repaired, and possesses good release properties from most commonly used composites. Perhaps the greatest advantage of this type of tooling is that the scale of the possible parts is limited by the size of the plating tank and the master production techniques.

2.3.3. Graphite/epoxy Tooling [1]

The problem of CTE mismatch with the use of metal tools has led to the increased popularity of graphite/epoxy tooling for autoclave molding processes. The tool making process is similar to that of electroforming of nickel in that it starts with the development of a master model that may, in some cases, be made from plaster or wood. This is then properly sealed and treated with mold release, and the carbon/epoxy material is laid up in the conventional manner. An autoclave curing cycle is applied to solidify the material. A high finish can be applied to the tooling by either polishing or applying a high-gloss gel coat to the master model.

Other than the ability to tailor CTE, additional advantages of graphite/epoxy tooling include lightness, ease of construction, and rapid response to heating profiles. The major limitation of the material is that it cannot be used when processing metals at temperatures greatly in excess of the glass transition temperature of the mold material, especially when high autoclave pressures are necessary. Since the surface is basically polymeric, graphite/epoxy tools tend to be somewhat less durable and resistant to scratches than conventional metallic tooling materials.

2.3.4 Elastomeric (or Rubber) Tooling [1]

Elastomeric tooling is employed to either intensify or redistribute pressure during a molding cycle. It is very useful in conditions where it is difficult to get a vacuum bag into the recesses of a complicated mold: for example, in the vertical elements of a stiffened panel. In such cases, the expansion of the elastomeric material generates high lateral loads as shown in Figure 4.6.

In its most basic form, elastomeric tooling may be applied as simply a rubber caul sheet between the vacuum bag and the part, or rubber pads that are used in areas of the part that are known to be difficult to consolidate. Elastomeric inserts are commonly used on parts where bridging or the composite or vacuum bag has caused problems of incomplete consolidations or bag rupture in tight radii.

The major disadvantages of elastomeric tooling are: low tool life?most elastomeric tools are seriously degraded after about 30 thermal pressurization cycles use for advanced composites; and low thermal conductivity, which slows the cure cycle and sometimes necessitates the redesign of the processing cycle. This may sometimes be alleviated by inserting metal plugs to reduce the rubber mass and increase conductivity.

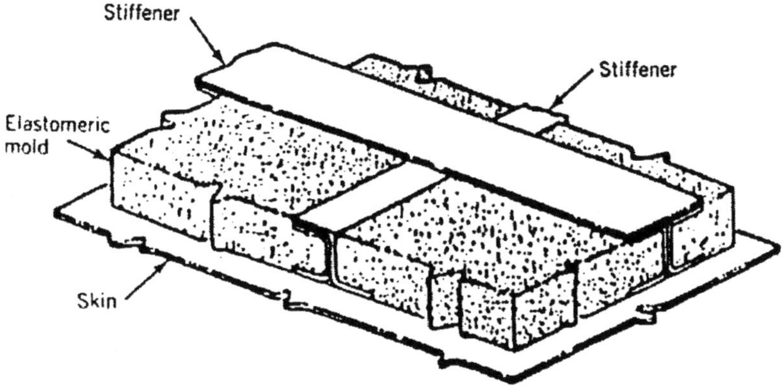

FIGURE 4.6 Example of elastomeric tooling for molding of stiffened panel (reproduced from "The autoclave processing of composites," by G. Dillon, P. Mallon and M. Monaghan, in *Advanced Composites Manufacturing,* by T. G. Gutowski, 1997, with permission from John Wiley and Sons).

2.3.5. Bulk Graphite and Ceramic Tooling [1]

The introduction of thermoplastic composites that must be processed considerably above their T_g has created the need for high temperature tooling. To date no thoroughly acceptable tooling material has been found. However, the need has sparked experimentation with various high temperature materials. Chief among them has been bulk or monolithic graphite and various ceramic tooling materials.

Bulk graphite tools have the advantages of low CTE, light weight, relatively easy fabrication, and high thermal conductivity and can be used to temperature as high as 2000°C. These tools are generally fabricated from blocks of monolithic graphite that are machined and joined together. The major drawback is that monolithic graphite tools tend to be extremely fragile, hence autoclave processing cycles can be quite low, perhaps less than 10.

2.4. Release Agent

Before the laying of the prepregs on the surface of the tool, a *release agent* needs to be applied. This is to facilitate the removal of the part after cure. Insufficient application of the release agent on the surface of the tool can result in the part sticking to the tool surface. Removal of the part that sticks to the tool may result in damage to both the tool surface and the part. Sometimes several coats of release agent may be required and baking of the release agent may be necessary. Release agents will be discussed further in the next section.

2.5. Lay-up of the Prepregs on the Tool to Make the Part

2.5.1. Determination of Number of Layers and Layer Orientation

Normally an engineering part requires a certain thickness to carry load. The thickness of composite laminates in aerospace applications is usually of several millimeters. Assuming that a part 3 mm thick is to be made, and assuming that a thickness of each layer after cure is 125 μm (thickness of a layer of prepreg of 150 μm was mentioned previously, some resin is bled out during the autoclaving process), one would require about 24 layers. The number of layers and the orientation of each of the layers can be tailored to meet the mechanical load requirement. (Discussion of the effects of different fiber orientation on the properties of the laminates can be found in Reference [2].) The order in which the layers are stacked on each other is called a stacking sequence. Figure 4.7 shows a typical stacking sequence. Normally the stacking sequence is written in a laminate code such as $[0/90/45/-45]_s$. This means that if one starts from the bottom, then the bottom layer has fibers that are oriented along the

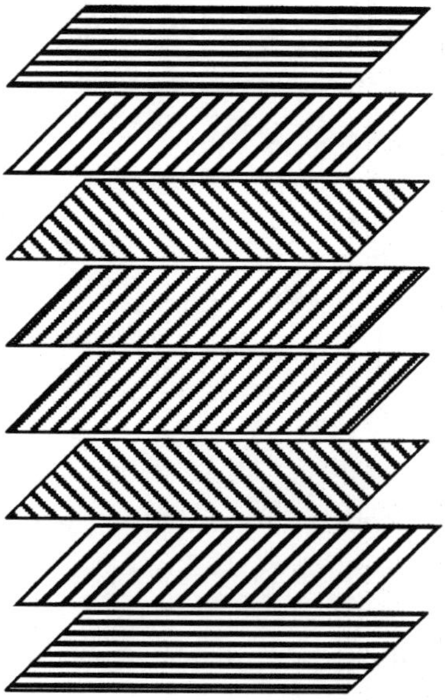

FIGURE 4.7 A typical lay-up sequence $[0/90/45/-45/]_s$.

axis of the part to be made, followed by a layer having fibers oriented at 90° with the axis of the part, then by a layer with fibers oriented at 45° etc. The subscript s indicates symmetry. In other words, the code can be written in full as [0/90/45/–45/–45/45/90/0].

After the prepregs are cut, they can either be laid on the tool to make the part right away or they can be kitted away for later laying up. Labeling of the kit is extremely important to avoid mix-up problems later on.

2.5.2. Laying Up

The laying up of the prepregs on the surface of the mold consists of not only the laying of the fiber prepregs on the mold, but also the placement of ancillary materials for the following purposes:

- To facilitate the removal of the part after cure (without the problem of the part sticking to the mold)
- To allow the compaction of the stack of prepregs using vacuum
- To prevent excess resin from running within the plane of the stack of fibers, which can distort the orientation of the fibers
- To provide an escape path for volatiles such as water vapor or gases that are generated during the curing process
- To provide materials that can absorb excess resins that ooze out of the laminate during the curing and molding process
- To obtain good surface finish on the part

The result of the laying up process is a stack of the prepregs along with many other layers within a vacuum bag as shown in Figure 4.8. The steps included in the laying up process are as follows.

2.5.2.1. Preparation of the Mold Surface

As discussed in Section 2.3 (Tool Preparation) of this chapter, the selection of the mold material depends on many factors. To assure good surface finish for the part, the mold surface needs to be well polished. Poor mold surface finish will result in poor surface finish for the part. Thermoset resins such as epoxies are basically glue. They will stick to the mold materials, which normally are made of metals, composites or ceramics. When the part sticks to the mold, removal of the part from the mold may result in damage to either the mold or the part. Molds are usually very expensive to prepare, therefore it is essential to prevent sticking. For this purpose, release agents are applied, either in liquid form that can be spread over the surface of the mold, or in the form of a release film.

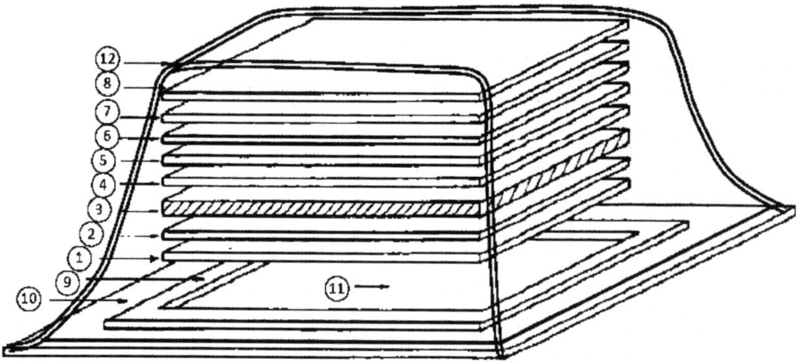

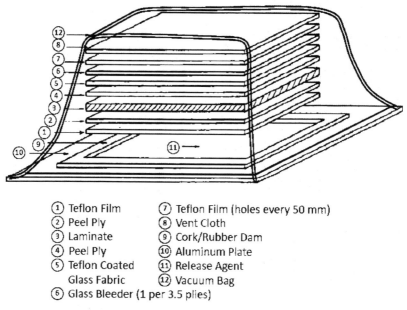

(1) Teflon Film
(2) Peel Ply
(3) Laminate
(4) Peel Ply
(5) Teflon Coated Glass Fabric
(6) Glass Bleeder (1 per 3.5 plies)
(7) Teflon Film (holes every 50 mm)
(8) Vent Cloth
(9) Cork/Rubber Dam
(10) Aluminum Plate
(11) Release Agent
(12) Vacuum Bag

FIGURE 4.8 The lay-up assembly of different layers.

Usually several coats of the release agent are applied, each coat being allowed to dry before the next one is put on. Release films are usually made of low surface tension polymer such as Teflon. Sometimes the release film is applied on top of the liquid release agent. After the release film, a peel ply may be applied to add to the release action, and also to provide a texture for the surface of the part.

In order to prevent resins to flow out of the prepregs in the plane of the stack of prepregs, dams made of ceramic or metal blocks are placed along the periphery of the stack.

2.5.2.2. Laying Up the Stack of Prepregs

After the release agents and peel ply have been placed on the surface of the mold, stacks of the prepregs are positioned. This has been done by hand, thus the term hand lay-up (HLU). One may ask why such high tech parts as composite components for airplanes are made by such a rudimentary technique as HLU instead of by machines, which perform with greater consistency. The reason is the low volume production for aircraft components: airplanes are more or less custom-made for different customers, so aircraft companies do not mass produce them. Because parts that look similar may have differences in detailed configurations, it is not economical to build machines to do the laying up. This apparent disadvantage has an advantage in that the HLU process is so versatile that it can be used to build most, if not all, the composite parts for the aircraft industry. The exception is the manufacturing of tubes or other components having surfaces of revolution. Recently the development of fiber placement machines allow for automation of the hand lay-up process. These are virtually robots with many degrees of freedom (the Cincinnati Milatron, for example) which require huge capital investment, affordable only by large companies.

It is essential that the stack of prepregs is well packed. As discussed in Chapter 3, fibers in the prepregs may be wavy while one needs to make laminates with straight fibers. To assure good packing, the usual practice is to perform *debulking*. In this process, after about 3–5 layers of prepregs have been laid, a vacuum bag of thick film is placed around the mold and the stack of prepregs. Vacuum is then applied for about 10 minutes onto the partially laid prepregs for consolidation. The process is repeated after each stack of 3–5 layers are laid until the whole laminate is laid up. This is a time consuming process but it is essential to assure good quality laminates. After the laminate has been laid and debulked, another peel ply is applied on top for part removal purpose.

2.5.2.3. Placement of Bleeder Materials

Bleeder materials are used to absorb any resin that may ooze out during the curing and molding process. Usually prepregs contain more resin than what remains in the final laminate. Some of the resins in the prepregs will flow out and absorb into the bleeder material. The bleeder material is placed outside the release fabrics.

2.5.2.4. Placement of Breather Materials

Breather materials are perforated films of polymer of high temperature resistance. The holes allow volatiles such as water vapor or gases that are formed during the curing process to escape. The breather materials are placed outside the bleeder materials.

2.5.2.5. Placement of a Caul Plate

A caul plate is sometimes used to aid in the consolidation. The weight of the caul plate delivers some pressure to the laminate. The caul plate is placed outside the bagging assembly but inside the vacuum bag.

2.5.2.6. Placement of Vacuum Bag

The final layer that goes on top of the whole assembly is the vacuum bag. A hole is made in the vacuum bag to incorporate the vacuum valve. It is then sealed. The vacuum bag is used to compact the bed of prepregs, and also to aid in the removal of volatiles that may be produced during the curing process. Vacuum is kept on at all times during the autoclaving process.

The assembly of all layers is shown in Figure 4.9. The whole lay-up assembly is then placed inside the autoclave for curing. For making a small laminate such as that of coupons for testing, a plate of about 300 mm × 300 mm may be used and the whole assembly may not be too large or too heavy. However, the manufacturing of a large structure such as parts of a wing of an aircraft with dimensions on the order of tens of meters can require a whole bagging assembly (including the tool) of several tons in weight. The handling of such an assembly requires careful planning.

2.6. Curing and Consolidation of the Part

The resin in the stack of layers of composite in the bag mentioned in the previous section is a viscous liquid. This resin needs to be transformed into a solid to make a useful composite, which requires heat to activate the chemical reaction between the molecules (as discussed in Chapter 2). During this transformation of the resin, it is important to assure that the fibers maintain their orientation and that no resin rich area or

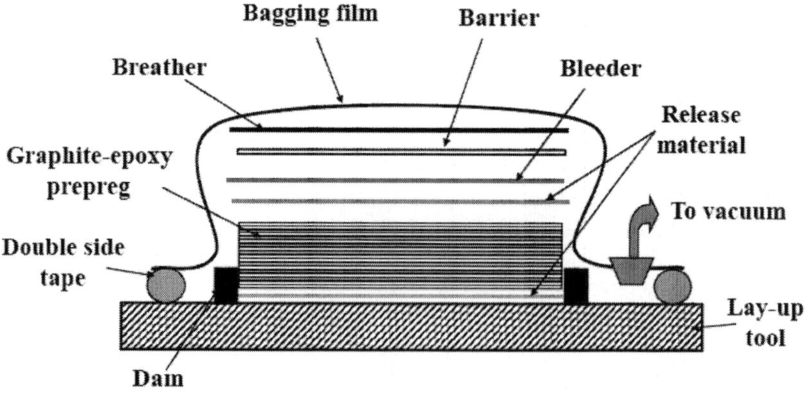

FIGURE 4.9 Assembly of the bagged composite.

other defects will exist. Sufficient amounts of pressure need to be applied for this purpose. Also, during the transformation of the resin from the liquid to solid state, volatiles such as water vapor or other gases may be generated. These need to be removed from the material in order to avoid the occurrence of voids after the resin has become solid.

Heating of the resin needs to follow a very carefully planned schedule. This is due to many phenomena that occur during the process:

a. Heating will kick start the chemical reaction between the bonds that have not reacted. In the case of epoxy, these are the bonds between the epoxy molecules and the molecules of the curing agents (such as amines or anhydrides). The reaction between these bonds is exothermic, i.e. heat is generated as a result of the reaction. When this occurs, the temperature of the resin can increase. This increase in temperature can accelerate quickly. If not properly handled, the temperature of the resin can pass the degradation temperature of the resin and damage such as burning may occur.
b. The percentage of the amount of chemical bonds that have formed represents the degree of cure of the resin. At the beginning when no chemical bond has formed, the degree of cure is 0. When all chemical bonds have formed, the degree of cure is 1. When only a portion of the bonds have formed, the material is partially cured and the degree of cure is between 0 and 1.
c. The viscosity of the liquid resin depends on the temperature and also on the degree of cure. The greater the temperature, the smaller the viscosity of the liquid is (before curing becomes significant). At the same time, the greater the degree of cure, the higher the viscosity. Higher temperature does speed up the degree of cure. As such higher temperature provides two competing effects on the viscosity. The value of the viscosity has influence on the flow of the resin. Flow of resin is important for the squeezing of excess resin out of the laminate to assure high fiber volume fraction. Flow of resin is also essential for the squeezing out of bubbles of water vapor or volatiles. Therefore, it is important to apply pressure when the viscosity is low. At the same time, in order to reduce processing time, it is good to have the resin cure as quickly as possible. These two competing effects have influence on the determination of the heating and pressurizing schedule.

2.6.1. Resin Kinetics

As presented in Chapter 2, the curing of the resin happens due to the chemical reaction between the epoxy molecules and the curing agent molecules. This reaction is fairly complex and depends on the combination of the type of epoxy molecule and the type of curing agent molecule. For the purpose of illustration, DGEPBA epoxy and TETA amine curing agent are used. The chemical formulae for these molecules were presented in Figure 2.8 and Table 2.5 and are repeated for convenience in Figure 4.10.

For the epoxy-amine resin system, curing takes place due to the reaction between the epoxy group (the ring CH_2–O–CH at the end of the molecule) and the amine groups (segments of the molecule containing N atom) of the amine molecule. The amine groups are classified into two categories: primary amine and secondary amine. The reason for this is because it takes less energy to activate a reaction involving the primary amine than that of a secondary amine.

Refer now back to the presentation for the chemical reaction between the epoxy and amine curing agents in Section 3.2.2 of Chapter 2. In the DETA molecule shown above, the segment NH_2 is called the primary amine and the segment NH is called the secondary amine. Also when the primary amine has finished its reaction, one hydrogen atom leaves the segment and the remaining part of the segment contains NH. This remaining segment becomes a secondary amine. With the application of catalysts and/or heat, reaction between segments of the epoxy molecules may occur. This is called homopolymerization or Etherification. One group of researchers [3] proposed that the activation energies for the different amine reactions with an epoxy are: primary amine: 83 kJ/mole and secondary amine: 131 kJ/mole. From each of these reactions there are different amounts of heat that are generated. One group obtained the

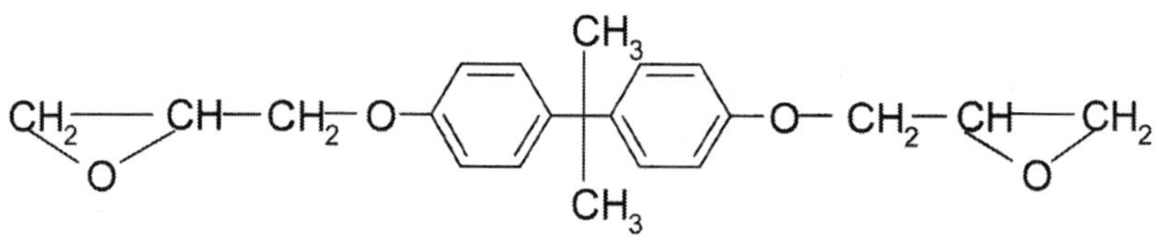

DGEBPA epoxy molecule

$$H_2N-CH_2-CH_2-NH-CH_2-CH_2-NH_2$$

Diethylenetriamine (DETA) molecule

FIGURE 4.10 Chemical formulae of epoxy and amine molecules.

amounts to be [4,5]: primary amine: 61.4 kJ/mole; secondary amine: 72 kJ/mole; and etherification: 101 kJ/mole.

Initially, some heating is required to get the reaction started and primary amines are normally the first involved. When some reactions have taken place, there will be heat generated from these reactions and the temperature of the material will increase. Apart from the involvement of the primary amines, secondary amines and even etherification may also participate into the reactions. This means that the reactions may not consume all of the primary amines before the secondary amines and etherification may get into the action. Besides this, the heat capacity of the material may change as it changes its state. This complicates determining the heat that may be generated during the course of the reaction. Too much heating from an external source may cause a runaway reaction which can become dangerous because the high temperature may cause burning. In the case where reaction takes place in an enclosed container, pressure may build up and this may lead to explosion.

In spite of the complexities involved, it is essential to control and quantify the process for engineering purposes. As mentioned in Chapter 2, the degree of cure of the resin is the percentage of the epoxy ends that have been consumed in the chemical reactions. This can be represented by the ratio of the amount of heat that has been generated over the total amount of heat that can be generated when the material is completely cured as:

$$\alpha = \frac{H_t}{H_T} \tag{4.1}$$

where,

α = the degree of cure
H_t = heat generated up to a certain time t
H_T = total heat generated at complete cure

The rate of cure is given as:

$$\frac{d\alpha}{dt} = \frac{1}{H_T}\frac{dH_t}{dt} \tag{4.2}$$

Differential scanning calorimeters have been used to measure the heat generated during curing of polymeric materials. Many efforts have been made to model the rate of cure of the resin. As can be seen the rate of cure depends on the temperature and the degree of cure. It also depends on the mechanism responsible for the curing reaction, such as whether primary amines, secondary amines, or etherification are involved. The general form for the rate of cure can be written as:

$$\frac{d\alpha}{dt} = f(\alpha, T) \tag{4.3}$$

Meticulous DSC measurements need to be carried out to determine the mathematical model for a particular resin system. One general form for the equation can be:

$$\frac{d\alpha}{dt} = K\alpha^m(1-\alpha)^n \tag{4.4}$$

where,

$K = Ae^{-E/RT}$
A = the frequency factor
E = activation energy
m, n = orders of the reactions

When $m = 0$, the model is called nth order and the reaction is fastest at the beginning when $\alpha = 0$. When $n = 0$, the model is called autocatalytic and the reaction is 0 at the beginning. Figure 4.11 shows the two extreme cases. The real case is when m and n have numerical values and the reaction is fastest after a certain amount of cure has already taken place.

Equation (4.4) shows the basic form for the model. The model can get

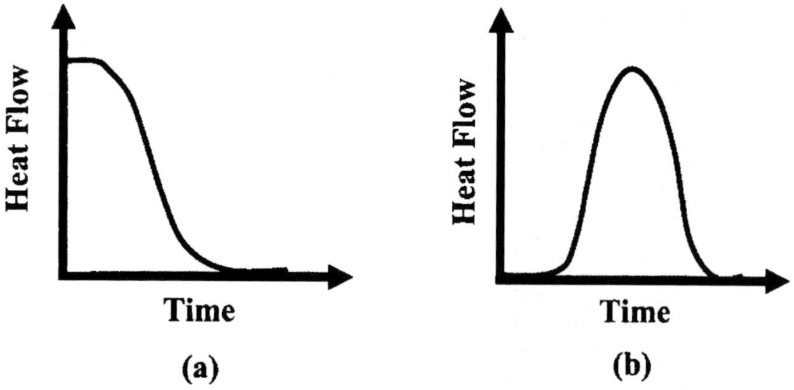

FIGURE 4.11 Two cases of cure rates: (a) nth order and (b) $m, n \neq 0$.

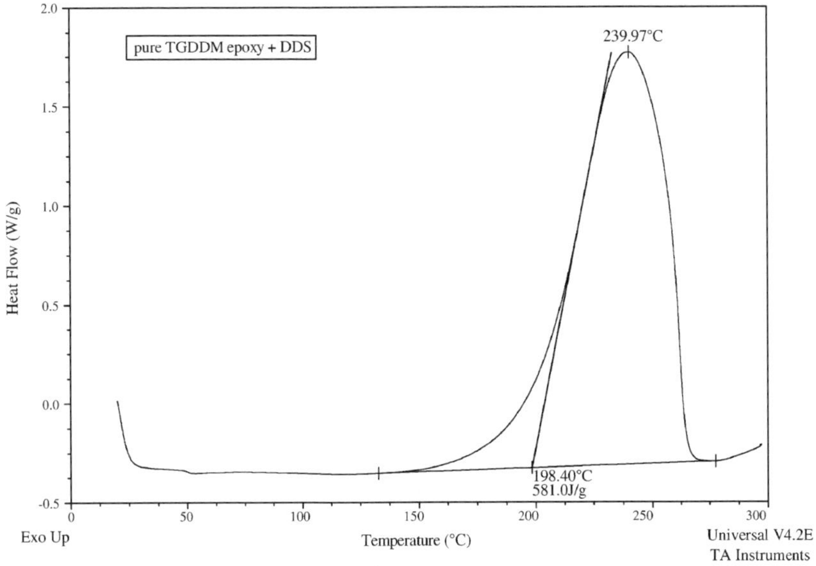

FIGURE 4.12 Heat flow versus time curve for TGDDM/DDS epoxy resin.

complicated where the mechanism of cure changes as curing progresses. One example would be:

$$\frac{d\alpha}{dt} = (K_1 + K_2\alpha)(1-\alpha)^{n1} + K_3(1-\alpha)^{n2} \tag{4.5}$$

In Equation (4.5), three activation energies E_1, E_2, E_3 (associated with K_1, K_2, K_3) are involved. Figure 4.12 shows an example of a more realistic heat flow versus time curve for an epoxy resin.

The above shows that the model for the rate of cure can get complicated. One needs to formulate a model for each resin system. The reason for this is because commercial resin systems usually contain blends of different resins. Also fillers such as accelerators or inhibitors are usually added to speed up or to retard the curing reaction, or only to kick up the curing reaction after a certain degree of cure.

2.6.2. Heat Transfer and Energy Balance

In order to determine the temperature of the composite material during the manufacturing process, a volume element of the material can be considered. For this volume element, there is heat transferred across the faces of the element and there is also heat generated due to chemical reaction. By neglecting energy transfer by convection as would be the case for a process such as autoclaving, the energy equation may be expressed as:

$$\frac{\partial(\rho C_p T)}{\partial t} = \frac{\partial}{\partial x}\left(K_x \frac{\partial T}{\partial X}\right) + \frac{\partial}{\partial y}\left(K_y \frac{\partial T}{\partial Y}\right) + \frac{\partial}{\partial z}\left(K_z \frac{\partial T}{\partial Z}\right) + \frac{dH}{dt} \quad (4.6)$$

where,

ρ and C_p = the density and specific heat of the composite
K_x, K_y, K_z = the thermal conductivities in the x, y and z directions
T = the temperature of the composite

The coefficients of thermal conductivity of the material are usually determined to be those along the principal material direction. They need to be transformed into the coordinate system of the laminate. Figure 4.13 shows the relative orientation between the two coordinate systems. The relation between the coefficients of conductivity along the two coordinate system is given as:

$$\begin{bmatrix} k_{xx} \\ k_{zz} \\ k_{xz} \end{bmatrix} = \begin{bmatrix} m^2 & n^2 & mn \\ n^2 & m^2 & -mn \\ -mn & mn & m^2 - n^2 \end{bmatrix} \begin{bmatrix} k_{11} \\ k_{33} \\ k_{13} \end{bmatrix} \quad (4.7)$$

where,

k_{ij}, $i,j = x,z$ are the off-axis coefficients of thermal conductivity
k_{ij}, $i,j = 1,3$ are the on-axis coefficients of thermal conductivity

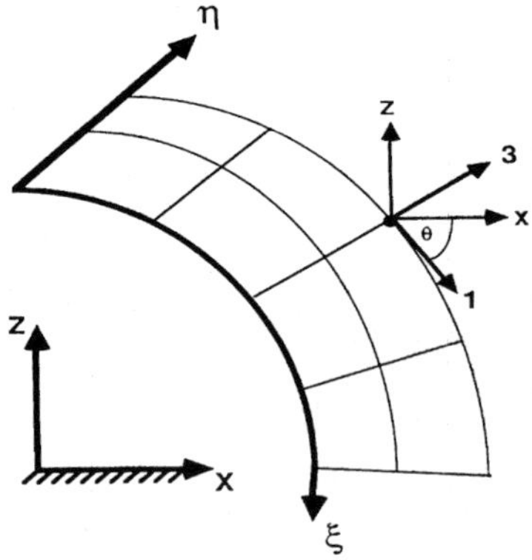

FIGURE 4.13 Coordinate systems for different coefficients of thermal conductivity.

TABLE 4.2 Thermal Properties for Glass/Polyester and Graphite/Epoxy [6].

	ρ (kg/m³)	C_p (kJ/(W·°C)	k_{33} (kW/(m·°C)	k_{11}/k_{33}
Glass/polyester	1.89×10^3	1.26	2.16×10^{-4}	2
Graphite/epoxy	1.52×10^3	0.94	4.46×10^{-4}	5

The thermal properties of a few composite systems are given in Table 4.2 [6].

From Equation (4.2), the rate of heat generation by chemical reaction *dH/dt* can be written as:

$$\frac{dH}{dt} = \frac{d\alpha}{dt} H_T \tag{4.8}$$

where H_T is the total heat of reaction and α is the degree of cure predicted by the kinetic model.

The effective total heat of reaction for the composite, H_T, is based on the following rule of mixture: $H_T = w_r H_r + w_f H_f$, where H_r and H_f are the heats of reaction of the resin and fiber, respectively. w_r and w_f are the weight fractions of the resin and fiber, respectively. The fiber heat of reaction is assumed to be zero. Only the resin heat of reaction and the resin weight fraction are required. The total heat of reaction for two composite systems is shown in Table 4.3 [6]. The reaction rate for each material system is unique due to the inherent differences in the overall order of the reaction kinetics. The reaction rate for one glass/polyester is given by:

$$\frac{d\alpha}{dt} = A \exp\left(-\frac{\Delta E}{RT}\right)\alpha^m (1-\alpha)^n \tag{4.9}$$

where,

$A = 3.7 \times 10^{22}\ \text{min}^{-1}$
$\Delta E = 1.674 \times 10^5$ J/mol
$m = 0.524$
$n = 1.476$

The reaction rate for graphite/epoxy (Hercules 3501-6 resin) follows the equation [8]:

$$\begin{aligned} \frac{d\alpha}{dt} &= (K_1 + K_2\alpha)(1-\alpha)(B-\alpha) \quad \alpha \le 0.3 \\ \frac{d\alpha}{dt} &= K_3(1-\alpha) \quad \alpha \succ 0.3 \end{aligned} \tag{4.10}$$

$$K_1 = A_1 \exp\left(-\frac{\Delta E_1}{RT}\right)$$

$$K_2 = A_2 \exp\left(-\frac{\Delta E_2}{RT}\right)$$

$$K_3 = A3 \exp\left(-\frac{\Delta E_3}{RT}\right)$$

$$A_1 = 2.101 \times 10^9 \text{ min}^{-1}$$

$$A_2 = -2.014 \times 10^9 \text{ min}^{-1}$$

$$A_3 = 1.960 \times 10^6 \text{ min}^{-1}$$

$$\Delta E_1 = 8.07 \times 10^4 \text{ J / mol}$$

$$\Delta E_2 = 7.78 \times 10^4 \text{ J / mol}$$

$$\Delta E_3 = 5.66 \times 10^4 \text{ J / mol}$$

For the case of autoclave manufacturing, it can be assumed that there are no variations of temperature along the in-plane directions (x and y) and Equation (4.6) can be written as:

$$\frac{\partial(\rho C_p T)}{\partial t} = \frac{\partial}{\partial z}\left(K_z \frac{\partial T}{\partial z}\right) + \frac{d\alpha}{dt} H_T \tag{4.11}$$

2.6.2.1. Thin Laminates

The solution for Equation (4.11) can be simplified when the laminate is thin. For thin laminates (thickness of about 3 mm or less), the tempera-

TABLE 4.3 Heat of Reaction for Two Composite Materials [6].

	Glass/polyester	Graphite/epoxy
w_r	46%	42%
w_f	54%	58%
H_{resin} (kJ/kg)	168.6	473.6
H_T (kJ/kg)	77.5	198.9

ture may be considered to be constant across the thickness of the laminate without large error. For that situation, Equation (4.11) can be simplified to be:

$$\frac{\partial(\rho C_p T)}{\partial t} = \frac{d\alpha}{dt} H_T \tag{4.12}$$

The determination of the development of the degree of cure and rate of cure for the case of thin laminates can be obtained using the incremental method and is illustrated in the following example.

Example 4.1

Consider a carbon/epoxy composite where the resin kinetic follows the equation:

$$\frac{d\alpha}{dt} = K\alpha^m (1-\alpha)^n$$
$$K = Ae^{-E/RT} \tag{a}$$

with,

$A = 1.27 \times 10^5$ sec^{-1}
$E = 63400$ J/mol
$m = 0.9$
$n = 2.1$

Values of the physical properties for many resins, fibers and composite systems are given in Reference [9].

Composite density: $\rho = 1580$ kg/m^3

Composite specific heat: $C_p = 870$ J/(kg·K)

$\rho C_p = 1.3746 \times 10^6$ J/(K·m^3)

Composite conductivity along the thickness direction: $K_z = 0.69$ W/m·K

$H_T = 150$ J/g (or 2.37×10^8 J/m^3)

R: Universal gas constant = 8.31 J/(mol·K)

It is desired to manufacture a laminate of eight layers with thickness of 0.150 mm per layer. The bottom of the laminate is an aluminum mold 12.7 mm thick and the top of the laminate has bagging materials (bleeder, breather layer, etc.). The whole assembly is placed inside an autoclave subject to a temperature cycle. Figure 4.14 shows the configuration.

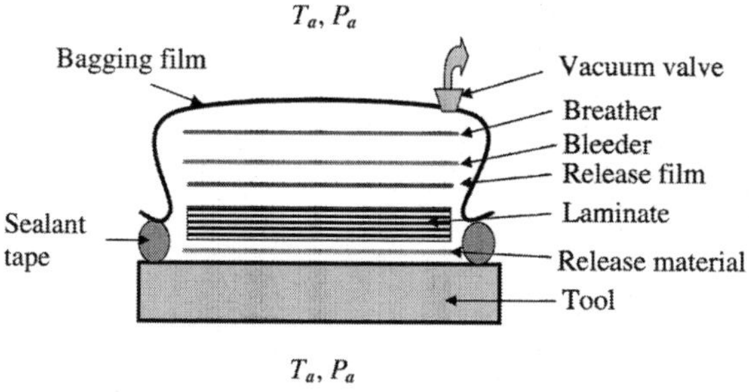

FIGURE 4.14 Configuration for the laminate curing.

Two schedules of temperature will be considered.

- *Linear increase in temperature:* In this schedule, the temperature of the autoclave is increased linearly from room temperature (20°C) at a rate of 5°C per minute.
- *Two-step temperature increase:* In this schedule, the autoclave temperature is increasing from room temperature (20°C) at a rate of 5°C per minute for 18 minutes. It is then held constant at 110°C for 10 minutes. Then it is increased by 5°C per minute for 14 minutes to a maximum of 180°C and held there for 60 minutes. (Figure 4.15 shows the temperature schedule.)

Determine the development of degree of cure and rate of cure as functions of time.

$$K_{Al} = 237 \text{ W/K(m}$$
$$L = 1.2 \text{ mm (laminate thickness)}$$
$$t_1 = 12.7 \text{ mm (thickness of mold)}$$
$$t_2 = 1 \text{ mm (thickness of bagging material on top of laminate)}$$
$$\rho C_p L = 1.65 \times 10^3 \text{ J/(K·m}^2\text{)}$$
$$K_{Al} = 237 \text{ W/m·K}$$
$$K_{\text{glass cloth}} = 0.26 \text{ W/m·K}$$

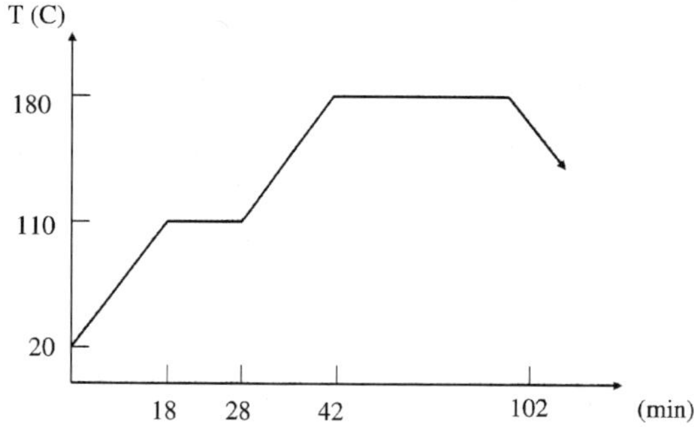

FIGURE 4.15 Schedule of boundary temperatures.

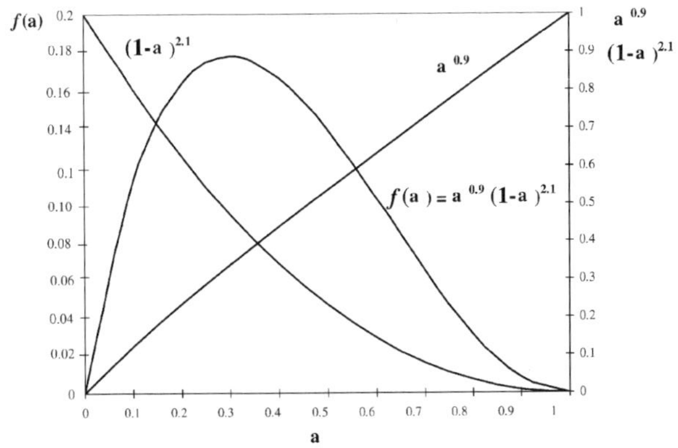

FIGURE 4.16 Function $f(\alpha)$ versus α.

Before proceeding to obtain the solution for this problem, it is informative to examine the influence of the curing order and the activation energy on the curing process. For this, we will examine the variation of the $f(\alpha)$ function and of the frequency factor K.

Define a function $f(\alpha) = \alpha^{0.9}(1-\alpha)^{2.1}$. A plot of $f(\alpha)$ versus α is shown in Figure 4.16. In Figure 4.17, the portion $(1-\alpha)^{2.1}$ represents the nth order of the reaction. It is maximum at the beginning and decreases to zero at full cure. This portion $\alpha^{0.9}$ may represent the segments involving primary amines, which are most active at the beginning but decrease in activity as curing proceeds. The portion may represent the autocatalytic segment of the reaction. It is zero at the beginning but increases as the curing proceeds. The combination of these two types of reactions gives rise to the curve of $f(\alpha)$, which has a local maximum with the fastest rate of reaction occurring at about 30% cure.

Denote another function $g(T) = e^{-E/RT} = e^{-63400/(8.31)T} = e^{-7629/T}$. A plot of $g(T)$ versus T is shown in Figure (4.17). Note that T has to be expressed in °K.

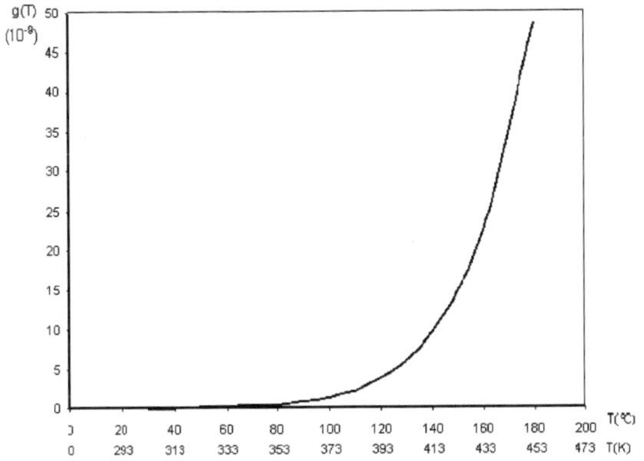

FIGURE 4.17 Plot of $g(T)$ versus T.

It can be seen that $g(T)$ increases very slowly with T at low T (less than 373 K or 100°C). Above 100°C, $g(T)$ increases sharply with T.

Solution

Assuming constant coefficients, Equation (4.12) can be written as:

$$\rho C_p \frac{\partial T}{\partial t} = H_T \frac{\partial \alpha}{\partial t} = H_T A e^{-E/RT} (\alpha)^{0.9} (1-\alpha)^{2.1} \qquad \text{(b)}$$

Equation (b) is a coupled equation and it takes numerical procedure to obtain a solution. However in order to examine the effect of different elements on the solution, an approximate procedure is used. In this procedure, the time scale is divided into small increments. For each time increment, the term on the right-hand side of Equation (b) will be calculated using the temperature of the material at the beginning of the increment. It will be treated as a constant during the increment of time. This term is the heat generated by the chemical reaction. It is the amount of heat per unit volume of the material. Instead of leaving this term in the governing Equation (b), this term will be considered as a heat flux entering into the element from one face (lower face). This equivalent heat flux Q_c is determined as:

$$Q_c = \frac{d\alpha}{dt} H_T L = A e^{-E/RT} \alpha^{0.9} (1-\alpha)^{2.1} H_T L \qquad \text{(c)}$$

In addition to this heat flux, there are also heat fluxes due to heat input from the autoclave through the aluminum mold, and the heat flux from the laminate across the bagging material on the top.

The problem can be formulated as shown in Figure 4.18 with heat flux Q_1 from the bottom face and Q_2 to the top.

The governing Equation (b) can then be written as:

$$\rho C_p L \frac{dT}{dt} = \text{ net heat flux to the sample } = Q_c + Q_1 + Q_2 \qquad \text{(d)}$$

where

Q_c is heat generated by the chemical reaction. Even though this is heat generated per unit volume, it was converted to an equivalent heat flux into the system. This is given by Equation (c) above. It can be seen that this is a function of temperature T. In the incremental procedure adopted below, this is calculated using the temperature at the end of the previous increment, and is considered as a constant during the time increment.

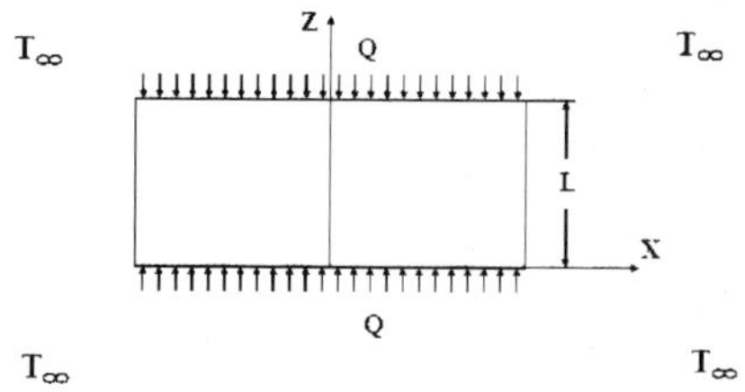

FIGURE 4.18 Equivalent heat transfer problem.

Q_1 is heat flux from the mold to the composite. This can obtained by the equation:

$$Q_1 = \frac{K_1}{t_1}(T_\infty - T) \tag{e}$$

where,

K_1 = the thermal conductivity of the mold
t_1 = the thickness of the mold
T_∞ = the autoclave temperature

Q_2 is heat flux from the mold across the bagging material to the atmosphere in the autoclave on the top side of the laminate. This can be obtained by the equation:

$$Q_2 = \frac{K_2}{t_2}(T_\infty - T) \tag{f}$$

where,

K_2 = the thermal conductivity of the bagging material
t_2 = the thickness of the bagging materials

Substituting Equations (e) and (f) into Equation (d) yields:

$$\rho C_p L \frac{dT}{dt} = Q_c + \frac{K_1}{t_1}(T_\infty - T) + \frac{K_2}{t_2}(T_\infty - T) \tag{g}$$

which can be written as:

$$a\frac{dT}{dt} + bT + c = 0 \tag{h}$$

where,

$$a = \rho C_p L$$

$$b = \frac{K_1}{t_1} + \frac{K_2}{t_2} \tag{i}$$

$$+c = -\left[Q_c + T_\infty\left(\frac{K_1}{t_1} + \frac{K_2}{t_2}\right)\right]$$

Assume that the solution of Equation (h) takes the form:

$$T = c_1 e^{-mt} + c_2$$

Substituting this into Equation (h) yields:

$$-mc_1 a e^{-mt} + bc_1 e^{-mt} + bc_2 + c = 0$$

yielding:

$$m = \frac{b}{c} \quad \text{and} \quad c_2 = -\frac{c}{b}$$

Applying the initial condition: at $t = 0$, $T = T_o$, where T_o is the initial temperature of the lower surface of the laminate, we have:

$$T_o = c_1 + c_2$$

$$c_1 = T_o - c_2 = T_o + \frac{c}{b}$$

The solution is then:

$$T = \left(T_o + \frac{c}{b}\right)e^{-mt} - \frac{c}{b} \qquad \text{(j)}$$

Substituting the expressions for a, b and c yields:

$$T = \left[T_o - \frac{Q_c + T_\infty\left(\dfrac{K_1}{t_1} + \dfrac{K_2}{t_2}\right)}{\dfrac{K_1}{t_1} + \dfrac{K_2}{t_2}} \right] e^{-mT} + \frac{Q_c + T_\infty\left(\dfrac{K_1}{t_1} + \dfrac{K_2}{t_2}\right)}{\dfrac{K_1}{t_1} + \dfrac{K_2}{t_2}} \qquad \text{(k)}$$

and

$$m = \frac{\dfrac{K_1}{t_1} + \dfrac{K_2}{t_2}}{\rho C_p L} \qquad \text{(l)}$$

For this problem, the following quantities have been determined:

$$Q_c = L\, H_T\, d\alpha/dt = [3.6 \times 10^{10}\ \text{J/(sec}\cdot\text{m}^2)]e^{-7629/T}\, \alpha^{0.9}(1-\alpha)^{2.1} \qquad \text{(m)}$$

$$\frac{K_1}{t_1} = \frac{K_{AL}}{t_1} = \frac{237\ \text{W/K.m}}{12.7\ \text{mm}} = 18661\ \text{W}/(\text{m}^2.K)$$

$$\frac{K_2}{t_2} = \frac{K_{glass\ cloth}}{1\ \text{mm}} = \frac{0.26\ \text{W/K.m}}{0.001\ \text{mm}} = 260\ \text{W}/(\text{m}^2.K)$$

$$\frac{K_1}{t_1} - \frac{K_2}{t_2} = 18921\ \text{W}/(\text{m}^2 K)$$

$$m = \frac{18921}{1.65 \times 10^{+3}} = 11.47\ \text{sec}^{-1}$$

Rate of cure:

$$\frac{d\alpha}{dt} = (1.27 \times 10^5/\text{sec})e^{-7629/T}(\alpha)^{0.9}(1-\alpha)^{2.1} \qquad \text{(n)}$$

Equation (k) can be written as:

$$T - T_\infty = \left[T_o - \frac{Q_c + 18401 T_\infty}{18401}\right]e^{-11.47t} + \frac{Q_c}{18401} \qquad \text{(o)}$$

The determination of the degree of cure at a time increment $(n+1)$ is determined from the degree of cure at the previous step (n) following the equation:

$$\alpha^{n+1} = \alpha^n + \left(\frac{d\alpha}{dt}\right)^{n+1} \Delta t \qquad \text{(p)}$$

Initial conditions:

$T_o = 273$ K

$\alpha_o = 0.1$

Increment 1: Time increment from 0 to 5 minutes

During this time increment, the temperature of the surrounding T_∞ increases from 20°C to 45°C (293–318 K). An average temperature of 305.5 K for this increment is used for T_∞.

We have:

$T_\infty =$ 305.5°C
$\alpha =$ 0.1

Q_c is calculated based on the average temperature of the increment, which is 305.5 K. From Equation (m): $Q_c = 0.05$ W/m^2.

Temperature of the composite is calculated from Equation (o) as:

$$T - 305.5 = \left[273 - \frac{0.05 + 18921(305.5)}{18921}\right] e^{-(11.47)/(300)} + \frac{0.05}{18921} = 2 \times 10^{-6}$$

$$T = 305.5K$$

It can be seen that the increase in temperature due to heat generation by chemical reaction is very small and the temperature of the part is controlled mainly by the autoclave temperature for this case. The rate of cure as given by Equation (*n*) can be calculated as:

$$\frac{d\alpha}{dt} = (1.27 \times 10^5 / \text{sec}) e^{-7629/3055} (0.1)^{0.9} (1 - 0.1)^{2.1} = 1.82 \times 10^{-7} / \text{sec}$$

The degree of cure at the end of this increment is:

$$\alpha_1 = 0.1 + (1.2 \times 10^{-7})(300) = 0.1$$

Increment 2: Time increment from 5 to 10 minutes

During this time increment, the temperature of the surrounding T_∞ increases from 45°C to 70°C (318–343 K). An average temperature of 330.5 K for this increment is used. In order to start the process, a degree of cure $\alpha_1 = 0.1$ is assumed. We have:

$T_\infty =$ 330.5°C
$\alpha_1 =$ 0.1
$Q_c =$ 0.34 W/m^2

The rate of cure as given by Equation (*n*) can be calculated as:

$$\frac{d\alpha}{dt} = (1.27 \times 10^5 / \text{sec}) e^{-7629/3305} (0.1)^{0.9} (1 - 0.1)^{2.1} = 1.2 \times 10^{-6} / \text{sec}$$

The degree of cure at the end of this increment is:

$$\alpha_2 = 0.1 + (1.2 \times 10^{-6})(300) = 0.1$$

TABLE 4.4 Progression of Cure of the Example (linear increase in autoclave temperature).

Time (minutes)	T_a (°C)	T_∞ (°C)	T_∞ (K)	Q_c (W/m^2)	T (K)	Rate of Cure (sec^{-1})	Degree of Cure (α)
0	20	20	293	0	293	1.2×10^{-8}	0.1
5	45	32.5	305.5	0.05	305.5	1.2×10^{-7}	0.1
10	70	57.5	330.5	0.34	330.5	1.2×10^{-6}	0.1
15	95	82.5	355.5	1.72	355.5	6.1×10^{-6}	0.102
20	120	107.5	380.5	7.21	380.5	2.55×10^{-5}	0.110
25	145	132.5	405.5	26.1	405.5	9.17×10^{-5}	0.138
30	170	157.5	430.5	89.2	430.5	2.46×10^{-4}	0.211
35	180	175	448	217	448	7.64×10^{-4}	0.440
40	180	180	453	247	453	8.71×10^{-4}	0.702
45	180	180	453	99.8	453	3.70×10^{-4}	0.813
50	180	180	453	42.9	453	1.89×10^{-4}	0.941
55	180	180	453	4.33	453	1.53×10^{-5}	0.946
60	180	180	453	3.6	453	1.27×10^{-5}	0.950

Proceed with the same procedure for subsequent time increments. Tables 4.4 and 4.5 show the values of the temperature of the composite part and its degree of cure at different time points for linear increase and a two-step increase in temperatures of the autoclave.

TABLE 4.5 Progression of Cure of the Example (two-step increase in autoclave temperature).

Time (minutes)	T_a (°C)	T_∞ (°C)	T_∞ (K)	Q_c (W/m^2)	T (K)	Rate of Cure (sec^{-1})	Degree of Cure (α)
0	20	20	293	0	293	1.2×10^{-8}	0.1
5	45	32.5	305.5	0.05	305.5	1.2×10^{-7}	0.1
10	70	57.5	330.5	0.34	330.5	1.2×10^{-6}	0.1
15	95	82.5	355.5	1.72	355.5	6.1×10^{-6}	0.102
18	110	102.5	375.5	5.52	375.5	1.95×10^{-5}	0.106
23	110	110	383	8.39	383	2.96×10^{-5}	0.114
28	110	110	383	8.86	383	3.13×10^{-5}	0.123
33	110	110	383	9.27	383	3.27×10^{-5}	0.132
38	110	110	383	9.62	383	3.40×10^{-5}	0.142
40	120	115	388	13.0	388	4.6×10^{-4}	0.148
45	145	132.5	405.5	31.1	405.5	1.1×10^{-4}	0.181
50	170	157.5	430.5	102	430.5	3.6×10^{-4}	0.289
52	180	175	448	231	448	8.2×10^{-4}	0.387
60	180	180	453	265	453	9.4×10^{-4}	0.837
70	180	180	453	32.7	453	1.15×10^{-4}	0.906
80	180	180	453	11.1	453	3.92×10^{-5}	0.930
90	180	180	453	6.14	453	2.17×10^{-5}	0.942

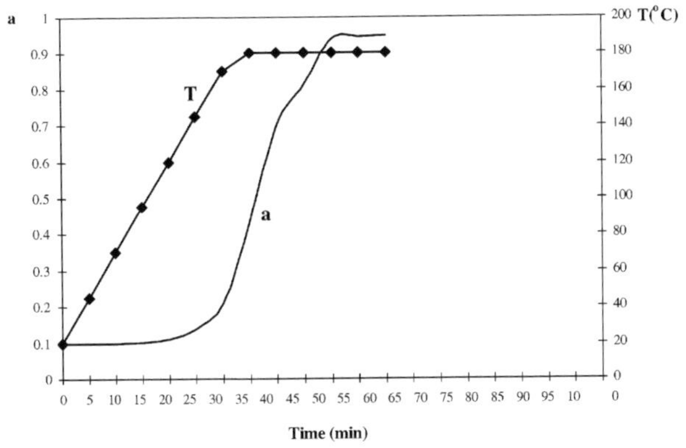

FIGURE 4.19 Variation of temperature and degree of cure as a function of time (linear increase in autoclave temperature).

Figures 4.19 and 4.20 show the variations of the temperatures and degrees of cure as functions of time for a linear increase and a two-step increase in temperature of the autoclave. It can be seen that the temperature of the part follows closely the temperature of the autoclave. The degree of cure reaches to 80% after 45 minutes but it takes a long time for it to reach full cure.

It can be seen that linear increase in temperature makes the resin cure very quickly (up to 30 minutes of low cure degree) while a two-step temperature increase allows more time for the resin to have a low degree of cure (up to 50 minutes of low-cured degree). Lower degree of cure ensures low viscosity, which is required for resin flow and for the elimination of voids as discussed in Section 2.6.5. Note that fairly coarse increments of time were used in the example for illustration purpose. Finer time increments should provide better results.

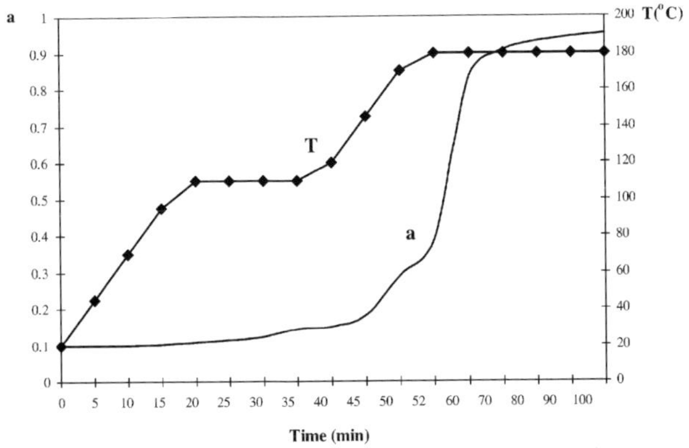

FIGURE 4.20 Variation of temperature and degree of cure as a function of time (two-step increase in autoclave temperature).

2.6.2.2. *Thick Laminates*

Thick laminates refer to laminates that are about 6 mm or more thick. For the thick laminates, the dependence on thickness of the temperature is significant and the simplification made for thin laminates does not apply. Equation (4.11) rather than (4.12) has to be solved. The solution procedure is complex and numerical method using a computer is necessary. Bogetti and Gillespie [7] have worked out the solution for a few thick composites. They examined thick laminates made of glass/polyester and graphite/epoxy. The variation of the modulus of the resin and resin shrinkage during cure were taken into account. The temperature distributions and stresses were calculated using a computer program. Figure 4.21 shows the temperature distribution across the thickness of laminates with different thicknesses. For laminates with thickness less than 2.54 cm, at 164 minutes into the curing process, the temperature at the mid-thickness region is larger than the temperature at the bottom and top of the laminate. However when the thickness is 5.08 cm, the temperature at the mid-thickness region is smaller than that on the surfaces. This is because for thinner laminates, heat flow from the mold can enter quickly into the mid-thickness area. This helps to kick-start the reaction in all parts of the laminate. Heat generated in the mid-thickness area cannot dissipate to the surrounding environment as quickly as heat generated

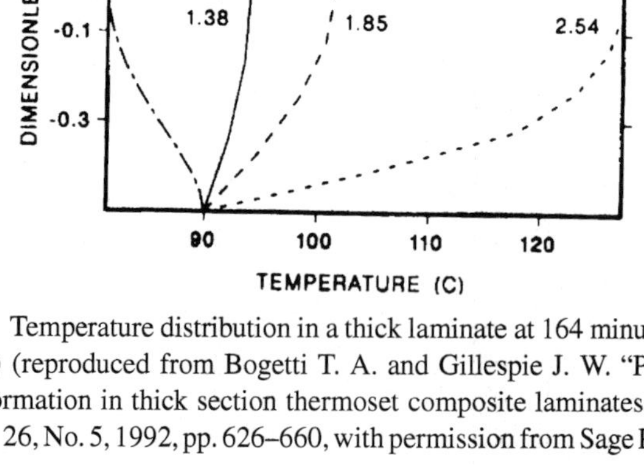

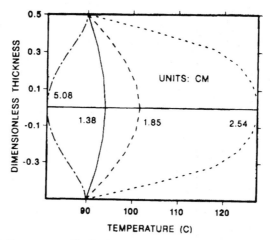

FIGURE 4.21 Temperature distribution in a thick laminate at 164 minutes (glass/polyester laminate) (reproduced from Bogetti T. A. and Gillespie J. W. "Process induced stress and deformation in thick section thermoset composite laminates," *J. Composite Materials,* Vol. 26, No. 5, 1992, pp. 626–660, with permission from Sage Publications).

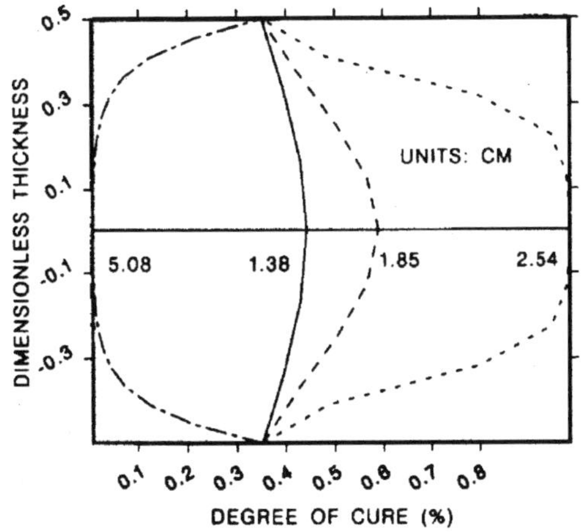

FIGURE 4.22 Degree of cure distribution in a thick laminate at 164 minutes (Glass/polyester laminate) (reproduced from Bogetti T. A. and Gillespie J.W. "Process induced stress and deformation in thick section thermoset composite laminates," *J. Composite Materials,* Vol. 26, No. 5, 1992, pp. 626–660,with permission from Sage Publications).

from areas close to the surface. As such, the mid-thickness region is hotter than regions close to the surface of the laminate.

On the other hand, for thicker laminates (5.08 cm thick), it takes longer for the heat from the tool to reach the mid-thickness region because the mid-thickness region has not had the chance for the reaction to take place at 164 minutes in to the cure. Temperature close to the surface of the laminate is therefore larger than that in the mid-thickness region.

Figure 4.22 shows the degree of cure development of the laminates studied by Bogetti and Gillespie [7]. The degree of cure follows the same distribution as the temperature distribution.

2.6.3. Viscosities

The viscosity of the resin as a function of time depends on the extent of the reaction of the components. An empirical relationship was used by Lee et al. in 1982 as [8]:

$$\mu = \mu_{\infty} \exp\left(\frac{U}{RT} + k\alpha\right) \tag{4.13}$$

where α is the degree of cure, μ_{∞} is a constant, U is the activation energy for viscosity, and k is a constant that is dependent on the temperature. To

TABLE 4.6 Parameters for Equation (4.13).

Author	System	μ_o (Pa·s)	E (J/mol)	K (Pa·s)	Range of Validity
Lee et al. [8]	Hercules 3501-6	7.93×10^{-14}	9.08×10^{4}	14.1 ± 1.2	$\alpha < 0.5$
Dusi et al. [13]	Fiberite 976	1.06×10^{-6}	3.76×10^{4}	18.8 ± 1.2	$\alpha < 0.2$

characterize the viscosity as a function of temperature it is necessary to determine the value of k by fitting a least- squares curve to the μ versus α data. The degree and rate of cure can be determined from isothermal scans and, in general, the behavior is fit into an Arrhenius-type equation. This treatment is usually empirical, but the information is key to understanding the viscosity of the resin and thus the bleed-off and fiber volume fraction distribution that results from a given autoclave processing cycle. The values for these parameters for two resin systems are listed in Table 4.6.

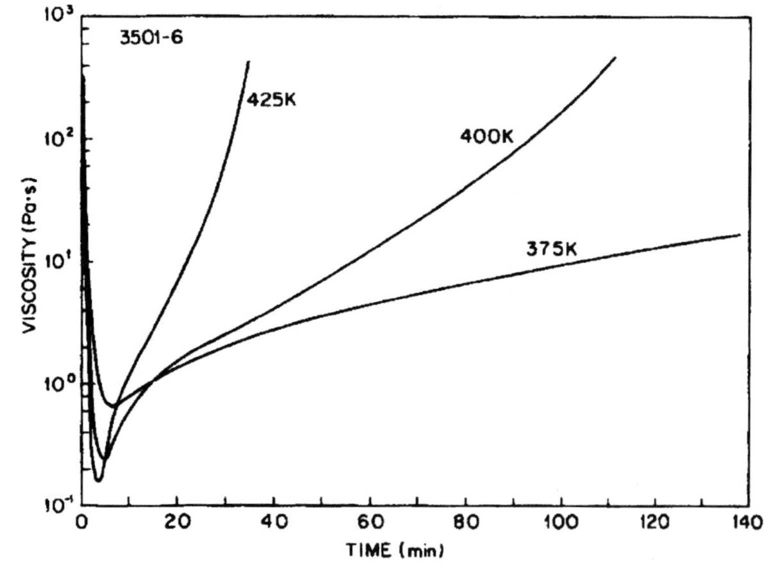

FIGURE 4.23 Isothermal viscosity for Hercules 3501-6 resin (reproduced from Lee W. I., Loos A. C. and Springer G. S., "Heat of reaction, degree of cure, and viscosity of Hercules 3501-6 resin," *Journal of Composite Materials,* Vol. 16, 1982, pp. 510–520, with permission from Sage Publications).

Substituting the values for Hercules 3501-6 into Equation (4.13) yields:

$$\mu = 7.93 \times 10^{-14} \exp\left(\frac{9.08 \times 10^4}{8.31T} + 14.1\alpha\right) \text{ in Pa - sec} \quad (4.14)$$

The variation of the degree of cure as a function of time at different temperatures for the above resin is shown in Figure 4.23.

Calado and Advani [9] gave a comprehensive table of many cure kinetic models for epoxies.

Example 4.2

Values of the temperature and degree of cure of example 4.1 are given in Tables 4.4 and 4.5 for two different schedules of heating. Using these values, determine the variation of the viscosity of the resin as a function of time, assuming that the viscosity follows Equation (4.14).

Table 4.7 shows the values of the viscosity for different times for a linear increase in temperature, and Table 4.8 shows the values for a two-step cure cycle.

Figure 4.24 shows the variation of viscosity for the two heating schedules. For the two- step heating schedule, long time at low viscosity allows more time to apply pressure. Note that after the dwell period at low viscosity, the viscosity increases rapidly.

TABLE 4.7 Viscosity for a linear increase in autoclave temperature.

Time (minutes)	T_a (°C)	T_∞ (°C)	T_∞ (K)	T (K)	Degree of Cure (sec^{-1})	Viscosity (Pa·s)
0	20	20	293	293	0.1	5138
5	45	32.5	305.5	305.5	0.1	1084
10	70	57.5	330.5	330.5	0.1	72.6
15	95	82.5	355.5	355.5	0.102	7.33
20	120	107.5	380.5	380.5	0.110	1.09
25	145	132.5	405.5	405.5	0.138	0.28
30	170	157.5	430.5	430.5	0.211	0.16
35	180	175	448	448	0.440	1.52
40	180	180	453	453	0.702	46.5
45	180	180	453	453	0.813	223
50	180	180	453	453	0.941	1356
55	180	180	453	453	0.946	1455
60	180	180	453	453	0.950	1540

TABLE 4.8 Viscosity for a Two-Step Temperature Cycle.

Time (minutes)	T_a (°C)	T_∞ (°C)	T_∞ (K)	T (K)	Degree of Cure (sec^{-1})	Viscosity (Pa·s)
0	20	20	293	293	0.1	5138
5	45	32.5	305.5	305.5	0.1	1084
10	70	57.5	330.5	330.5	0.1	72.6
15	95	82.5	355.5	355.5	0.102	7.33
18	110	102.5	375.5	375.5	0.106	1.50
23	110	110	383	383	0.114	0.95
28	110	110	383	383	0.123	1.08
33	110	110	383	383	0.132	1.22
38	110	110	383	383	0.142	1.42
40	120	115	388	388	0.148	1.06
45	145	132.5	405.5	405.5	0.181	0.50
50	170	157.5	430.5	430.5	0.289	0.48
52	180	175	448	448	0.387	0.71
60	180	180	453	453	0.837	310
70	180	180	453	453	0.906	820
80	180	180	453	453	0.930	1150
90	180	180	453	453	0.942	1362
100	180	189	453	453	0.951	1546

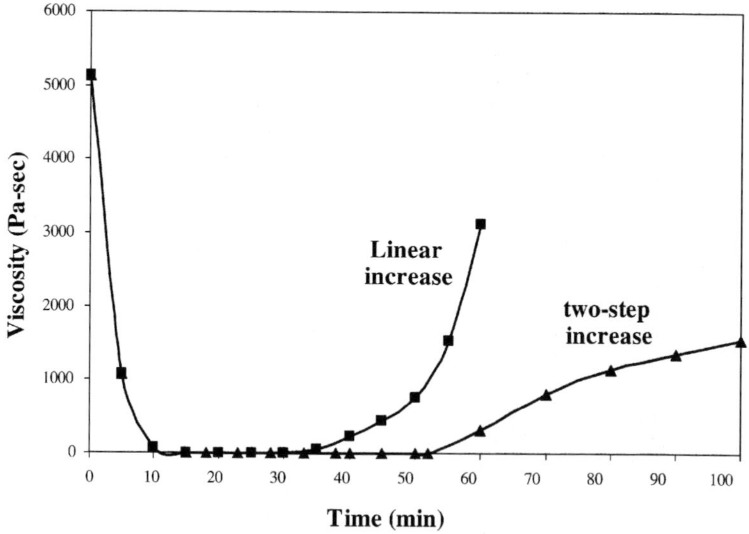

FIGURE 4.24 Variation of viscosity for two heating schedules.

2.6.4. Resin Flow and Consolidation

Usually the prepregs contain more resin than the final cured composite. During curing, the excess resin is oozed out. The excess resin flows from the bottom layers to the top layers. The bleeder materials absorb the excess resin. In order for the excess resin to flow from the mold side to the bleeder side, pressure needs to be applied. In addition, fibers are wavy and the fiber bed is similar to a springy network. It is necessary to flatten these fibers so that the composite layers are well consolidated. The combined springy fiber network and liquid resin is shown schematically in Figure 4.25.

This can be modeled as spring and dash pot as shown in Figure 4.26, where the stress P_a represents the pressure in the autoclave. The equilibrium condition can be written as:

$$P_a = P_r + P_f \tag{4.15}$$

In order to fully model the consolidation behavior of composites, both the resin flow and fiber deformation must be taken into account. The requirements for the resin pressure and the pressure required to consolidate the fibers are shown following.

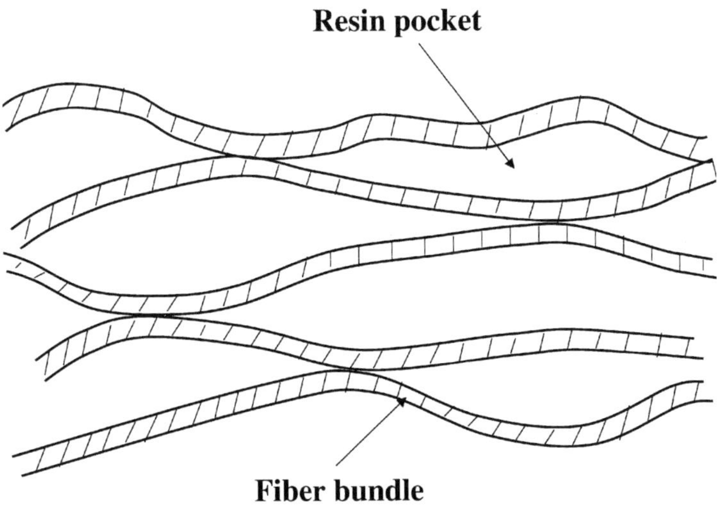

FIGURE 4.25 Springy fiber network surrounded by resin.

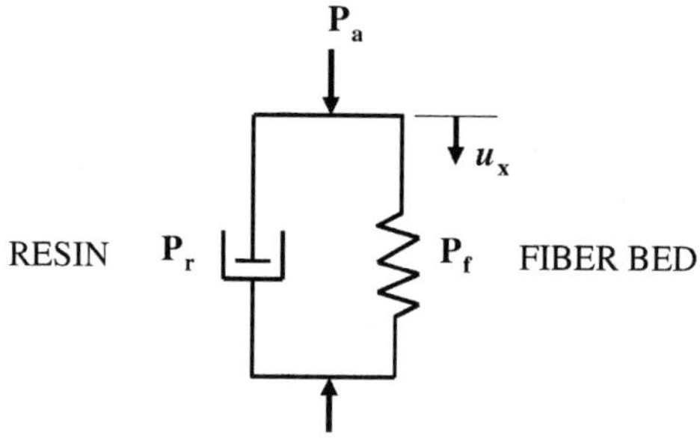

FIGURE 4.26 Model of fiber bed and resin.

2.6.4.1. Resin Flow

Flow through a fiber assembly is treated as flow through porous media, using the general form of Darcy's law, which states that the flow rate is dependent on the applied pressure, the fluid viscosity, and the permeability of the fiber network. Darcy's law can be written in vector form as:

$$\begin{bmatrix} u_x \\ u_y \\ u_z \end{bmatrix} = \frac{1}{\mu} \begin{bmatrix} K_{xx} & K_{xy} & K_{xz} \\ K_{yx} & K_{yy} & Kyz \\ K_{zx} & K_{zy} & K_{zz} \end{bmatrix} \begin{bmatrix} \frac{\partial p}{\partial x} \\ \frac{\partial p}{\partial y} \\ \frac{\partial p}{\partial z} \end{bmatrix} \tag{4.16}$$

where u_i are the components of the superficial velocity vector $\boldsymbol{u}$,K_{ij} are components of the permeability tensor K of the porous medium, $\partial p/\partial x$, $\partial p/\partial y$, $\partial p/\partial z$ are components of the pressure gradient vector $\Delta\mathbf{P}$, and μ is the fluid viscosity.

For the matrix of the coefficients of permeability, when the coordinate axes coincide with the principal material directions, only the diagonal values (K_{xx}, K_{yy}, K_{zz}) are non-zero. For a unidirectional ply, which is transversely isotropic, these directions correspond to the fiber axis and any two mutually perpendicular transverse axes. Generally the permeability is much higher in the direction along the fibers than transverse or through the thickness. The permeability is also a function of the fiber volume function V_f, fiber radius r_{fi} and fiber architecture. A number of studies have been carried out to determine the permeability of fibrous preforms. In general, the dependence of permeabilities on the material parameters mentioned above, and on the fiber volume fraction have been treated using the Kozeny-Carman equation, which can be written as:

$$K = \frac{r_f^2}{4k_o}\frac{(1-v_f)^3}{v_f^2} \tag{4.17}$$

where k_o is the Kozeny constant. The value of k_o varies with the fiber architecture and the direction of the flow. For flow parallel to the fibers, values of 0.5–0.7 have been reported, while for transverse flow, values of k_o = 11–18 have been quoted. The measured values of the Kozeny constant show significant variation and are sensitive to experimental conditions. In practice, when the fiber volume fraction approaches its theoretical limit, lateral or through-the-thickness flow can shut off as the fibers are forced into contact with one another along their entire lengths.

Figure 4.27 shows a curve of axial permeability values for aligned AS4 fibers with the Kozeny-Carman equation. In this figure, the symbol S_{xx} is

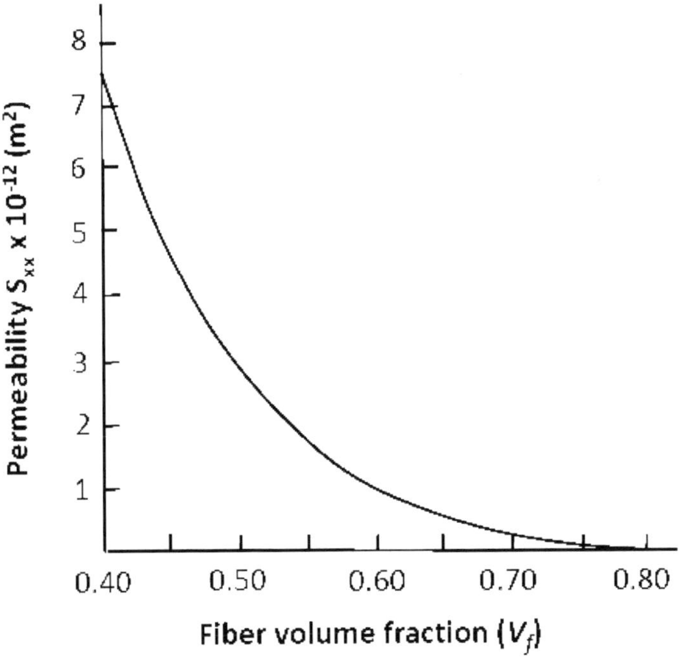

FIGURE 4.27 Axial permeability of aligned AS4 fibers with Kozeny-Carman equation (k_{xx} = 0.7) (corn oil and silicone liquid).

used to denote the permeability. This symbol has the same meaning as the symbol K_{xx}. Note that the permeability has the unit of length squared.

For most cases of practical importance, flow across the thickness of the laminate is considered. In this case, Equation (4.16) can be simplified to be:

$$u_z = \frac{K_{zz}}{\mu}\frac{dp}{dz} \tag{4.18}$$

The velocities in Equations (4.16) and (4.18) are dependent on viscosity, which varies greatly during an autoclave curing cycle. Most resins employed in aerospace material prepregs are in a B-staged or partly reacted condition. Their initial viscosities are therefore relatively high. As the material is heated, the viscosity drops dramatically and reaches a minimum at a temperature that is generally prescribed as a "hold" temperature in the autoclave curing cycle (Figure 4.23). This is to ensure that complete wet-out is achieved and that excess resin is effectively squeezed out of the plies and into the bleeder material.

This is also the temperature at which the resin is most susceptible to void formation (void formation is discussed in Section 5.5 later in this chapter). Thus the pressure developed in the resin at this time is absolutely critical. Continued heating leads to the initiation of cure and the resin viscosity increases progressively until the resin solidifies. The resin state changes during this process can be conveniently represented in a time-temperature-transformation (TTT) curve as shown in Figure 4.28.

In general, the resin in a typical epoxy prepreg system goes from the ungelled glass region (in the freezer) to a liquid (viscosity drops) through gelation (as the reaction proceeds) and into the vitrification regime (when solidification occurs). If the resin solidified before the voids are either compressed to negligible size or squeezed out of the laminate, then there will be holes in the solidified resin.

2.6.4.2. Consolidation

Consolidation involves the squeezing of the resin out of the laminate (into the bleeder materials) and also the flattening of the fiber network. Dave et al. [11] modeled the squeezing out of the resin by a system of water (representing the liquid resin) and spring (representing the fiber network) in a container (Figure 4.29). On top of the spring and water is a constant weight simulating the pressure applied by the autoclave, similar to the case of a piston in a cylinder. The piston has a relief valve which is

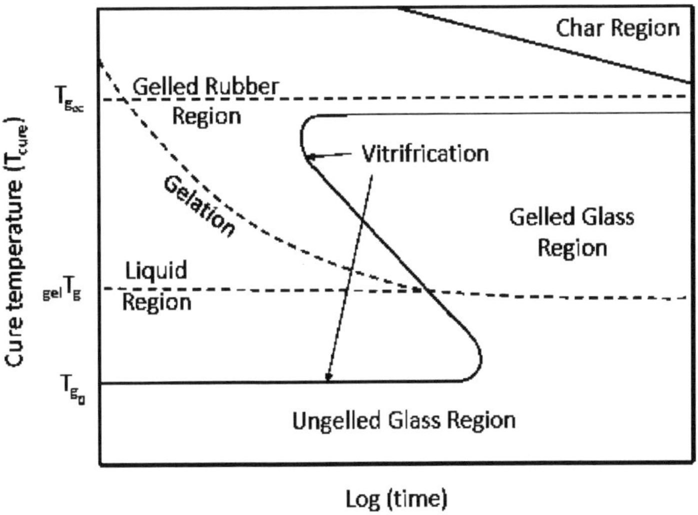

FIGURE 4.28 The time temperature transformation diagram for isothermal cure (reproduced from Enns J. B. and Gillham J. K., *Journal of Applied Polymer Science,* Vol. 28, 1983, pp. 2567–2591, with permission from Wiley Interscience).

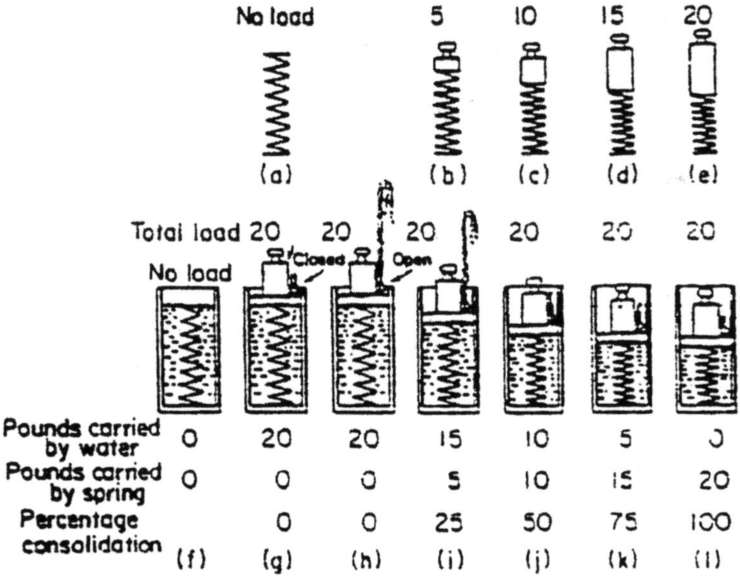

FIGURE 4.29 Conceptual representation of the consolidation of the fiber/resin system (reproduced from "A model for resin flow during composite processing. Part I: General mathematical development," by R. Dave, J. L. Kardos, and M. P. Dudukovic, *Polymer Composites,* Vol. 8, No. 1, 1987, 29–38, with permission from Society of Plastic Engineers).

closed at the beginning. The liquid (water) absorbs most of the load. At a certain instant of time during the curing process, the relief valve is opened. Liquid is allowed to leak out of the container. The spring is compressed and the spring takes up more and more of the load. At the end of the process, the spring takes up all of the load and the pressure in the liquid diminishes to almost nil.

Gutowski [1] proposed another model that considers the waviness of the fiber network. The formulation to calculate the stresses developed in the fiber network due to compression was presented in Chapter 3. As resin flows out of the composite, the fiber volume fraction increases. This decreases the spacing between the fibers to the point where significant fiber-fiber contact occurs. When significant contact exists, the fiber network becomes load-bearing and can have a significant effect on the overall flow behavior of the composite. One of the most important consequences of the chain of events is reduction in resin pressure, which can lead to voids in the matrix and a subsequent loss in the structural properties of the composite. This effect may be included in consolidation models by describing the load-carrying behavior of the fiber bundle. For transverse compression this relationship can be given as a function between the applied stress and the fiber volume fraction (repeated from Chapter 3).

$$\sigma_b = \frac{3\pi E}{\beta^4} \frac{1 - \sqrt{\dfrac{V_f}{V_o}}}{\left(\sqrt{\dfrac{V_a}{V_f}} - 1\right)^4} \tag{4.19}$$

where V_a is the maximum attainable fiber volume fraction, V_o is the initial fiber volume fraction corresponding to zero load, and β is a constant that describes the shape of the fiber.

Example 4.3

Continue from the case considered in Examples 4.1 and 4.2. Assuming that the pressure applied in the autoclave is 80 psi (547 kPa). Determine:

a. Stress absorbed by the fiber network.
b. Pressure in the resin.
c. Speed of flow of the resin out of the fiber network at the period of lowest viscosity (Figure 4.24).

d. Whether this speed is sufficient for the resin to flow out of the laminate before gelling.

The fiber volume fractions are:

Initial fiber volume fraction $V_o = 0.50$
Allowable fiber volume fraction: $V_a = 0.85$
Final volume fraction $V_f = 0.68$.
Fiber modulus: $E = 234$ GPa
Fiber aspect ratio: $\beta = 300$

Solution

a. Stress in the fiber bundle can be obtained using Equation (4.19):

$$\sigma_b = \frac{3\pi E}{\beta^4} \frac{1-\sqrt{\frac{V_f}{V_o}}}{\left(\sqrt{\frac{V_a}{V_f}}-1\right)^4} = \frac{3\pi(234 \text{ GPa})}{300^4} \frac{1-\sqrt{\frac{0.68}{0.50}}}{\left(\sqrt{\frac{0.85}{0.68}}-1\right)^4} = 232.7 \text{ kPa}$$

b. Pressure in the resin:

From Equation (4.15),

$$P_a = P_r + P_f$$

$$P_r = P_a - P_f = 547 - 232.7 = 314.3 \text{ kPa}$$

c. Speed of flow of resin:

Using Equation (4.17) for permeability transverse to fiber direction with $r_f = 3.5$ µm and $k_o = 18$, we have:

$$K = \frac{r_f^2}{4k_o} \frac{(1-v_f)^3}{v_f^2} = \frac{(3.5\times10^{-6} m)^2}{4\times18} \frac{(-0.68)^3}{(0.68)^2} = 1.2\times10^{-14} m^2$$

Using Darcy's law, Equation (4.18)

$$u_z = \frac{K_{zz}}{\mu} \frac{dp}{dz}$$

The thickness of the laminate is 1.2 mm and the viscosity at the lowest point in Figure 4.23) is about 1 Pa·sec. We have:

$$u_z = \frac{1.2\times10^{-14} m^3}{1 \text{ Pa-sec}} \frac{314.3 \text{ kPa}}{1.2 \text{ mm}} = 3.14\times10^{-3} \text{ mm / sec}$$

d. Is there sufficient time for resin to flow out of the laminate?

Time required for resin to flow out of the laminate:

$$t = \frac{\text{thickness}}{u_z} = \frac{1.2 \text{ mm}}{3.14\times10^{-3} \text{ mm / sec}} = 382 \text{ sec} = 6.37 \text{ minutes}$$

Examining Figure 4.24, the duration where viscosity is low is about 25 minutes (from 10–35 minutes) for the linear increase in temperature and the duration is about 40 minutes (from 10–50 minutes) for the two-step increase. As such both heating schedules are adequate for the flow.

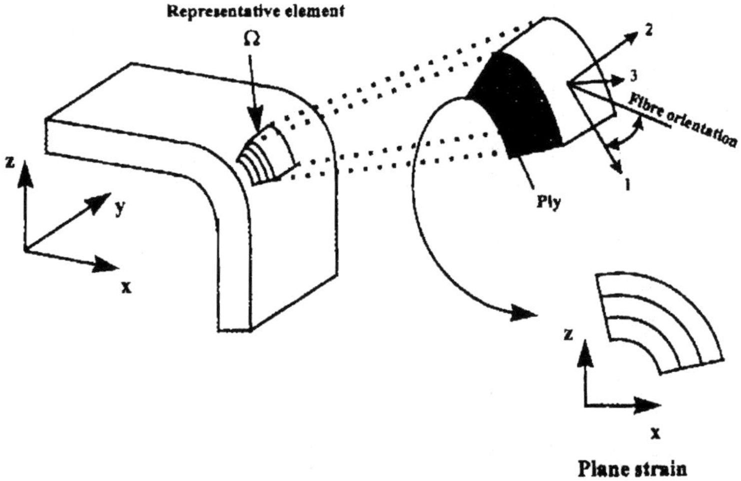

FIGURE 4.30 Angled piece (reproduced from Hubert P., Vaziri R. and A. Poursartip, "A two dimensional flow model for the process simulation of complex shape composite laminates," *Inter. J. for Numerical methods in Engineering,* Vol. 44, 1999, pp. 1–26, with permission from Wiley Interscience Publications.

For the manufacturing of curved pieces such as an angle piece, apart from the compression, there is also the possibility for shear flow. Figure 4.30 shows the configuration of an angled piece.

The stress-relation in this case needs to consider not only normal stresses but also shear stresses as shown in Figure 4.31.

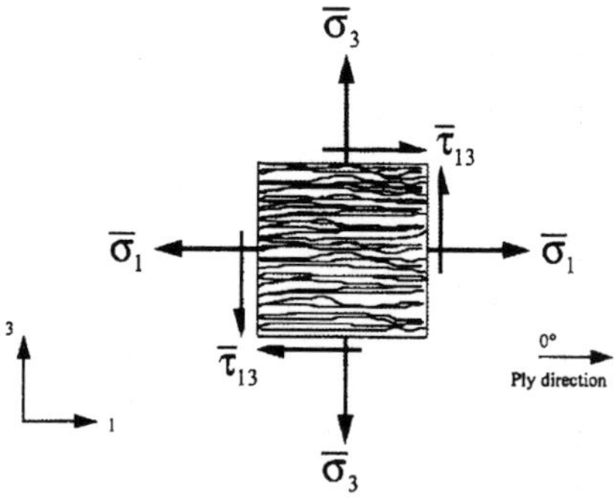

FIGURE 4.31 Fiber bundle subjected to three-directional loading (reproduced from Hubert P., Vaziri R. and A. Poursartip, "A two dimensional flow model for the process simulation of complex shape composite laminates," *Inter. J. for Numerical Methods in Engineering,* Vol. 44, 1999, pp. 1–26, with permission from Wiley Interscience Publications).

A simplified model was employed by Hubert et al. [12] to describe the stress-strain relation for the fiber bundle in Figure 4.31 as:

$$\begin{bmatrix} \bar{\sigma}_1 \\ \bar{\sigma}_3 \\ \bar{\tau}_{13} \end{bmatrix} = \begin{bmatrix} E_1 & 0 & 0 \\ 0 & E_2 & 0 \\ 0 & 0 & G_{13} \end{bmatrix} \begin{bmatrix} \varepsilon_1 \\ \varepsilon_3 \\ \gamma_{13} \end{bmatrix} \tag{4.20}$$

where E_1, E_2, G_{13} are the elastic constants describing the fiber bundle compaction behavior.

2.6.5. Void Suppression and/or Removal [16]

2.6.5.1. Void Formation

For autoclave processing, the prepregs are removed from the freezer and normally left at room condition for some time to thaw out. The resin in the prepreg may contain a certain amount of moisture, which may have diffused inside the resin material. During processing, upon initial heating, the resin first becomes a liquid. The vapor (or any other gases that may be contained within the resin) may grow to bubbles of a certain size, depending on the pressure in the resin. If the pressure in the resin (surrounding the bubble) is not high enough, these moisture or gas bubbles may grow to significant size. To avoid this there are two things that should happen. One is that if the pressure in the resin is high enough, then the moisture or gas may not be able to form bubbles of significant size. The other is that if the pressure in the resin is high enough, then whatever bubbles have already formed can be squeezed out of the material system while the resin is still a liquid. In either case, the pressure in the resin has to be sufficiently large to prevent large bubbles from remaining inside the resin after solidification has occurred. If bubbles do remain, they will appear as voids.

Let first consider the synergistic effect that water has on void stabilization. It is likely that a distribution of air voids occurs at ply interfaces because of pockets, wrinkles, ply ends, and particulate bridging. The pressure inside these voids is not sufficient to prevent their collapse on subsequent pressurization and compaction. However, as water vapor diffuses into the voids or when water vapor voids are nucleated, there will be, at any one temperature, an equilibrium water vapor pressure (and therefore partial pressure in the air-water void) that, under constant volume conditions, will cause the total pressure in the void to rise above that of a pure-air void. When the void equals or exceeds the surrounding resin hydrostatic pressure plus the surface tension forces, the void becomes stable and can even grow. Equation (4.21) expresses this relationship:

$$P_g - P_r = \frac{\gamma_{LV}}{m_{LV}} \tag{4.21}$$

where P_g and P_r are the gas and resin pressures, respectively, γ_{LV} is the liquid resin void surface tension, and m_{LV} is the ratio of volume to its surface area. Thus the difference in pressure is counterbalanced by the surface tension forces. When the temperature rises for a constant-volume system, P_g will rise faster than P_r, whereas γ_{LV} will decrease slightly. In addition to P_g increasing according to the perfect gas law, the partial pressure of water in the void can rise exponentially because of the temperature effect on the water vapor pressure.

In order to prevent the potential for pure water void growth by moisture diffusion in a laminate at all times and temperatures during the curing cycle, Kardos [14] developed a recommendation for minimum resin pressure at any point as:

$$P_{\min} \geq 4962 \exp\left(-\frac{4892}{T}\right)(RH)\% \tag{4.22}$$

where $P_{\min}$(atm) = minimum pressure in the resin required to prevent void growth by moisture diffusion at any time during cure. $(RH)_o(\%)$ = relative humidity to which the resin in the prepreg is equilibrated prior to processing, and $T(K)$ = temperature at any time during the curing cycle.

Equation (4.22) was derived from the requirement that void growth by moisture diffusion at any temperature cannot occur if the pressure within the void is greater than the saturated vapor pressure at that temperature. A plot of Equation (4.22) for two relative humidities (50% and 100%) yields a void stability map, which is shown in Figure 4.32. It is evident from this map that vacuum can be applied without encouraging void growth, if such application is coordinated with the temperature of the system.

At the same time, it is also clear that in order to prevent void formation, the resin pressure must be kept high as the temperature increases.

For a relative humidity of 52% and a temperature of 110°C (383 K), Equation (4.22) yields a minimum pressure of 0.732 atm (10.76 psi or 73.5 kPa). Considering the resin pressure of 340 kPa as determined in Example 4.3, the minimum pressure can be easily satisfied.

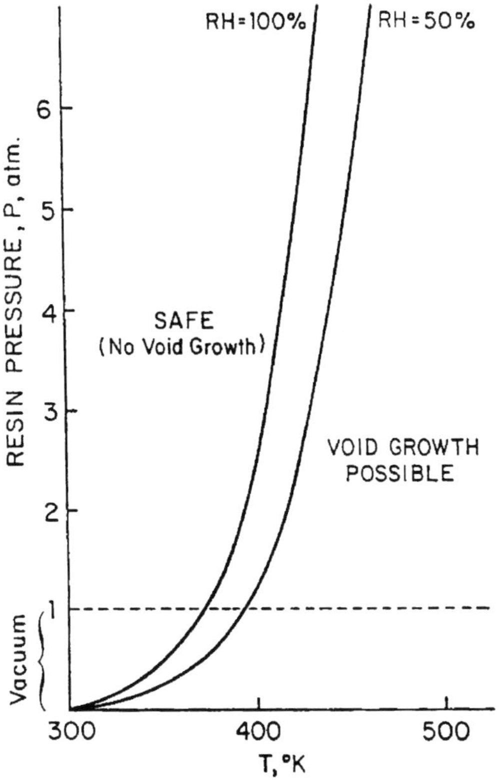

FIGURE 4.32 Void stability map for pure water void formation in Narmco 5208 epoxy matrix (reproduced from "The processing science of reactive polymer composites," by J. L. Kardos in *Advanced Composites Manufacturing,* T. G. Gutowski, ed. ,1997, with permission from Wiley Interscience Publications).

2.6.6. Curing Cycle

The above considerations (determination of temperature, degree of cure, rate of cure, viscosity, resin flow, void formation) can be taken into account to develop the curing cycle. Figure 4.33 shows a typical cure cycle for a carbon/epoxy composite laminate. The cure cycle consists of the temperature cycle, pressure cycle and vacuum. The cycle is called a two-step cure cycle since the temperature cycle has two holds (the horizontal parts of the cycle). Note that the first hold temperature is to allow sufficient time for the low viscosity period in which the resin can wet the fibers, and also to allow time for the resin to flow to the bleeders.

2.6.7. Autoclave [1]

An autoclave may be described as a large pressure vessel with an integral heating capacity. Figure 4.34 shows a typical configuration of an industrial autoclave.

For aerospace applications, the composite parts are quite large, and hence the autoclave has to be larger. For example, the dimensions shown in Figure 4.34 can be $D = 30$ m and $L = 50$ m.

Commonly used aerospace thermoset materials (e.g., high temperature cure epoxies) require cure temperature and pressure on the order of 175°C and 600 kPa (about 80 psi), respectively. Hence the autoclave must be strong at high temperatures. For some composite materials, the pressure and temperature requirements may be even higher. For example, thermoplastic composites (e.g., PEEK, PEl) and higher temperature

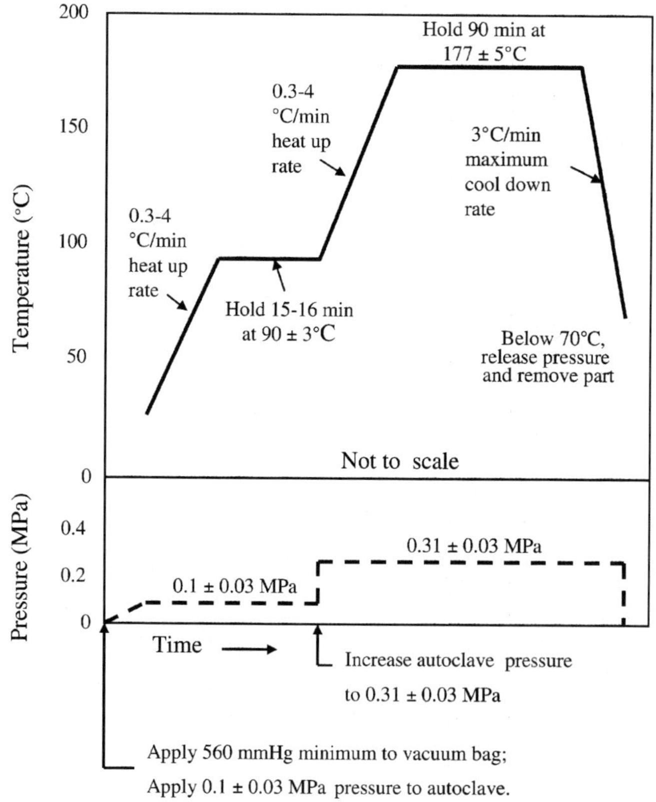

FIGURE 4.33 A typical cure cycle for epoxies.

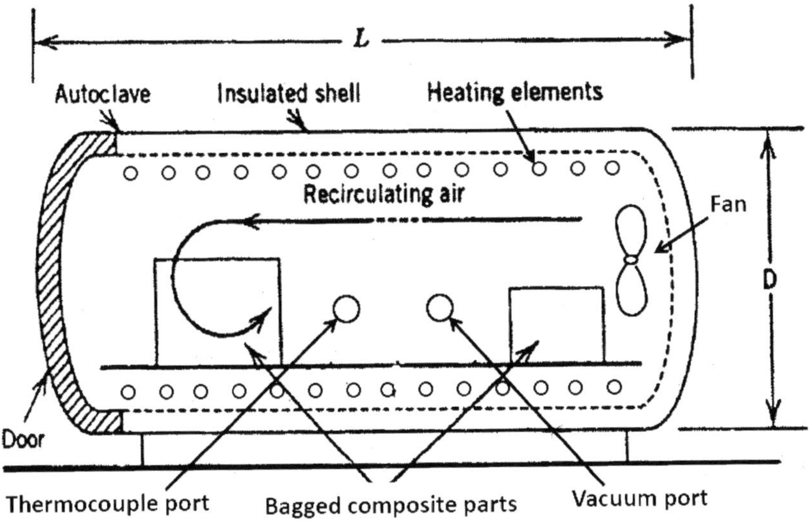

FIGURE 4.34 Schematic drawing of an aerospace autoclave.

thermoset materials (polyimides, PMR15) may require temperatures in the range of 300–400°C and pressures to 689 kPa (100 psi) or higher. For these applications, special autoclaves must be constructed to meet the special requirements but at a significantly higher cost. Fortunately from a cost point of view, most applications are at the mid-range in temperature and pressure.

2.6.7.1. Advantages of the Autoclave

One of the major advantages of the autoclave is its ability to process a wide variety of materials. In general, any polymeric material can be processed as long as its cure cycle falls within the temperature and pressure limitations for the autoclave. A typical cure cycle is shown in Figure 4.33. A second major advantage of the autoclave is that it represents an extremely flexible method to apply pressure to a part. Usually parts are laid-up on one-sided tooling and then wrapped in a plastic bagging material. Pressure is then applied onto the part, which presses it against the tool. The pressure on the part is further intensified by simultaneously pulling a vacuum in the bag. This technique allows the autoclave to be used for the processing of a variety of part shapes. Because of these advantages, the autoclave is used extensively for the production of aerospace advanced composite parts.

2.6.7.2. *Disadvantages of the Autoclave [1]*

Chief among the shortcomings are its sluggish temperature and pressure response, and relatively poor temperature control. To some extent, all of these problems are related to the large size of the autoclave and the methods of heating and pressurization. Note, however, that the cure cycles of many of the materials developed for aerospace applications are rather slow, on the order of hours. Hence the generally sluggish heating and cooling rates of a large autoclave and associated tooling do not severely limit the process for such materials. Temperature control can be addressed by an improved internal gas circulation system along with temperature sensing and control. Pressure buildup is related to temperature control since temperature is used to heat the gas and increases the pressure.

There are, however, basic fundamental limits to the performance of industrial autoclaves. For example, the time Δt to achieve a given adiabatic temperature rise ΔT for a given power is controlled by the mass M and heat capacity at constant pressure C_p as:

$$\Delta t \approx \frac{Mc_p \Delta T}{\text{power}} \tag{4.23}$$

In general, the Mc_p term for the autoclave and the tooling can be very large, much larger than that of the part. Thus the thermal inertia of the equipment and tooling represents a limitation on the heating and cooling rates for the process. As a result, large autoclaves are not suitable for the rapid processing of material with potentially short thermal cycles.

Currently the most popular pressurizing medium for autoclave is nitrogen. Although air is still sometimes used because of the risk of auto-ignition, its use is generally confined to low temperature cure systems. Either way, the pressurizing medium is compressible, while the volume of the autoclave is larger. Both of these result in relatively slow pressurization rates.

A primary concern for autoclave operators is safety. Autoclaves have to be built according to the ASME pressure vessel code. It is not an exaggeration to say that a heated autoclave is effectively a bomb, so safety procedures are rigorous. Industrial autoclaves are generally fitted with several relief valves, and inspections are carried out frequently to ensure that flaws have not developed in the structure. It is common practice in the aerospace industry to house only the door of the autoclave in the plant interior. Shell structures are on the outside.

Because of their size, temperature and pressure requirements, autoclaves are usually expensive. For example, even small lab scale autoclaves (approximately $D = 0–5$ m, $L = 1$ m) currently cost on the order of US $200,000. Peripheral equipment such as air or nitrogen lines, cooling elements, heaters, and monitoring equipment can also add to the cost. Large-scale industrial autoclaves and auxiliary equipment can cost in the range of US 1 million dollars.

2.7. Resin Shrinkage, Out-of-Dimensions and Residual Stresses

During the process of curing the resin, the part has to go through a temperature cycle. There are many phenomena that may affect the dimensions of the resin and of the part. These are as follows.

- There is shrinkage of the resin due to chemical reaction. The change of the state of the resin from liquid to solid is accompanied by some shrinkage of the resin.
- The increase and decrease in temperature during the cycle create thermal expansion and contraction in both resin and fibers.
- The combined chemical shrinkage and the thermal expansion (or contraction) will result in changes in dimensions of the part.
- Friction at the interface between the part and the mold and the mismatch between thermal expansion (or contraction) of the mold and deformation of the part can result in residual stresses in the part.

Figure 4.35 shows the schematic of the development of modulus and shrinkage of the resin during the curing process. This development can be divided into three stages. Stage 1 is from the liquid state until the degree of cure reaches the gel point where the modulus starts to be developed. Stage 2 extends from the gel point until the modulus is fully developed, and stage 3 is the solid phase. The shrinkage development can also be classified into three similar stages. However the limit of these stages may not coincide with the limits for the stages for the modulus.

2.7.1. Shrinkage and Modulus Development

Epoxy resins are known to exhibit shrinkage upon curing. The average shrinkage (percent change in volume in solid state as compared to liquid state of the same mass of resin) is about 5%–8%. The usual method to determine shrinkage is to measure the volume of the material in the liquid state and that in the solid state. One single number for the shrinkage is usually given. The PVT method [15] may be used to determine the de-

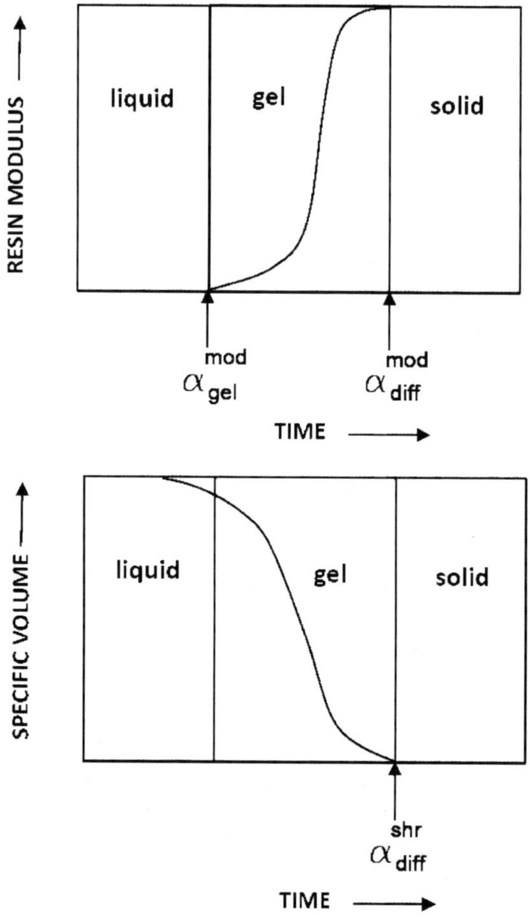

FIGURE 4.35 Shrinkage and modulus development.

gree of shrinkage as the resin changes state. Recently the ultrasonic method (PSM) developed at Concordia University [16] provides monitoring of the development of shrinkage and modulus of the resin during the curing process.

The history of shrinkage and modulus development for one epoxy system (Shell Epon 828 and Epicure 3046) determined using the PSM method is given in Figures 4.36 and 4.37. It can be seen that the shrinkage modulus developments of the resin depend on the amount of curing agent. There is a discontinuity in the modulus curve due to loss of the ultrasonic signal used for the detection of the development.

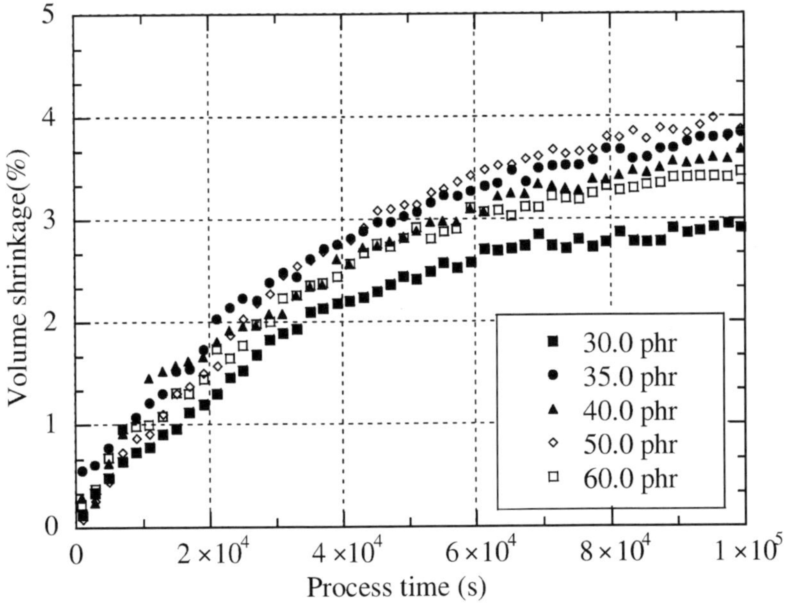

FIGURE 4.36 Shrinkage development in Shell Epon 828 and Epicure 3046 using the PSM method. Different curves show the effect of different amounts of curing agent [16].

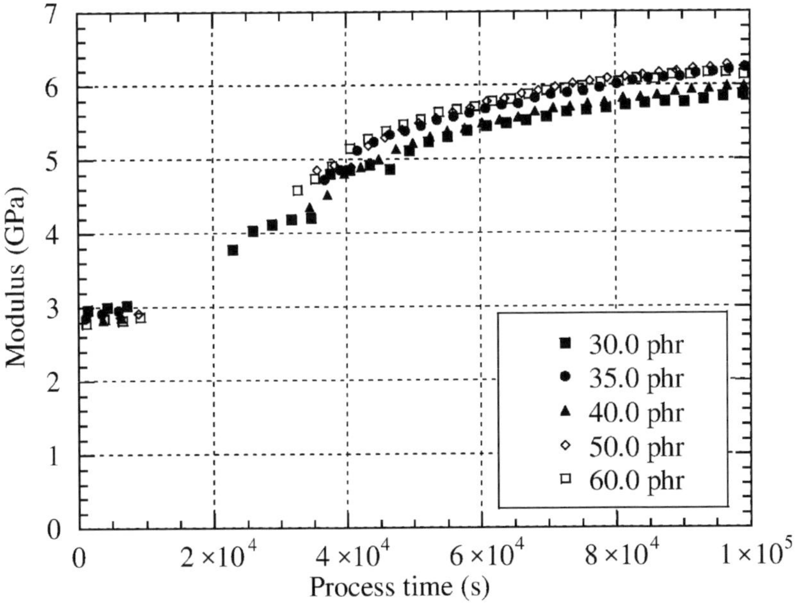

FIGURE 4.37 Modulus development of Shell Epon 828 and Epicure 3046 using the PSM method. Different curves show the effect of different amounts of curing agent [16].

Correlation between shrinkage and the degree of cure is shown in Figure 4.38. The shrinkage seems to follow the degree of conversion at the beginning of the process. After a certain amount of time, however, the conversion is faster than the shrinkage. Toward the latter part of the process, shrinkage catches up with the degree of conversion to reach full value at about the same time. The combined information on the shrinkage and modulus development can be used to determine the change in dimensions of the part and the residual stresses.

2.7.2. Determination of Changes in Dimensions of the Part and Residual Stresses

Due to resin shrinkage, coefficient of expansion (or contraction) of the part and of the tool, the dimensions of the final part may not follow the di-

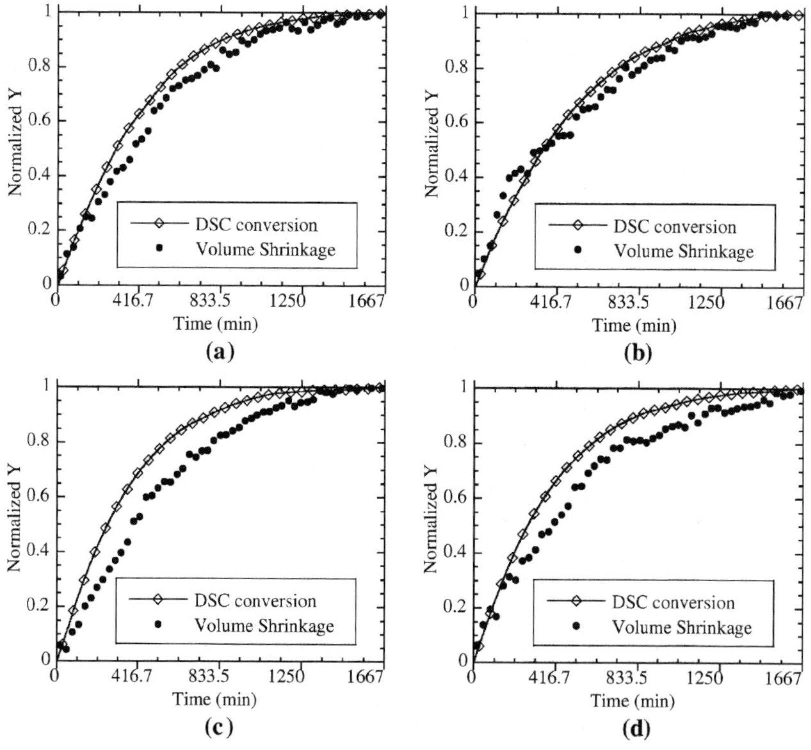

FIGURE 4.38 Correlation between degree of cure (DSC) and shrinkage development using the PSM method—Epon 828 and Epicure 3046: (a) 30 phr, (b) 40 phr, (c) 50 phr, and (d) 60 phr [16].

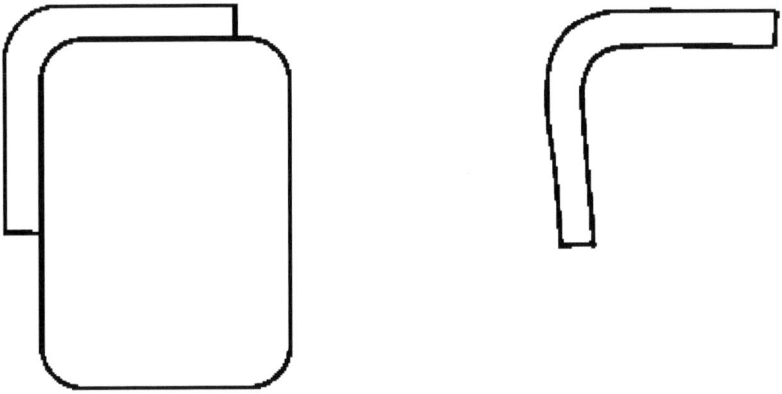

FIGURE 4.39 Out-of-dimensions due to shrinkage.

mensions of the tool. The result is a part with out-of-dimensions. Figure 4.39 shows an example of the spring-in phenomenon on parts with curved portions. This, together with possible friction between the part and the tool during the cool down cycle, may create residual stresses in the part. For large parts or for parts with complex geometry, this can present a very serious problem.

Numerical analysis (finite element analysis) is normally used to calculate the deformation of the part and its residual stresses, provided that information on the history of shrinkage and modulus development is available [6]. For the simple case of a flat plate, closed form solution is available. For this case, the situation can be represented as shown in Figure 4.40.

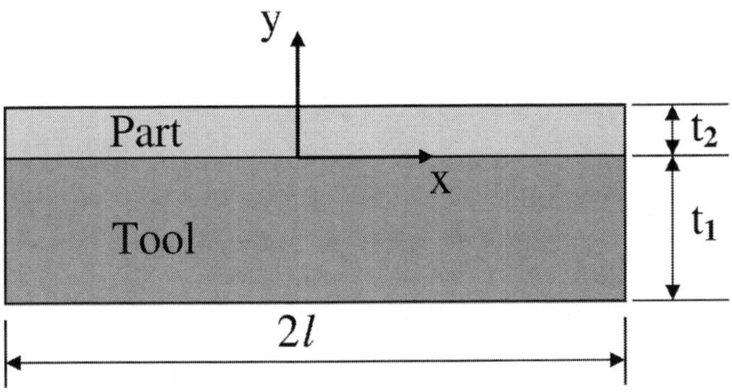

FIGURE 4.40 Flat composite part of a flat tool.

For the case of a flat part on a flat tool, the mismatch is due to thermal strain along the axial direction. For stress-free boundary conditions at the top surface of the part and at the bottom surface of the tool, the axial displacement and shear stress in the tool and part are given as [17]:

For the tool (denoted with subscript 1):

$$u_1 = \sum_{n=1}^{\infty} [\{D_{1n}^f \sin(k_n^f x) \cosh[\beta_{1n}^f (y + t_1)]\}] + \varepsilon_1^{ther} x \tag{4.24}$$

$$\tau_1 = \sum_{n=1}^{\infty} \{G_1 D_{1n}^f \beta_{1n}^f \sin(k_n^f x) \sinh[\beta_{1n}^f (y + t_1)]\} \tag{4.25}$$

For the part (denoted with subscript 2):

$$u_2 = \sum_{n=1}^{\infty} [\{D_{2n}^f \sin(k_n^f x) \cosh[\beta_{2n}^f (y - t_2)]\}] + \varepsilon_2^{ther} x \tag{4.26}$$

$$\tau_2 = \sum_{n=1}^{\infty} \{G_2 D_{2n}^f \beta_{2n}^f \sin(k_n^f x) \sinh[\beta_{2n}^f (y - t_2)]\} \tag{4.27}$$

where D_{1n}^f and D_{2n}^f can be found by applying the displacement and shear stress compatibility conditions at the interface (i.e., at $y = 0$), namely,

$$D_{1n}^f = (-1)^{(n+1)} \frac{2(\varepsilon_2^{theo} - \varepsilon_1^{theo})}{l(k_n^f)^2} \times \frac{G_2 \beta_{2n}^f \sinh[-B_{2n}^f t_2]}{\{G_2 \beta_{2n}^f \cosh[\beta_{1n}^f i_1] \sinh[-\beta_{2n}^f t_2] - G_1 \beta_{1n}^f \sinh[\beta_{1n}^f t_1] \cosh[-\beta_{2n}^f t_2]} \tag{4.28}$$

$$D_{2n}^f = (-1)^{(n+1)} \frac{2(\varepsilon_2^{theo} - \varepsilon_1^{theo})}{l(k_n^f)^2} \times \frac{G_1 \beta_{1n}^f \sinh[-B_{1n}^f t_1]}{\{G_2 \beta_{2n}^f \cosh[\beta_{1n}^f i_1] \sinh[-\beta_{2n}^f t_2] - G_1 \beta_{1n}^f \sinh[\beta_{1n}^f t_1] \cosh[-\beta_{2n}^f t_2]} \tag{4.29}$$

and

$$G = G_{xy}$$

$$\beta_n^f \sqrt{\frac{E_{xx}}{G_{xy}}} \frac{\pi(2n-1)}{2l}$$

$$k_n^f = 2n - 1$$

and ε^{ther} is the free thermal strain (due to CTE and cure shrinkage) in the axial direction.

Using the material properties for the tool and the part as given in Table 4.9, deformation and stresses were calculated by Arafath et al. [17]. One set of results on the variation of the residual stresses for different laminate lay-up and thicknesses is shown in Figure 4.41.

It can be seen that by assuming the development of the modulus as shown in Table 4.9, the residual stresses in the part first develop at the in-

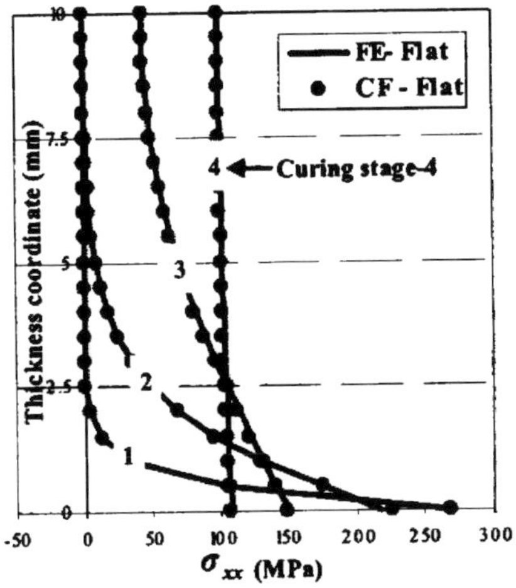

FIGURE 4.41 Variation of residual stresses for a flat composite plate across its thickness (reproduced from Arafath A. R. A., Vaziri R. and Poursartip A. "Closed form solutions for process induced stresses and deformations of flat and curved composite parts," *Proc. 6th Canadian International Conference on Composites, CANCOM 2007, Winnipeg, Canada, August 2007,* with permission from Canadian Association for Composite Structures and Materials).

TABLE 4.9 Thermo-mechanical Properties of Aluminum and Composite Materials [17].

Cure Stage	E_{11} (GPa)	E_{22}, E_{33} (MPa)	ν_{12}, ν_{13}	ν_{23}	G_{12}, G_{13} (MPa)	G_{23} (MPa)	CTE_1 $\mu\varepsilon/°C$	CTE_2, CTE_3 $\mu\varepsilon/°C$
Aluminum								
	69.00	69×10^3	0.0	0.0	2.6×10^4	2.6×10^4	23.6	23.6
Composite								
1	124.0	0.177	0.0	0.0	0.0665	0.058	3.74×10^2	29.5
2	124.0	1.77	0.0	0.0	0.665	0.58	3.74×10^2	29.5
3	124.0	17.7	0.0	0.0	6.64	5.8	3.74×10^2	29.5
4	124.0	175.0	0.0	0.0	66.3	57.6	3.74×10^2	29.5

terface between the part and the tool at stage 1. The stress then spreads toward the upper part of the laminate until the whole laminate feels the residual stress when the modulus is fully developed.

Bogetti and Gillespie [7] also obtained the residual stresses for glass/polyester. Figure 4.42 shows the residual stresses for laminates up to 2.54 cm thick. The transverse stress is compressive at the center of the laminate and tensile at regions close to the surface.

Figure 4.43 shows the residual stresses for laminates thicker than 2.54 cm. Thicker laminates show tensile residual stresses at the mid-thickness region and compressive stresses at regions closer to the surface of the laminate. This is the result of the temperature distributions as shown in Figures 4.21 and 4.22.

2.8. Out-of-Autoclave (OOA) Manufacturing Processes

Manufacturing of composites using the autoclave has been the workhorse for the composite industry for a long time. However, due to the large equipment required (large size autoclave) and the associated long heating time required to bring the environment in the autoclave to proper

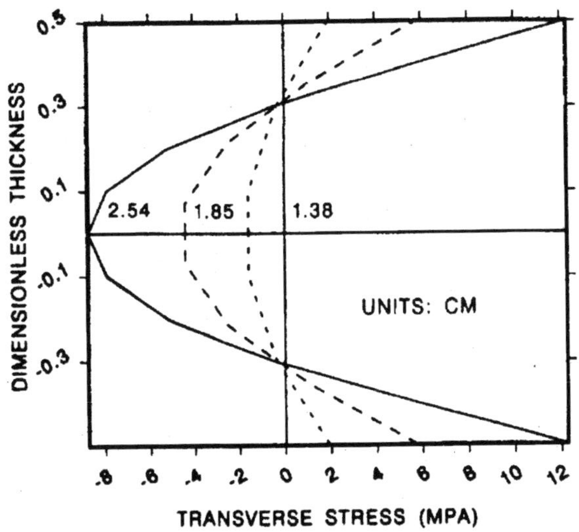

FIGURE 4.42 Residual process induced in-plane transverse stress distribution (t 2.54 cm) (reproduced from "Process induced stress and deformation in thick-section thermoset composite laminates," by Bogetti T. A. and Gillespie J. W., *Journal of Composite Materials,* Vol. 26, No. 5, 1992, pp. 626–660, with permission from Sage Publications).

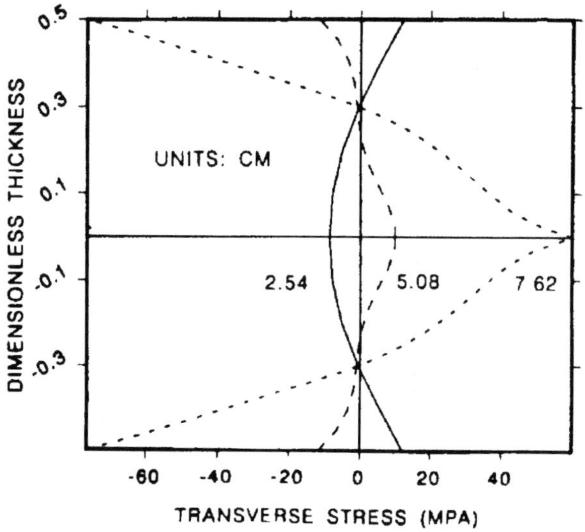

FIGURE 4.43 Residual process induced in-plane transverse stress distribution (t 2.54 cm) (reproduced from "Process induced stress and deformation in thick-section thermoset composite laminates," by Bogetti T. A. and Gillespie J. W., *Journal of Composite Materials,* Vol. 26, No. 5, 1992, pp. 626–660, with permission from Sage Publications).

temperature and the associated cost, there have been attempts to cure composites without using the autoclave. The term *out-of-autoclave manufacturing process* refers to those processes that do not require an autoclave. The main differences between OOA and autoclave processes are that the maximum compaction pressure for OOA is less than 1 atm (about 6 atm for autoclave processes) and that the curing temperature is typically less than 100°C for OOA (about 180°C for autoclave process). These processes are still in the experimental stage and have not yet reached widespread industrial applications.

3. REFERENCES

1. Gutowski T.G. *Advanced Composites Manufacturing,* John Wiley and Sons, 1997.
2. Hyer M.W. *Stress Analysis of Fiber reinforced Composite Materials,* McGraw Hill, 1998.
3. DeBakker. C. J. St. John N. A. and George G. A. M, "Simultaneous differential scanning calorimetry and near-infra-red analysis of the curing of Tetraglycidyldiaminodiphenylmethane with diaminodiphenylsulphone," *Polymer,* 34(4), 1993, pp. 716–725.

4. Cole K.C. "A new method to modeling the cure kinetics of epoxy/amine thermosetting resins. 1. Mathematical development," *Macromolecules,* 1991, 24(11), pp. 3093–3097.
5. Cole K. C., Hechler J. J. and Noel D. " A new method to modeling the cure kinetics of epoxy/amine thermosetting resins. 2. Application to a typical system based on bis[4-(diglycidylamino)phenyl]methane and bis(4-aminophenyl) sulfone," Macromolecules, 1991, 24(11), pp. 3098–3110.
6. Bogetti T. A. "Process induced stress and deformation in thick-section thermosetting composites," Ph.D thesis, Department of Mechanical Engineering, University of Delaware, 1989.
7. Bogetti T. A. and Gillespie J. W. "Process-induced stress and deformation in thick-section thermoset composite laminate,s" *J. Composite Materials,* 1992, Vol. 26, No. 5, pp. 626–660.
8. Lee W. I., Loos A. C. and Springer G. S. "Heat of reaction, degree of cure, and viscosity of Hercules 3501-6 resin," *Journal of Composite Materials,* 1982, Vol. 16, pp. 510–520.
9. Calado V. M. A. and Advani S. G. "Thermoset resin cure kinetics and rheology" in *Processing of Composites,* Dave R. S. and Loos A. C., eds., Hanser, Cincinnati, 1999.
10. Roylance D. "Reaction kinetics for thermoset resins", in *Manufacturing International,* 1988 Proc. , Vol. IV, The Manufacturing Science of Composites, 1988, pp. 7–11.
11. Dave R., J. L. Kardos, and M.-P. Dudu Kovic "A model for resin flow during composite processing. Part I, general material development." *Polymer Composites,* Vol. 8, No. 1, 1987, pp. 29–38.
12. Hubert P., Vaziri R. and Poursartip A. "A two-dimensional flow model for the process simulation of complex shape composite laminates," *Int. J. for Numerical Methods in Engineering,* 1999, Vol. 44, pp. 1–26.
13. Dusi M. R.et al. *Journal of Composite Materials,* 1987, Vol. 21, pp. 243–261.
14. Kardos J. L., Dudukovic M. P. and Dave R. "Void growth and resin transport during processing of thermosetting matrix composites," *Advances in Polymer Science,* K. Dusek ed., 1986, Vol. 80, Springer-Verlag, Berlin,
15. Kinkelaar M. and Lee L. J. "Development of a dilatometer and its application to low-shrink unsaturated polyester resins," *J. of Applied Polymer Science,* 1992,Vol. 45, pp. 37–50.
16. Hoa S.V., Ouellette P., and Ngo T. D. "Determination of shrinkage and modulus development of thermoset resins,"accepted for publication *Journal of Composite Materials,* 2008.
17. Arafath A. R. A, Vaziri R. and Poursartip A. "Closed-form solutions for process- induced stresses and deformations of flat and curved composite parts," *Proceedings of the 6th Canadian International Conference on Composites, CANCOM 2007, Winnipeg, Canada August 2007.*

4. HOMEWORK

1. Determine the variation of the degree of cure for a carbon/epoxy laminate of Example 1 with the exception that the resin is 3501-6 with the kinetic equation given by the equation:

$$\frac{d\alpha}{dt} = (K_1 + K_2\alpha)(1-\alpha)(B-\alpha) \quad \alpha \le 0.3$$

$$\frac{d\alpha}{dt} = K_3(1-\alpha) \quad \alpha \succ 0.3$$

$$K_1 = A_1 \exp\left(-\frac{\Delta E_1}{RT}\right)$$

$$K_2 = A_2 \exp\left(-\frac{\Delta E_2}{RT}\right)$$

$$K_3 = A3 \exp\left(-\frac{\Delta E_3}{RT}\right)$$

$$A_1 = 2.101 \times 10^9 \text{ min}^{-1}$$

$$A_2 = -2.014 \times 10^9 \text{ min}^{-1}$$

$$A_3 = 1.960 \times 10^6 \text{ min}^{-1}$$

$$\Delta E_1 = 8.07 \times 10^4 \text{ J/mol}$$

$$\Delta E_2 = 7.78 \times 10^4 \text{ J/mol}$$

$$\Delta E_3 = 5.66 \times 10^4 \text{ J/mol}$$

$$B = 0.47$$

CHAPTER 5

Filament Winding and Fiber Placement

1. FILAMENT WINDING

1.1. Introduction

Filament winding is a process used to make composite structures such as pressure vessels, storage tanks or pipes. Composite pressure vessels offer light weight and high strength. Applications include oxygen tanks used in aircraft and by mountain climbers, compressed natural gas cylinders for vehicles, drive shafts for automobiles, and pipes for conducting corrosive liquids.

Filament winding is a comparatively simple operation in which continuous reinforcements in the form of rovings or monofilaments are wound over a rotating mandrel. Specially designed machines, traversing at speeds synchronized with the mandrel rotation, control the winding angles and the placement of the reinforcements. Structures may be plain cylinders or pipes or tubing, varying from a few centimeters to one or two meters in diameter. Spherical, conical, and geodesic shapes are within winding capability. End closures can be incorporated into the winding to produce pressure vessels and storage tanks.

A schematic of a simple filament winding setup is shown in Figure 1.2(b) and repeated here as Figure 5.1. Figure 5.2 shows a photo of a filament winding machine [repeat of Figure 1.2(c)].

The basic mechanism consists of pulling a roving (number of strands) of fibers from the creels. These are spread out using a bank of combs. The fibers then go through a bath of resin (for the case of wet winding). On exit from the bath of resin, the fibers are collimated into a band. The band

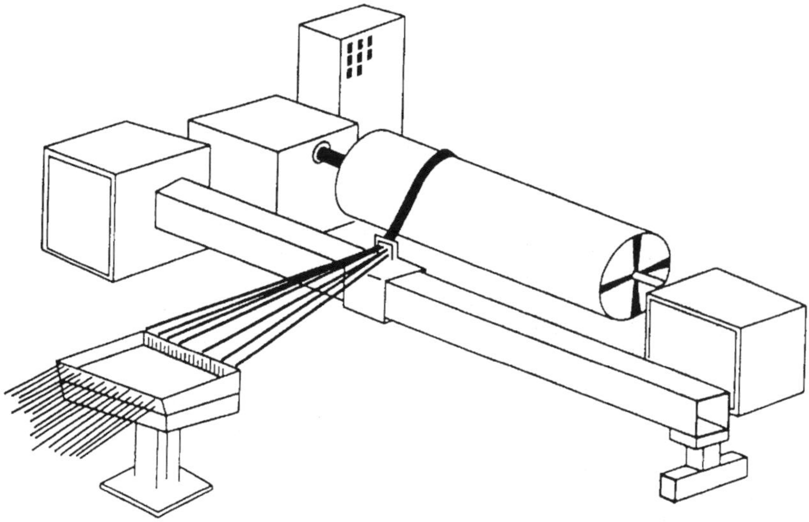

FIGURE 5.1 Schematic of the filament winding process (courtesy of Wiley Interscience).

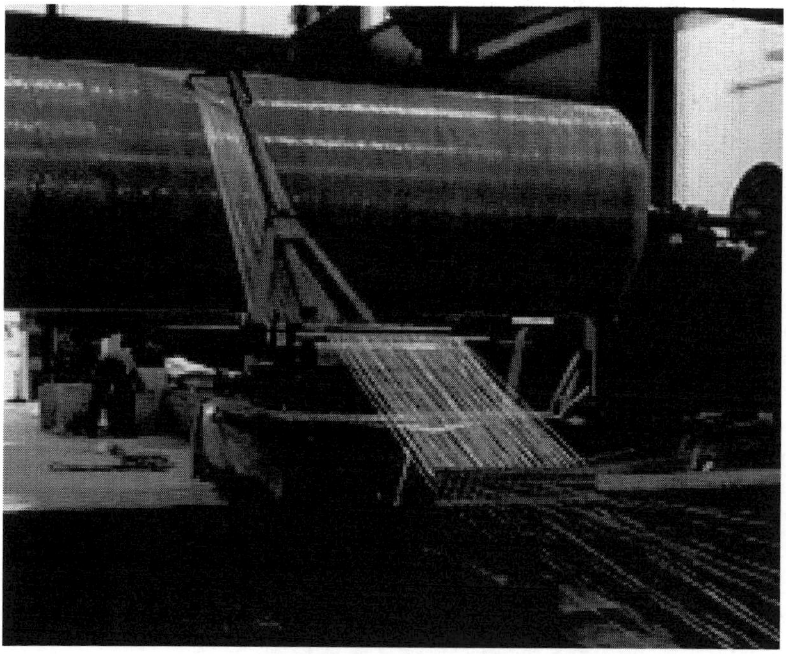

FIGURE 5.2 The placement of fiber band on the mandrel. (www.gilgwang.com/english/frp/grp01.html).

goes through a fiber feed and is then placed on the surface of a mandrel. The fiber feed traverses back and forth along the length of the mandrel. The mandrel is attached to a motor, which gives it rotational motion. The combined motion of the fiber feed and the rotation of the mandrel make the fiber bands spread over the surface of the mandrel. By covering the surface of the mandrel with many layers, one can build up the thickness of the part. The fiber orientation can be controlled by varying the speed of traverse of the fiber feed and the rotational speed of the mandrel. Filament winding is usually used to make a composite structure in the form of surfaces of revolution, such as cylinders or spheres. The surfaces of these structures are usually convex due to the need to apply tension on the tows while these tows are placed on the surface of the mandrel. If the surface is concave, bridging of the fibers over the concave surface may occur. As can be seen from these figures, the basic components of a filament winding system consist of a mandrel and devices to place the fiber tows on the surface of the mandrel to build up the thickness for the part.

1.2. The Winding Process

The operation of filament winding is the reverse of the conventional machining process of milling on a lathe. In milling, one starts with a cylindrical surface and one removes the material from the surface one strip at a time. In filament winding, one deposits the material on the surface of the mandrel one strip at a time. The most basic form of filament winding is a two-degrees-of-freedom operation. This consists of the rotation of the mandrel and the linear movement of the feed along the axis of the mandrel. Two-axis filament winding machines can be used to wind pipes. Filament winding machines with more degrees of freedom exist. The availability of the additional degrees of freedom can be useful in winding at the end of the part, such as heads of pressure vessels, or the winding of shapes more complex than straight cylinders such as those with variation in cross section (i.e. cones) or spheres. For example, for the case of a four-axis winding machine, the basic movements are mandrel rotation and feed traverse. To these are added a cross-slide perpendicular to the mandrel axis and a fourth axis of motion, rotation of the feed eye. These latter permit more accurate fiber placement at the ends. Winding machines with more degrees of freedom up to the level of multi-degrees-of-freedom robots are available. To illustrate the concept of filament winding, only the simple operation of machines with two degrees of freedom will be described in this chapter. Depending on the coordination between the rotational motion and the axial motion, different types of winding can be obtained. These are: polar, helical, circuit and pattern, layer, hoop, longitudinal, and combination.

1.2.1. Polar Winding

This is also called planar winding. In this process, the mandrel remains stationary while a fiber feed arm rotates about the longitudinal axis, inclined at the prescribed angle of the wind. The mandrel is indexed to advance one fiber bandwidth for each rotation of the feed arm. This pattern is described as a single circuit polar wrap (Figure 5.3) .The fiber bands lie adjacent to each other; a completed layer consists of two plies oriented at plus and minus the winding angle.

1.2.2. Helical Winding

In this process, the mandrel rotates continuously while the fiber feed carriage traverses back and forth. The carriage speed and mandrel rotation are regulated to generate the desired winding angle. The normal pattern is multi-circuit helical. After the first traverse, the fiber bands are not adjacent. Several circuits are required before the pattern repeats. A typical 10-circuit pattern is shown in Figure 5.4.

In the above configuration one needs to distinguish between the straight cylindrical part and the head (or dome). In the straight cylindrical part, the relation between the rotational displacement and axial displacement can be established. Refer to Figure 5.5. This figure shows the developed surface of the straight part of the cylinder. The dimension of the base is πD where D is the diameter of the mandrel. Let α be the winding angle (angle between fiber path and the axis of the cylinder), b be the band width of the fiber band, and L be the axial distance traveled by the

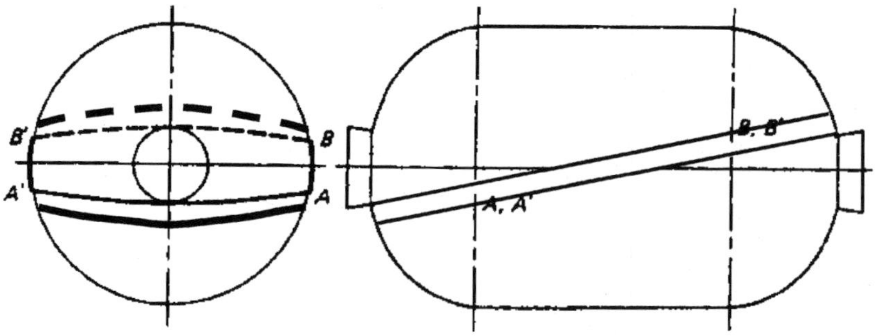

FIGURE 5.3 Planar winding.

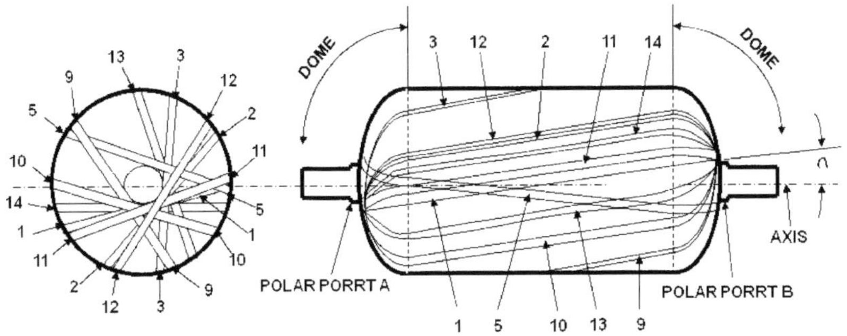

FIGURE 5.4 An example of a helical winding pattern.

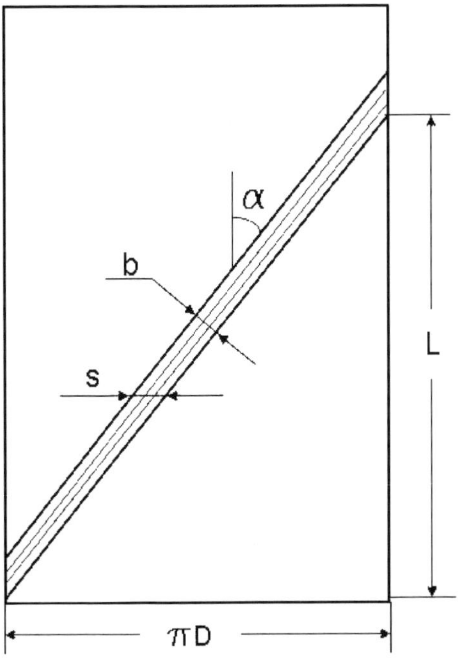

FIGURE 5.5 Developed envelope with fiber path.

fiber feed corresponding to one rotational revolution. The relation between the rotational distance and axial distance can be written as:

$$L = \frac{\pi D}{\tan \alpha} \tag{5.1}$$

If h represents the length of the straight part of the cylinder to be built, the number of revolutions required for the fiber feed to travel this distance is given as:

$$n = \frac{h}{L} = \frac{h \tan \alpha}{\pi D} \tag{5.2}$$

Equation (5.2) gives the number of revolutions. This can be a whole number or a decimal number. One needs to convert this into the number of degrees (by multiplying n by 360) in order to determine the number of degrees of revolution.

Since filament winding is a continuous process, the fiber feed has to reverse its motion to go back to the other end. Also it is essential that tension be maintained in the fibers to ensure good properties of the final product. One also needs to identify the location of the fiber feed (point A in Figure 5.6) and the point of separation between the fiber band and the surface of the mandrel (point B).

When the point B reaches the end of the straight part of the cylinder, this point will move over the surface of the head of the component to be built (i.e. a vessel). The fiber feed (point A) starts to go into reverse. It takes some time before point B touches the end of the straight part of the cylinder again (point B′). The number of degrees of rotation of the man-

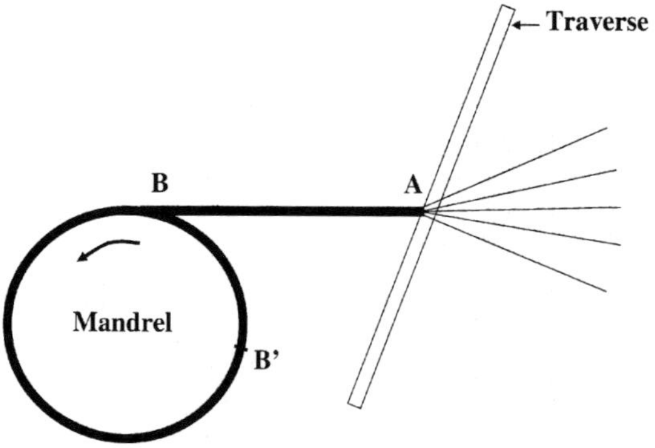

FIGURE 5.6 Relative position of the fiber feed (point A) and point of separation (point B) between fiber band and mandrel surface.

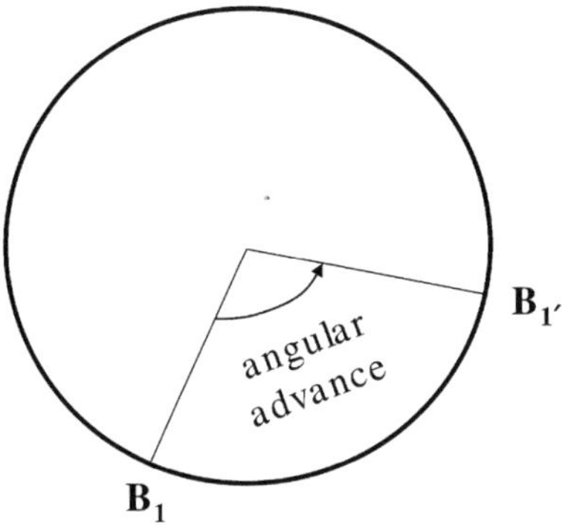

FIGURE 5.7 Relative angular position of the point of separation at the end and at the beginning of a circuit.

drel during this time is termed the *dwell angle*. A dwell angle exists at both ends of the cylinder.

1.2.3. Circuit and Pattern

When the point B has gone one complete cycle and returns to the same axial position along the length of the cylinder and goes in the same direction, a circuit has been completed. Due to the complexity of the motion, there is no guarantee that after one circuit, the point $B_{1'}$ at the end of one circuit will have the same angular position as its position at the beginning of the circuit (B_1). Figure 5.7 illustrates this point.

It takes a number of circuits before the point B can return to its position at the beginning. When this happens, one has a pattern. This can be illustrated in the following example.

Example 5.1

It is desirable to wind a 30 cm diameter by 100 cm long cylinder at a 30° wind angle. The fiber band width is assumed to be 0.6 cm and the dwell angle is 180°. Determine the number of circuits required to make a pattern.

Solution

First, define the *reference circle* as the circle of the cross-sectional area of the cylinder at one end, say, the left end. Assume that winding starts from a point B on that circle. In one circuit, the feed moves twice the length of the mandrel. This means forward once and backward once along the length of the mandrel. When a pattern is complete, a set of circuits has been made and the fiber path returns to the initial position.

Equation (5.2) gives the number of revolutions required for the fiber feed to move a distance h which is equal to the length of the cylinder. For a circuit, two cylinder lengths need to be traveled. The corresponding number of revolutions will therefore be:

$$2n = \frac{2h\tan\alpha}{\pi D} = \frac{(2)(100\text{ cm})\tan 30}{\pi(30\text{ cm})} = 1.23 \quad \text{(a)}$$

The corresponding number of degrees is:

$$(1.23)(360) = 441° \quad \text{(b)}$$

In addition to the number of degrees in Equation (b), one has to add two times the dwell angle in order to obtain the total number of degrees required to make a circuit. This gives:

$$\theta = 441 + 2(180) = 801° \quad \text{(c)}$$

If one subtracts the above number by a whole multiple of 360°, one would obtain the angular advance of the starting point (new point $B_{1'}$) as compared to starting point B_1 on the reference circle. This angular advance is 801 – (360)(2) = 81°. This is shown in Figure 5.7.

In order to make a pattern, one needs to have a multiple of the advance angles such that this multiple will be equal to a multiple of 360°. This can be expressed as:

$$(m)(81) = (n)(360) \quad \text{(d)}$$

where m and n are integers and should be as small as possible.

Equation (d) shows that m and n can be quite large before the equation is satisfied. This may not be practical. In order to reduce the numbers m and n, one needs to adjust the operation to make the advance angle a good whole number. One good whole number close to 81 is 90. This can be done by adjusting the dwell angle to be 180 + 9/2 = 184.5°.

(This can be done by adjusting the machine setting.) If this is done, Equation (d) becomes:

$$(m)(90) = (n)(360) \quad \text{or} \quad \frac{m}{n} = 4 \quad \text{(e)}$$

One can select $m = 4$ and $n = 1$. What this means is that it takes 4 times the advance angle (or 4 circuits) to make a pattern.

Note: In the pattern calculated above, the fiber band will go back exactly to the same position on the reference circle as at the beginning of the winding process. This may not be desirable because if one continues this process, the fiber will follow the same path as before and one may not be able to cover the whole surface of the mandrel. It is desirable to advance the position $B_{1'}$ one bandwidth distance along the circumferential direction after one pattern. This distance in angular value can be calculated to be (note that the circumferential coverage of a bandwidth b is $b/\cos\alpha$):

$$\Delta\theta = \frac{b}{\pi D\cos\alpha}(360) = \frac{0.6}{\pi(30)\cos(30)}(360) = 2.65^{\circ} \quad \text{(f)}$$

This advanced angular value is accumulated over 4 circuits. The value for each circuit is 2.65/4 = 0.66°. This angle is then divided by two dwell angles. The dwell angle is then adjusted to be: 184.5 + 0.66/2 = 184.8°.

1.2.4. Layer

A pattern may consist of fiber intersections (fiber crossovers—see Figure 5.4) at certain sections. Crossovers may occur at more than one section, depending on the wind angle. A layer is defined as a set of patterns that completely cover the surface of the mandrel with fibers.

From Figure 5.5, it can be seen that the relation between the circumferential coverage S and the bandwidth b can be written as:

$$S = \frac{b}{\cos\alpha} \quad (5.3)$$

In order to make a layer, the whole circumferential distance πD has to be covered. The number of circuits per layer C can be calculated as:

$$C = \frac{\pi D}{S} = \frac{\pi D\cos\alpha}{b} \quad (5.4)$$

Example 5.2

Continue with Example 5.1 and determine the number of circuits required to make a layer.

Solution

For $\alpha = 30°$, one has (from Equation 5.4):

$$C = \frac{\pi(30)\cos 30}{0.60} = 136$$

There are 136 circuits to make up a layer. Recall from Example 5.1 that it takes 4 circuits to make a pattern. The number of patterns per layer is then 136/4 = 34.

1.2.5. Hoop Winding

Hoop or circumferential layers are wound close to 90°. The feed advances one bandwidth per revolution. The layer is considered a single ply. Hoop layers may also be applied as doublers or localized stiffeners at strategic points along the cylinder.

1.2.6. Longitudinal Winding

Longitudinal winding applies to low angle wrap which is either planar or helical. For closed pressure vessels, the minimum angle is determined by the polar openings at each end.

1.2.7. Combination Winding

Longitudinals are reinforced with hoop layers. The customary practice with pressure vessels is to place the bulk of the hoop wraps in the outer layer. A balance of hoop and longitudinal reinforcement can also be achieved by winding at two or more helical angles.

1.2.8. Wet/dry Winding

In addition to the classification of different winding patterns, one also distinguishes the type of winding depending on whether fibers are wetted with liquid resin in-situ or prepregs are used. These are referred to as wet winding or dry winding. In wet winding, the resin is applied during the winding stage (Figure 5.1). The alternate dry winding method utilizes the pre-impregnated B-staged rovings. Wet winding tends to be messy due to the possible dripping of the wet resin. Dry winding is cleaner but the raw materials (prepregs) are more expensive than the tows.

1.3. End Closures

End closures for pressure vessels are either mechanically fastened to the cylindrical portion or are integrally wound. If end closures are fastened to the cylindrical portion, both the end closures and cylindrical portion need to have flanges. Integrally wound end closures can provide better pressure containment than mechanically fastened heads. The fiber path yields a balance of meridional and circumferential forces and is consistent with winding conditions so that no slippage occurs. The head contours and related polar bosses are critical in vessel design.

One common contour follows the geodesic isotensoid. This contour is normally adapted to helical winding. The fiber path is taken as tangential to the polar boss (Figure 5.8). The geodesic path is the path of shortest distance between two points on a curved surface. The reason to select the

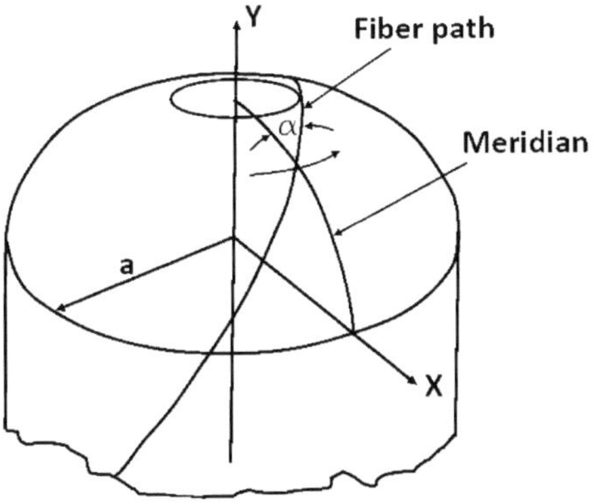

FIGURE 5.8 Geodesic path.

geodesic path is because if winding is made along this path, no slippage of the fiber relative to the surface of the mandrel will occur. For winding on frictionless surface, geodesic paths need to be followed. If one wants to wind a long non-geodesic path, some friction needs to be created between the fibers and the surface, either by rough surface or some form of adhesive applied to the fibers.

For the path to be geodesic (connecting the two points with the shortest distance on the curved surface), Equation (5.5) holds:

$$\mathrm{X} \sin \alpha = const \tag{5.5}$$

where X is the coordinate of the point and α is the angle between the fiber direction and the meridian at the point X.

At the tangency point, $\alpha = 90°$ and:

$$X \sin \alpha = X_o \tag{5.6}$$

where X_o is the boss radius.

1.4. Materials

1.4.1. Reinforcements

Nearly all filament winding for making pipes or low-pressure vessels is conducted with continuous E glass roving as reinforcement. A stronger but higher priced S glass roving is used less frequently, principally in the aerospace industry. Pressure vessels are not normally subjected to bending loads. As such the low modulus of glass fibers is not of concern.

However for pipes that are supported by saddles over long spans, deflection and ovalization of the cross section may be a concern. For these cases, thickness higher than what is required from internal pressure may be required. Graphite fibers and Kevlar fibers have a higher modulus than glass fibers but are more expensive. It is possible to have hybrid windings where different layers in the thickness of the structures can be wound using different type of fibers. Combinations of different types of fibers and different winding angles can provide the wound structures with unique capability to withstand internal pressure and also bending loads.

1.4.2. Resins

The major matrix systems for filament windings are based on epoxy, polyester or vinyl ester resins. The viscosity of the resin system should be sufficiently low so that wetting of the fibers can be done quickly and easily as the fibers are run through the bath of resin. However the viscosity should thicken rapidly after the fiber bands are deposited onto the surface of the mandrel to avoid resin dripping and to avoid resin from being squeezed from the pressure of subsequent layers. Normal curing is conducted either at room temperature or at elevated temperature without pressurization. Figure 5.9 shows the viscosity of a few resin systems used for filament winding.

Epoxy resins for filament winding are essentially the same as the laminating resins. The diglycidyl ether of bisphenol-A (DGEBA) is the most important resin type. Epoxy novolacs and cycloaliphatics are utilized to a lesser extent. Other available systems are based on brominated epoxy for improved ignition resistance, resorcinol diglycidyl ether for extended processability, and flexible epoxies for impact resistance and greater elongation.

Due to their lower cost and a balance of physical and chemical properties, polyestser and vinyl ester resins find extensive use in commercial practice. Their handling characteristics are readily adapted to filament winding. Processing viscosity is comparatively easy to control. As with the epoxies, no fundamental distinctions can be made between filament winding and laminating systems.

1.5. Mandrels

Invariably all pressure vessels or pipes made of composites have a liner. The function of the liner is to seal the liquid or gas inside the vessel or pipe. Normally the fibers provide the strength and stiffness for the

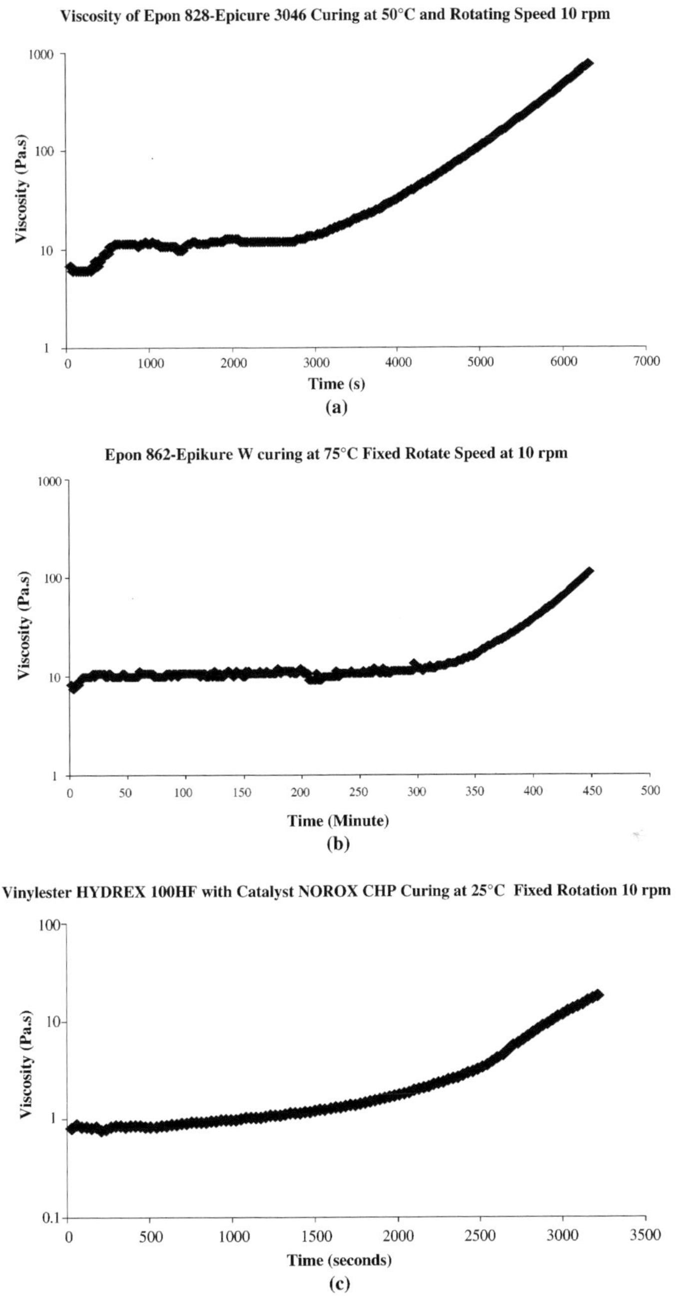

FIGURE 5.9 Viscosity of (a) Epon 828/Epicure 3046 at 50°C, 10 rpm rotating speed. 100 phr Epon: 47 phr 3046, (b) Epon 82-Epikure W at 75°C, 10 rpm rotating speed, (c) Vinylester at 25°C.

structure. In cases where there are cracks in the matrix, these cracks may not cause rupture in the vessel or pipe, but the fluid inside may leak or weep out of the container. In case of flammable fluid, this can be dangerous, even though the fiber network is sufficient to contain the pressure. In order to seal the fluid, a flexible liner is usually used. A liner can be a rubber bladder, a soft layer of thermoplastic such as PVC, or a thin layer of aluminum. When the liner is stiff enough, the liner may be used as a mandrel or over-winding of the fiber and resin.

In a situation where the liner is not stiff enough to withstand the compression due to winding force, or in cases where it is essential to take the mandrel out, strategies for mandrels should be developed. There are many requirements for mandrels. The important considerations for mandrels are:

- Mandrel should be sufficiently stiff to withstand the compression imposed by the winding force.
- The resin should not stick to the surface of the mandrel. Release agents need to be applied.
- The mandrel should be extractable from the part after curing.

Mandrels can be classified according the following categories:

1.5.1. Extractable Mandrels

For winding fiberglass pipes, steel tubes can be used as mandrels. The steel tube is made longer than the length of the pipe to be made, and a pin can be inserted into the steel tubes end. After the composite has cured, a winch can be used to extract the mandrel from the composite pipe. If machining can be done, the mandrel may have a slight taper along the length to facilitate extraction. A release agent should be placed on the outside of the mandrel to facilitate extraction. Figure 5.10 shows the mandrel for a pipe.

Inflatable mandrels made of a rubber bladder can also be used. The rubber bladder is inflated with air pressure to press against the wall of a form. The pressure provides stiffness for the mandrel. This may be used to supplement the stiffness of a thin liner.

1.5.2. Collapsible Mandrels

These mandrels consist of a segmented surface consisting of many pieces. These pieces are expanded to take the shape of the final form by collapsible linkages, similar to the operation of a collapsible umbrella. After the part is cured, the mandrel is collapsed to be removed.

1.5.3. Breakable Mandrels

Mandrels can be made out of plaster, which can be molded to take the form of the final part. After the part is cured, the plaster is broken to be removed. For some class projects where small tubes of composites are made, glass tubes such as the tube used for neon light can be used as a mandrel. The tube is broken to be removed after the part is cured.

1.5.4. Dissolvable Mandrels

Mandrels can be made of materials that can be dissolved in solution. One example is low melting alloy. These are high in density and tend to creep under moderate winding tension. They are limited to small vessels in the order of 0.3 m in diameter to 0.3 m in length. Another example is eutectic salts. These can be melted by moderately high temperature. These are better suited than the alloys and are applicable up to 0.6 m in diameter. With care, they can be flush molded, and they are easy to remove. Another example is soluble plaster. These have a long plastic stage and can be wiped to contour. They are easily washed out. Another example is a mixture of sand-polyvinyl alcohol (PVA). This material is an excellent choice for diameter up to 1.5 m and for limited quantities. It dissolves readily in hot water, but requires careful molding control. Low compressive strength is a limitation.

1.6. Material Handling and Process Controls

In filament winding, there are many machine components that are required for the handling of the fibers. Referring back to Figures 5.1 and 5.2, the process begins with the creels. Many creels are used to provide sufficient number of tows or yarns to make a fiber band. After the fibers leave the creels, they need to be tensioned so that the fibers are kept straight and taut. Tension is provided by guide eyes, drum eye brakes, scissor bars, and the drag through the resin tank. Typically, tension ranges from 1.1–4.4 N per end (an end is like a tow or yarn). Normal pro-

FIGURE 5.10 A mandrel to manufacture composite pipes.

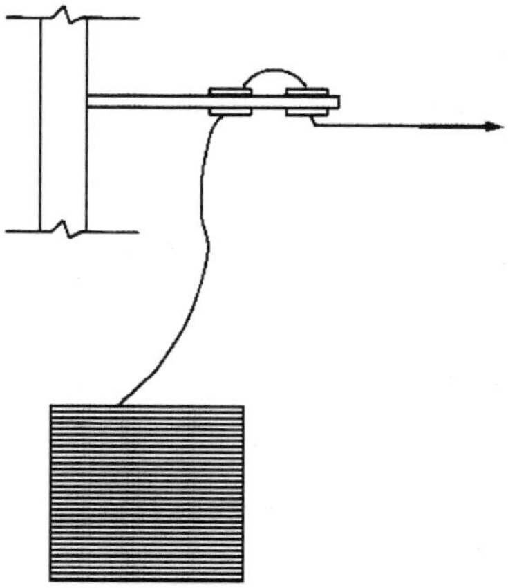

FIGURE 5.11 Example of a tensioning device.

cedure is to keep tension on the dry fibers to a minimum to prevent excessive abrasion and snarling. Figure 5.11 shows one example of the tensioner device. The friction between the fibers and the plates provides the tension force.

After the tensioner, the fiber tows or yarns are fed though a bath of resin for impregnation. Figure 5.12 shows an example of the bath of resin and the impregnation mechanism.

Before the fibers are deposited onto the mandrel, a band of fibers is formed. A device similar to that shown in Figure 5.13 is used to form the band. A uniform flat band will result in improved strength as well as a more uniform thickness. The thickness of a single layer can be calculated for a specific band density (ends/cm) and glass content. Wider band width can cover the surface of the mandrel more quickly. However with a constant amount of fibers, wider band means smaller thickness of the band. The band width has an influence on the number of circuits required to cover the mandrel surface as discussed earlier.

1.7. Netting Analysis for Pressure Vessels under Internal Pressure

The behavior of filament wound composites is analogous to that of other angle-plied laminates so that the analytical methods developed for laminates can be applied to filament wound structures. The netting analysis is a simplified procedure (as compared to laminate theory) used mainly to estimate fiber stresses in a cylindrical vessel subjected to internal pressure. This method is based on the assumption that only the rein-

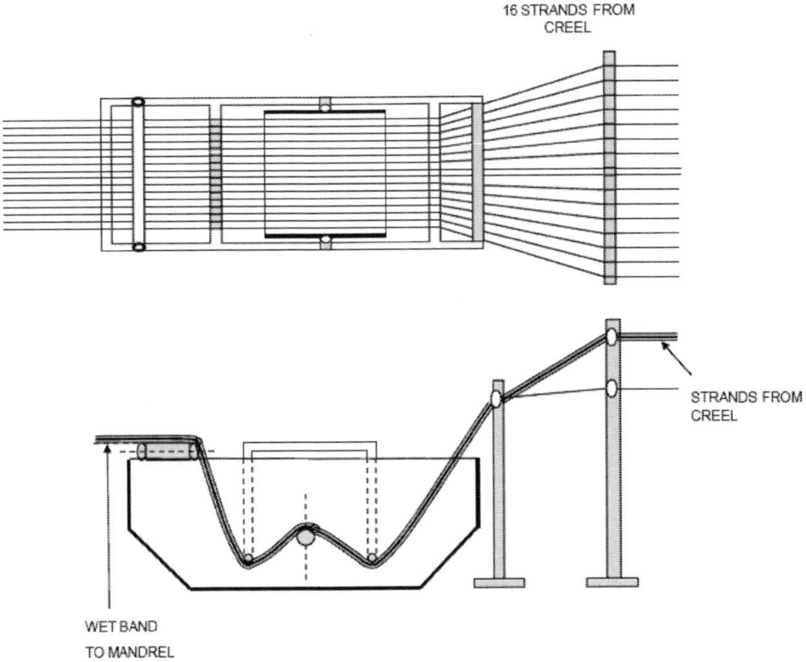

FIGURE 5.12 Impregnation tank with feed from double level.

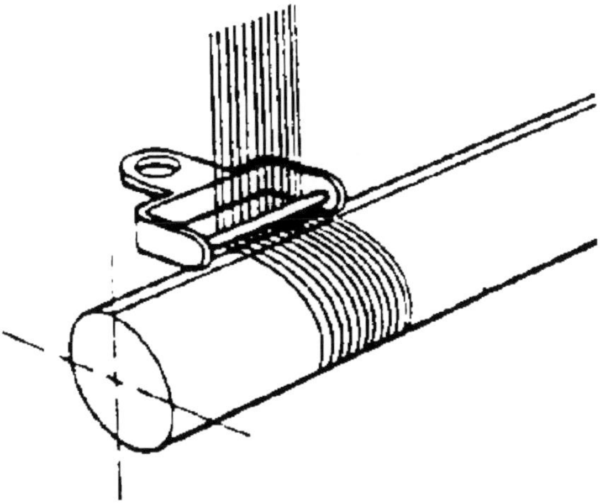

FIGURE 5.13 A band forming mechanism.

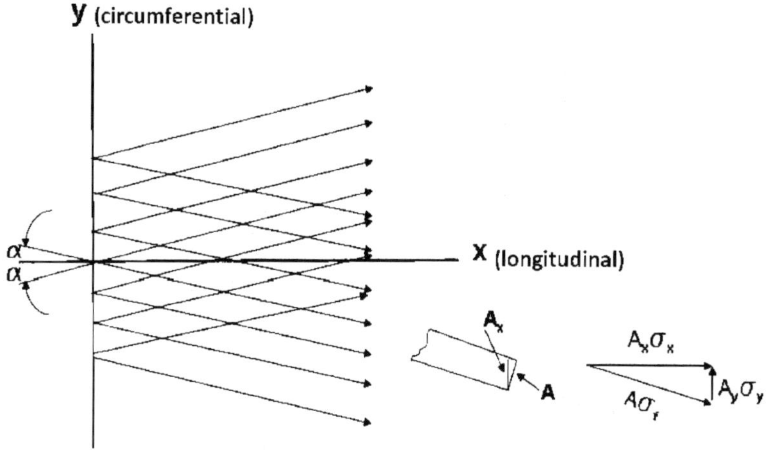

FIGURE 5.14 Netting analysis.

forcing fibers have a load carrying capability and that all fibers are uniformly stressed in tension. Figure 5.14 represents a two-layered system of parallel fibers from which the netting equations can be derived. In this figure, the x axis is along the length of the vessel whereas the y axis represents the circumferential direction.

The stress in each fiber is σ_f, acting on the cross-sectional area A of the fiber band. The force on the fiber is $A\sigma_f$. The component of this force along the x direction (along the length of the cylinder) is $A\sigma_f \cos\alpha$. This force is acting over an area $A_x = A/\cos\alpha$. Dividing the force along the x direction over the area normal to the x direction yields the stress σ_x as:

$$\sigma_x = \frac{F_x}{A_x} = \frac{\sigma_f A\cos\alpha}{A/\cos\alpha} = \sigma_f \cos^2\alpha \qquad (5.7)$$

Similarly, for the stress along the y direction (or hoop direction):

$$\sigma_y = \sigma_f \sin^2\alpha \qquad (5.8)$$

where α is the winding angle. It can easily be seen that:

$$\frac{\sigma_y}{\sigma_x} = \tan^2\alpha \qquad (5.9)$$

Winding with only one angle: In the case where filament winding is only with one angle, there is an optimal angle such that the structure may fail along the longitudinal and hoop directions simultaneously. From equilibrium conditions, it can be shown that for a thin-walled pressure vessel under internal pressure, the longitudinal and hoop stresses are given by:

$$\sigma_x = \frac{pr}{2t} \tag{5.10}$$

and

$$\sigma_y = \frac{pr}{t} \tag{5.11}$$

where

σ_x, σ_y = longitudinal and hoop stresses respectively
p = internal pressure
t = thickness
r = radius of the cylinder (for thin cylinder either inside or outside radius can be used)

From equations (5.10) and (5.11) one has:

$$\frac{\sigma_y}{\sigma_x} = 2 \tag{5.12}$$

Equating equations (5.10) and (5.11) yields:

$$\tan^2 \alpha = 2 \quad \text{or} \quad \alpha = 54.7^\circ \tag{5.13}$$

What the above indicates is that if the winding is done at 54.7°, the vessel will fail both along the longitudinal and hoop directions at the same time.

Winding at layers at two different angles: When winding consists of two layers at different angles, the following equations can be derived. Using equilibrium along the longitudinal direction:

$$\sigma_x (t_1 + t_2) = \sigma_{f1} t_1 \cos^2 \alpha_1 + \sigma_{f2} t_2 \cos^2 \alpha_2 \tag{5.14}$$

Using equilibrium along the hoop direction:

$$\sigma_y (t_1 + t_2) = \sigma_{f1} t_1 \sin^2 \alpha_1 + \sigma_{f2} t_2 \sin^2 \alpha_2 \tag{5.15}$$

The above two equations are general for winding with two layers at two different angles α_1 and α_2. These equations also allow for different materials to be used at the two layers, each with different thicknesses. For example, one layer can be made of carbon/epoxy (with fiber strength σ_{f1} of a certain thickness t_1 while the other layer can be made of glass/epoxy (with fiber strength σ_{f2} of another thickness t_2) etc.

When the outer layer is composed of a hoop layer only (i.e. $\alpha_2 = 90°$) (this is a common case where the inner layers are helical while the outer layer is hoop), the above equations reduce to:

$$\sigma_x(t_1 + t_2) = \sigma_{f1} t_1 \cos^2 \alpha_1 \tag{5.16}$$

and

$$\sigma_y(t_1 + t_2) = \sigma_{f1} t_1 \sin^2 \alpha_1 + \alpha_{f2} t_2 \tag{5.17}$$

If the fibers of the two different layers are made of the same material, i.e., $\sigma_{f1} = \sigma_{f2} = \sigma_F$, and utilizing also the relation in Equation (5.13), it can be shown that:

$$\frac{t_{\text{hoop}}}{t_{\text{longitudinal}}} = 2\cos^2 \alpha_1 - \sin^2 \alpha_1 = 3\cos^2 \alpha_1 - 1 \tag{5.18}$$

Winding with many layers: The above analysis can be extended to the case of winding with many layers. For this general case, equations (5.14) and (5.15) can be generalized as:

$$\sigma_x(t_1 + t_2 + t_3 + \ldots + t_n) =$$
$$\sigma_{f1} t_1 \cos^2 \alpha_1 + \sigma_{f2} t_2 \cos^2 \alpha_2 + \sigma_{f3} t_3 \cos^2 \alpha_3 + \ldots + \sigma_{fn} t_n \cos^2 \alpha_n \tag{5.19}$$

and

$$\sigma_y(t_1 + t_2 + t_3 + \ldots + t_n) =$$
$$\sigma_{f1} t_1 \sin^2 \alpha_1 + \sigma_{f2} t_2 \sin^2 \alpha_2 + \sigma_{f3} t_3 \sin^2 \alpha_3 + \ldots + \sigma_{fn} t_n \sin^2 \alpha_n \tag{5.20}$$

Example 5.3

A pressure vessel with internal diameter of 40 cm is subjected to an internal pressure of 7 MPa. It is to be wound using fibers at 90° and at +/–45°. Fiberglass with a strength of 1 GPa is used. If the thickness of the hoop (90°) layer is 2 mm, what would be the required thickness of the +/–45° layer?

Solution

From Equations (5.19) and (5.20) we have:

$$\sigma_x(t_1 + t_2) = \sigma_{f1}t_1\cos^2\alpha_1 + \sigma_{f2}t_2\cos^2\alpha_2$$

$$\sigma_y(t_1 + t_2) = \sigma_{f1}t_1\sin^2\alpha_1 + \sigma_{f2}t_2\sin^2\alpha_2$$

Let layer 1 be the 90° layer and layer 2 be the +/–45° layer.

$$\sigma_x(0.002 + t_2) = (10^9\ \text{Pa})t_2(0.5)$$

$$\sigma_y(0.002 + t_2) = (10^9\ \text{Pa})(0.002\ \text{m}) + (10^9)t_2(0.5)$$

From equilibrium:

$$\sigma_x = \frac{pr}{2t} \qquad \sigma_y = \frac{pr}{t} \quad \text{yielding} \quad \frac{\sigma_y}{\sigma_x} = 2$$

Substituting this relation to the above expression yields:

$$(10^9\ \text{Pa})(0.002\ \text{m}) + (10^9\ \text{Pa})(0.5t_2) = (10^9\ \text{Pa})t_2$$

This gives:

$$t_2 = 4\ \text{mm}$$

1.8. Fiber Motion [1,2]

During processing, the fibers may move causing a change in fiber tension and in fiber position. Springer et al. [1,2] developed a model for the determination of the stress and the position of the fiber. Assume that the fiber angle remains constant during processing. A fiber layer consists of a fiber sheet of thickness $\Delta\xi$ surrounded by the resin. The cross-sectional area of the fiber sheet is:

$$A_f = V_f A \tag{5.21}$$

where A is the cross-sectional area of the entire layer (resin + fiber sheet, $A = b\Delta h$), V_f is the fiber volume fraction, b is the width of the fiber band, and Δh is the band thickness.

Motions of the fiber sheet in the axial and hoop directions are not considered because the axial and hoop components of the fiber tension are in equilibrium. However, the radial position of the fiber sheet, r_f, may change for the following reasons:

1. Fiber tension in the curved fibers causes the fibers to move through the resin while the resin viscosity is low. The radial displacement of a fiber sheet relative to the resin is denoted as u_f.

2. Temperature changes in the mandrel and the composite and the chemical changes (shrinkage) of the resin may cause the mandrel and the composite to expand or to contract, causing radial displacement of the fiber, denoted by

$$u_{mc} = u_m + u_c$$

where

u_{mc} = the radial displacement due to changes in mandrel dimensions and chemical shrinkage of the resin.
u_m = the radial displacement due to changes in mandrel dimensions (thermal expansion or contraction).
u_c = the radial displacement due to chemical shrinkage of resin.

Thus, the instantaneous fiber position relative to the axis of the cylinder is:

$$r_f = R_f^o + u_f + u_{mc} \tag{5.22}$$

where R_f^o is the radial position of the fiber sheet at time t_o. u_f is obtained by assuming that the entire layer is deposited instantaneously at time t_o, so that the radial position of the entire sheet is the same at every point in a given layer. Furthermore, end effects are neglected. Thus, the stress in the direction of the fibers is:

$$\sigma_f = \frac{F}{A_f} \tag{5.23}$$

where F is the instantaneous fiber tension and A_f is the cross section of the fiber sheet. The circumferential component of the fiber tension is (from Equation 5.8):

$$\sigma_\theta \sigma_f \sin^2 \alpha \tag{5.24}$$

There is a pressure difference Δp across a fiber sheet (Figure 5.15). For a fiber sheet at the position r_f, force equilibrium gives:

$$\frac{\Delta p}{\Delta \xi} = \frac{dp}{dr} = \frac{\sigma_\theta}{r_f} \tag{5.25}$$

The relative velocity between the fibers and the resin is described by Darcy's law:

$$u_f = \frac{S}{\mu} \frac{dp}{dr} \tag{5.26}$$

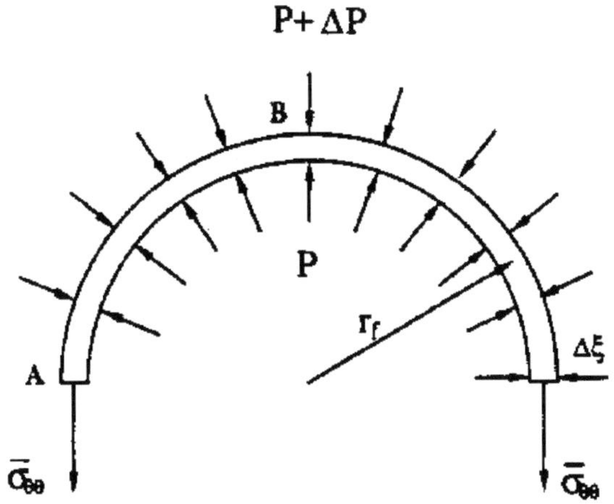

FIGURE 5.15 Pressures on the inside and outside of a fiber sheet.

where S is apparent permeability of the fiber sheet and μ is the resin viscosity. Then the change in fiber position Δu_f during a small time step Δt is:

$$\Delta u_f = -\frac{S\Delta t}{\mu}\frac{\sigma_f}{r_f}\sin^2\alpha \tag{5.27}$$

From geometric considerations, the change in length of the fiber band can be obtained by referring to Figure 5.16.

The change in length of the fiber can be expressed as:

$$\Delta L_f = AB - L_f$$

$$\begin{aligned} AB^2 &= L_f^2 + 2\pi(\Delta u_f)^2 + 2L_f[2\pi(\Delta u_f)]\sin\phi_o \\ &\approx L_f^2 + 2L_f[2\pi(\Delta u_f)]\sin\phi_o \end{aligned}$$

$$\Delta L_f = L_f\sqrt{1+\frac{4\pi\Delta u_f\sin\phi_o}{L_f}} - L_f \approx 2\pi\Delta u_f\sin\phi_o$$

From the lower part of Figure 5.16:

$$2\pi r_f = L_f\sin\phi_o$$

Substituting this into the above equation yields:

$$\frac{\Delta u_f}{r_f}\sin^2\alpha = \frac{\Delta L_f}{L_f} = \Delta\varepsilon_f \tag{5.28}$$

where ΔL_f and L_f are the elongation and original length of the fiber sheet, respectively. The change in fiber stress corresponding to the change in strain $\Delta\varepsilon_f$ is:

$$\Delta\sigma_f = \sigma_f^{t+\Delta t} - \sigma_f^t = E_f\Delta\varepsilon_f \tag{5.29}$$

where E_f is the longitudinal modulus of the fiber. These equations give the fiber stress at time $t + \Delta t$:

$$\sigma_t^{t+\Delta t} = \sigma_f^t\left[1 - \frac{E_f S\Delta t}{\mu}\frac{\sin^4\alpha}{r_f^2}\right] \tag{5.30}$$

Solutions of Equations (5.27) to (5.30) provide the fiber position and

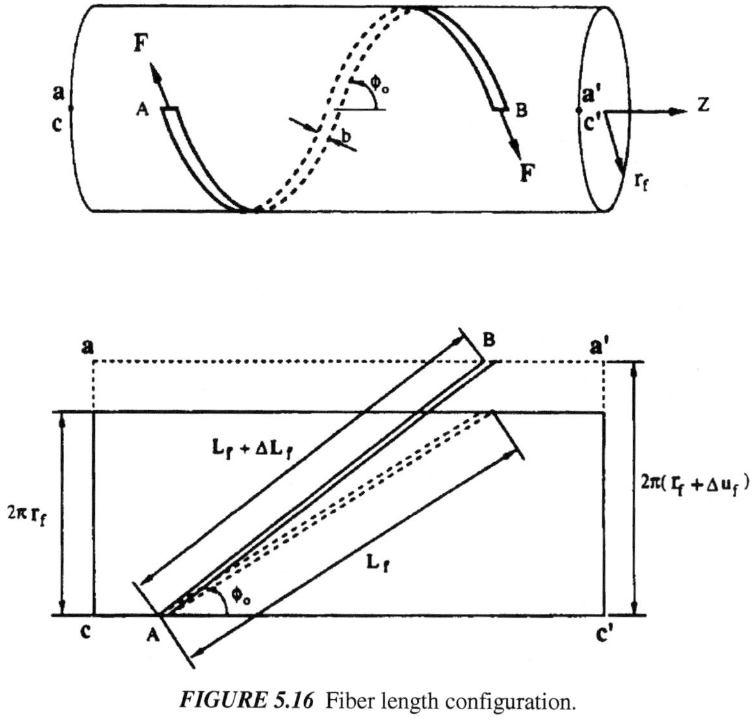

FIGURE 5.16 Fiber length configuration.

fiber stress (fiber tension) at time $t + \Delta t$. The corresponding initial conditions are:

At $t = t_o$,

$$u_f = 0$$

$$\sigma_f = \sigma_f^o = \frac{F_o}{A_f}$$

Example 5.4

It is desired to filament wind a cylinder using glass/epoxy at room temperature using a resin system with a viscosity of 600 cps. The diameter of the mandrel is 30 cm. The width of the fiber band is 10 mm and its height is 1 mm. The initial tension in the fiber band is 10 kPa. The fiber modulus is 70 GPa. The fiber volume fraction is 0.65. The fiber winding angle is +/–45°. Assuming that the permeability of the fiber network is 10^{-12} m^2, determine the position of the fiber and tension in the fiber as a function of time.

Solution

The solution utilizes Equations (5.27) and (5.30). Initial fiber tension is 10 kPa. For the first 4.5 hours, the viscosity is constant at 600 cps. The fiber tension as a function of time during this period is:

$$\sigma_f^t = \sigma_f^o \left[1 - \frac{E_f S \Delta t}{\mu} \frac{\sin^4 \alpha}{r_f^2}\right]$$

$$= (10 \text{ kPa}) \left[1 - \frac{(70 \text{ GPa})(10^{-12} \text{ m}^2)t}{600 \times 10^{-3} \text{ Pa - sec}} \frac{0.25}{(0.15 \text{ m})^2}\right] = 10 \text{ kPa } (1 - 1.3t)$$

$$\sigma_f^t = 10 - 13t \text{ kPa}$$

The fiber position at any time is given by:

$$\Delta u_f = -\frac{S \Delta t}{\mu} \frac{\sigma_f}{r_f} \sin^2 \alpha$$

$$= -\frac{(10^{-12} \text{ m}^2)t}{600 \times 10^{-3} \text{ Pa - sec}} \frac{(10 - 13t)10^3 \text{ Pa}}{0.15 \text{ m}}(0.5) = 5.55 \times 10^{-9}(10 - 13t) \text{ m}$$

It can be seen from the above results that the fiber tension decreases quickly due to the fiber motion even though the position of the fiber does not change very much. The fiber tension is reduced to nil in less than a second. It is essential that a tensioning device be used to take up this slack by maintaining tension in the fiber and avoiding fiber waviness.

2. FIBER PLACEMENT PROCESS

Fiber placement is a process similar to filament winding in which the fibers are placed onto the surface of the mandrel one strip at a time. Figure 5.17 shows a schematic for the fiber placement process. The differences between fiber placement and filament winding are as follows.

- In filament winding, the fiber tows are subject to tension while they are being placed onto the surface of the mandrel. In fiber placement, the fibers are pushed toward the surface of the mandrel. As such, flexible tows cannot be used for fiber placement. Instead, tapes with a certain degree of rigidity are used. Filament winding can be done using both wet winding and dry winding, whereas for fiber placement, only prepregs or tapes can be used (i.e. no wet fiber placement process).
- In fiber placement process, there is a pressure applicator that presses the fibers as they are being placed on the surface of the mandrel. This pressure applicator consolidates the fibers as they are being wrapped around the mandrel. With the pressure applicator, surfaces other than convex (as required for filament winding) can used.
- In the fiber placement process, usually heat is applied at the nip point (point where the fiber bands meet the surface of the mandrel). The application of heat allows the liquefication of the resin. Combination of heat and pressure provides the drive of flow and consolidation. Due to the presence of heat and pressure,

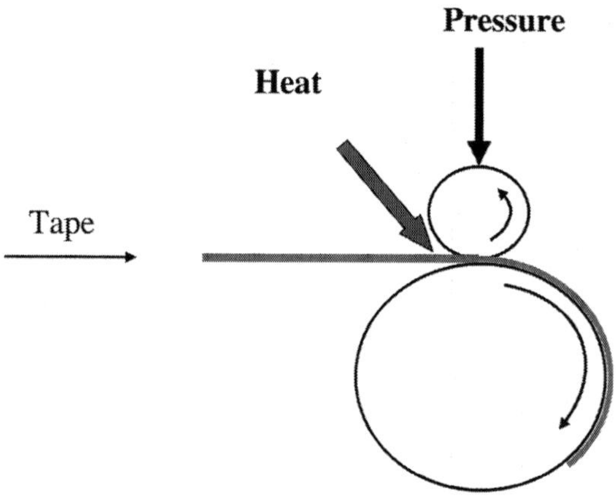

FIGURE 5.17 Schematic of the fiber placement process.

the resin systems used for the fiber placement process can be different from those used for filament winding.
- The fiber placement process can be applied to both thermoset and thermoplastic composites without significant change in the machine setup (except for the laying head).

3. REFERENCES

1. Calius E. P. and Springer G. S. A model of filament wound thin cylinders, *Internat. J. Solids Structures,* 1990, 26, pp. 271–297.
2. Lee S. Y. and Springer G. S. Filament winding cylinders: 1. Process model, *J. Compos. Mater.,* 24, pp. 1270–1298, 1990.

4. HOMEWORK

1. It is desired to filament wind a 300 mm diameter pipe with a winding angle of +/–54°. The length of the pipe is 1300 mm. The bandwidth of the fibers is 6 mm. Assume a dwell angle of 180° at each end. Assume also that there is no overlap or gap between adjacent bandwidths. Determine:

   - The number of circuits per pattern
   - The number of patterns per layer
   - The number of circuits per layer

   You are free to adjust the dwell angle and the bandwidth slightly to obtain your results.

2. A pipe of 300 mm in diameter is filament wound with fiberglass. Two types of wind angle will be used. One is 30° and the other is 45°. Assume that there is 1/3 of the thickness of the 30° type and 2/3 of the thickness of the 45° type. Assume also that the strength of the fiberglass used in this case is 1360 MPa. Determine the thickness of each type of winding needed to contain an internal pressure of 41 MPa.

CHAPTER 6

Pultrusion

1. GENERAL

Pultrusion is a manufacturing process that combines many steps in the manufacturing of composites into two steps. Refer to Figure 1.7(a) of Chapter 1. That figure shows four steps in the manufacturing process. Figure 6.1 shows a representation of the pultrusion process. For pultrusion, one goes directly from step *a* to step *d*, and one does not have intermediate products in between.

- Step *a*: Beginning from the left-hand side, the fiber tows drawn from fiber racks are routed through a series of guides. The fibers then traverse through a bath of low viscosity resin for impregnation.
- Step *d*: Upon exiting from the resin bath, fibers are collimated into an aligned bundle before entering into a heated die. While inside the die, the following takes place:

  —The resin flows and wets the fibers.
  —The ensemble of fibers and resin is compacted.
  —The resin cures and the fiber/resin system becomes solid.

  Upon exiting from the die, the composite structural component is pulled by a puller. It is then cut to length and is ready for shipment.

Since the pultrusion process combines many steps into one, the opportunity for checking the quality is reduced. Also the requirements for control of quality are increased to ensure good quality. However, pultrusion offers a fast production rate. The production rate can be in terms of meters per minute. It can also produce composite parts at low cost. This is because fibers in tow form are less expensive than in prepreg or woven form. The rule is that if more processing is involved in transforming the fiber, either by wetting with the resin or in changing the form from unidirectional tows to woven or braids, cost and time are involved. As such, pultrusion has been used to produce many low cost composite structural elements, if strict quality is not a requirement. However there are also limitations. One of the main limitations is that the fibers are mostly unidirectional.

By pultrusion, many lightweight, corrosion resistant and low electrical conductivity components can be made. These include standard shapes such as rods, angles, clips, I beams, panels, plates, and rebars for concrete reinforcement. Other components include side rails for ladders, fishing rods, tool handles, bus components, sign posts, and sucker rods for oil drilling rigs.

Due to the low cost of the process, E glass fibers are mostly used even though applications include fibers such as S glass, carbon, and Kevlar. Resins are usually low cost polyester or vinyl ester even though other resins such as epoxy, phenolic and thermoplastic have also been used.

There are two forms of pultrusion products. The first category consists of solid rod and bar stock produced from axial fiberglass reinforcements and polyester resins. These are used to make fishing rods and electrical insulator rods, which require high axial strength. The second category is structural profiles, which use a combination of axial

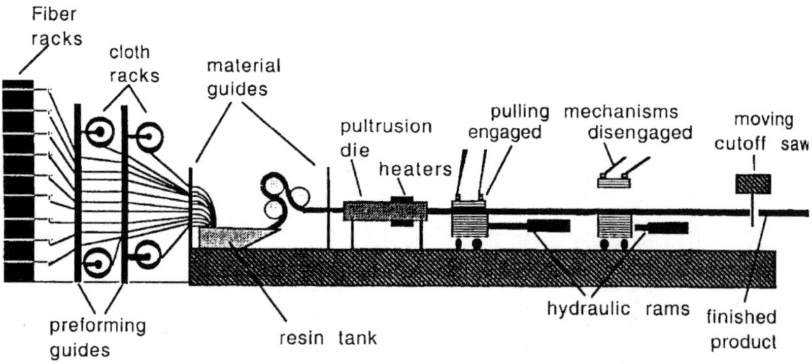

FIGURE 6.1 Schematic of the pultrusion process (reproduced from "Pultrusion of composites," by J. P. Fanucci, S. Nolet, and S. McCarthy, in *Advanced Composites Manufacturing* by T.G. Gutowski, 1997, with permission from John Wiley and Sons).

fibers and multidirectional fiber mats to increase a set of properties that meet the requirements of the application in the transverse and longitudinal directions.

2. MATERIALS

2.1. Fibers

2.1.1. Unidirectional Rovings

Fiberglass is the most commonly used fiber for pultrusion. Unidirectional fibers are the least expensive reinforcements available. Most pultruded profiles are used in structural applications where loading is highly oriented along the length of the profile. Unidirectional loading minimizes friction drag in the die, provides the highest pulling strength possible, and simplifies the design and fabrication of forming guides at the entrance of the die.

The combination of cost, design applicability, and manufacturing ease make unidirectional rovings the most widely used reinforcement in pultrusion processing. Unidirectional filaments in the form of rovings, tows or threads are the building blocks for virtually all other reinforcement forms.

For most practical applications, the use of all-unidirectional rovings is unrealistic. The highly orthotropic behavior of the unidirectional composite results in transverse properties that are much too low. Parts constructed this way might have unacceptable resistance to crushing or splitting parallel to the fiber direction. Some means of providing strength in the transverse direction is often mandatory.

2.1.2. Woven and Nonwoven Broad Goods

Low cost commercial pultrusions made of unidirectional glass rovings often include inexpensive forms of nonwoven broad goods called *continuous strands* and *chopped strand mat.* The random orientation of these materials provides some degree of off-axis strength and stiffness enhancement at very low cost. Woven materials used in the pultrusion process must be placed between more stable forms such as layers of unidirectional rovings. Figures 6.2 and 6.3 show the incorporation of mats along with unidirectional rovings.

When more complicated laminates are required, cloth, random mats, and preplied fabrics can be added to obtain transverse and off-axis rein-

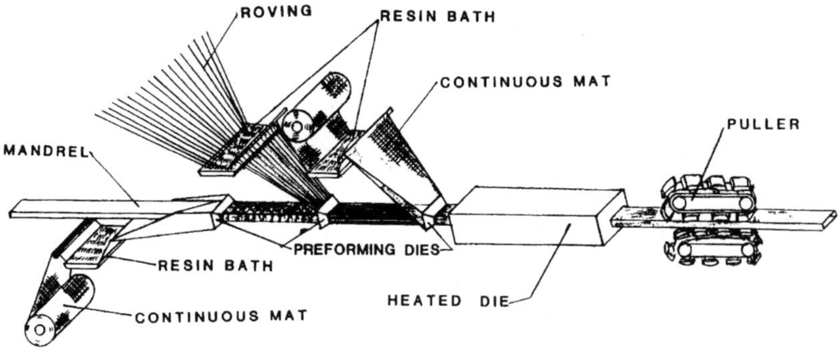

FIGURE 6.2 Introduction of mats into the pultruded products (reproduced from *Handbook of Pultrusion Technology,* by R. W. Meyer, with permission from Springer).

forcing. Fiber tension is not a severe problem for commonly pultruded constructions composed primarily of unidirectional tows. Problems begin to occur in more sophisticated applications when broad goods and other off-axis reinforcements are included in the laminate. If not properly handled, these laminates tend to distort and deform as they are folded during assembly outside the die, and can be further distorted when dragged along the tooling surfaces inside the die.

2.2. Resins

The necessary characteristics for a resin to be used to make pultruded products are that it have low viscosity and that gel time and cure time

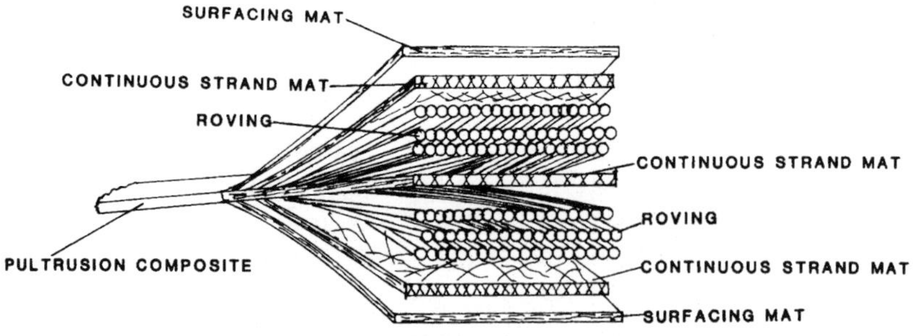

FIGURE 6.3 Exploded view of materials in pultrusion (reproduced from *Handbook of Pultrusion Technology,* by R. W. Meyer, with permission from Springer).

are short to allow for the high rate of production. If, for example, a rate of production of 20 cm/minute is desired, for the length of a die of 100 cm long, the duration of the resin inside the mold is 5 minutes. The resin should flow through the interstices of the tows, wet the fibers, gel and cure during this time. When the resin gels, it also shrinks, which helps to release the composite from the die wall. This, in turn, will reduce the pulling force. The resin normally used to make pultruded products is polyester resin, due to its low cost and low viscosity. Table 6.1 shows the viscosity of polyester resin along with gel time and cure time. When better corrosion resistance is required, vinyl ester resins are used. When a combination of superior mechanical and electrical properties is required, epoxy resin is used. Epoxy resins are expensive materials in a number of aspects. The resins are three to six times more expensive than polyesters and have a number of process-related costs not found with polyesters. Because they are cured by a stepwise reaction rather than an addition reaction, as with polyester resins, their reaction rate is very slow. The gelation of epoxy resins occurs at a later stage of reaction, and it is critical that the exotherm developed be contained within the die. This dictates a slow reaction rate, which results in a high labor. Because the epoxy begins to react slowly as soon as it is mixed, the pot life is short. The resin scrap rate is potentially higher if viscosity buildup affects wet-out to the extent that the bath must be recharged. The die temperature profiles used for epoxy are typically hotter than polyesters, and the drip-off at the entrance must often be discarded rather than recirculated to the bath. Because of the tendency for the epoxy resin to bond strongly to the die wall, epoxy products often display surface defects, such as exposed fibers, chipping, or loss of dimension, all of which increase finished-product scrap rate. These additional costs place epoxy resins in a class in which the end-use requirements must justify the high price [1].

TABLE 6.1 General Properties of Polyester Resins Used for Pultrusion.

Property	Value
Viscosity at 25°C (cP)	500–2000
Specific gravity	1.1
Gel time (minutes)	3–8
Cure time (minutes)	5–18
Peak exotherm (max temperature during the process)	415–470°F (213–243°C)

3. COMBINATION OF OTHER PROCESSES WITH PULTRUSION

As part of the process of automation, other types of processes have been combined with pultrusion to produce parts with enhanced properties. These include in-line filament winding or in-line braiding, together with the pultrusion process.

3.1. In-Line Winding

In situations such as the case of pultruded rods used for the reinforcement of concrete, it is necessary to provide roughness on the surface of the putruded rods to enhance the mechanical interlock between the reinforcement rod and concrete. Filament wound strips can be placed on the outer surface of the pultruded rods for this purpose. At the exit of the pultrusion machine, spools of unidirectional tows loaded onto two counter-rotating disks can be circumferentially wound around the exited pultruded rod.

3.2. In-Line Braiding

In in-line braiding, a vertical braider is positioned in front of the pultrusion die. As the braider pays off the material, the material is drawn forward through the braiding ring and laid down on the mandrel. The resulting fiber angle is a function of the braider speed and pultrusion line speed. Impregnation of the thin walled braids is accomplished via continuous resin transfer molding, called *direct resin injection.* The impregnated braid is drawn into the pultrusion die and the resin is polymerized. The cured tube is manufactured continuously.

4. FACTORS AFFECTING THE PULTRUDABILITY OF A COMPOSITE COMPONENT

Two important characteristics affecting the pultrudability of a composite product are: the pulling force required to move the product steadily through the system and the pulling speed. The pulling force has to be sufficient to overcome the resistance at the different stages of the pultrusion process, and the pulling speed determines the productivity of the process. These are discussed below.

4.1. Pulling Forces in Pultrusion

Controlling the buildup of pulling loads and developing ways to deal with their inevitable presence are major concerns for all pultruders. Preform compaction and the effects of fiber packing in the die contribute the most to pulling force generation during pultrusion. This is particularly true in the processing of epoxies where fiber and filler must be kept high to prevent sloughing or resin adhesion to die surfaces, and to maintain good surface finish.

One way to analyze the parameters that contribute to the pulling force is to examine the contribution of the resistance from each step of the process. As can be seen from Figure 6.1, the resistance should be considered from the four different steps: Force required to collimate the fiber tows from the creel to the entrance to the die, force required to compact the fiber tows into the cavity in the die, force required to overcome the viscosity of the liquid resin, and force required to overcome the friction between the wall of the die and the solid composite product. This can be written as:

$$F_{total} = F_{col} + F_{compaction} + F_{viscous} + F_{friction} \tag{6.1}$$

4.1.1. Force Due to Collimation

The collimation force F_{col} depends on the loading of the fibers relative to the size (diameter) of the die. There is a limit as to the difference between the diameter of the collimated fiber bundles to the diameter of the die. Within limits, the larger the amount of fibers, the larger the F_{col}. One other aspect is the ease with which the fiber bundles are introduced into the die. Figure 6.4 shows the arrangement of a tapered entrance into the die. The taper configuration facilitates the introduction of the fibers into the die and reduces the F_{col}.

4.1.2. Force Due to Compaction

In Equation (6.1), the compaction force F_{comp} is a force along the pull (axial) direction of the process. However, compaction is occurring along the radial direction of the die, which is normally the pull direction. Normally, the greater the number of fibers that are squeezed into the cavity of the die, the larger will be the compaction. If too much fiber is put in, the compaction will be too large and pulling may not be possible. However, if too few fibers are put into the die, insufficient compaction will occur and voids may appear. In addition, there is resin shrinkage, which can

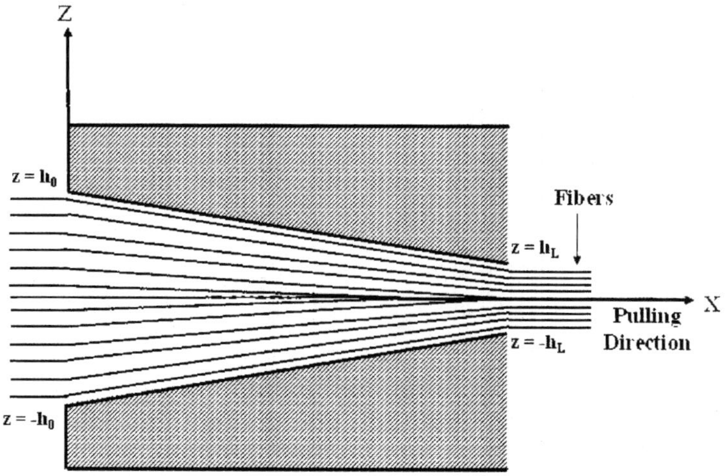

FIGURE 6.4 Tapered entrance facilitates the introduction of fibers into the die.

also affect the compaction pressure. One way to get an estimate of the pressure in the material inside the die is to use Equation (3.13) in Chapter 3, repeated here as:

$$\begin{bmatrix} e_1 \\ e_b \end{bmatrix} = \begin{bmatrix} F_{11} & F_{1b} \\ F_{b1} & F_{bb} \end{bmatrix} \begin{bmatrix} \sigma_1 \\ \sigma_b \end{bmatrix} \tag{6.2}$$

where,

e_1, σ_1 = the strain and stress along the pull direction
e_b, σ_b = the strain and stress in the radial direction
F_{ij} = the compliance coefficients of the fiber bundles

The strain e_b is governed by the geometry of the die. If R_o is the radius of the die entrance with an initial volume fraction V_o, and R_f is the die radius at the position corresponding to the fiber volume fraction V_f, then the strain e_b is calculated to be:

$$e_b = \frac{R_f - R_o}{R_o} \tag{6.3}$$

The axial strain depends on the degree of waviness of the fibers in the fiber bundle. Refer to Figure 3.19 (Chapter 3), assuming that the length of the stretched fiber can be calculated by:

$$\frac{l^2}{4} = \frac{L^2}{4} + a \tag{6.4}$$

where,

l = the length of the stretched fiber over one cycle
L = the length of the unstretched fiber over one cycle
a = the amplitude of the waviness as shown in Figure 3.19 (Chapter 3)

The strain e_1 can be calculated to be:

$$e_1 = \frac{l - L}{L} = \sqrt{1 + \frac{4}{\beta^2}} - 1 \tag{6.5}$$

where $\beta = L/a$. If $\beta = 200$, one has $e_1 = 0.00005$.

Recall the expressions for F_{ij} from Chapter 3 as:

$$F_{11} = \frac{4}{\pi}\frac{1}{E}\varsigma^2[1 + 2(\varsigma - 1)^1] \tag{6.6a}$$

$$F_{1b} = F_{b1} = -\frac{16}{\pi^3}\frac{\beta^2}{E}\varsigma(\varsigma - 1)^3 \tag{6.6b}$$

and

$$F_{bb} = \frac{\beta^4}{3\pi E}(\varsigma - 1)^4 \tag{6.6c}$$

One can calculate the stresses with information on the strains.

Example 6.1

A pultrusion machine having a radius R_o = 12.7 mm and R_f = 6.35 mm is used to pultrude fiberglass/polyester rod. The desired volume fraction is 0.60. Determine the contribution to pull force due to compaction. Assume $\beta = 200$ and $E = 70$ GPa, $V_o = 0.4$ and $V_a = 0.785$.

From Equation (6.3), $e_b = \dfrac{6.35 - 12.7}{12.7} = -0.50$

$$\varsigma = \sqrt{\frac{V_a}{V_f}} = \sqrt{\frac{0.785}{0.60}} = 1.144$$

The compliance coefficients can be calculated using Equations (6.6) as

$$F_{11} = \frac{4}{\pi}\frac{1}{E}\varsigma^2[1+2(\varsigma-1)^2] = \frac{4}{\pi}\frac{1}{70\text{ GPa}}(1.144)^2[1+2(1.144-1)^2] = 0.025\text{ GPa}^{-1}$$

$$F_{1b} = F_{b1} = \frac{16}{\pi^3}\frac{\beta^2}{E}\varsigma^2(\varsigma-1)^3 = -\frac{16}{\pi^3}\frac{200^2}{70\text{ GPa}}(1.144)(1.144-1)^3 = -1.009\text{ GPa}^{-1}$$

$$F_{bb} = \frac{\beta^4}{3\pi E}(\varsigma-1)^4 = \frac{200^4}{3\pi(70\text{ GPa})}(1.144-1)^4 = 1043\text{ GPa}^{-1}$$

Equation (6.2) can be inverted to be written as:

$$\begin{bmatrix}\sigma_1\\ \sigma_b\end{bmatrix} = \frac{1}{F_{11}F_{bb}-F_{1b}^2}\begin{bmatrix}F_{bb} & -F_{1b}\\ -F_{1b} & F_{11}\end{bmatrix}\begin{bmatrix}e_1\\ e_b\end{bmatrix} \tag{6.7}$$

For this particular case:

$$\begin{bmatrix}\sigma_1\\ \sigma_b\end{bmatrix} = \frac{1}{26.08-1.02}\begin{bmatrix}1043 & 1.009\\ 1.009 & 0.025\end{bmatrix}\begin{bmatrix}0.00005\\ 0.5\end{bmatrix}$$

$$= \frac{1}{25.06}\begin{bmatrix}0.552\\ 0.0125\end{bmatrix} = \begin{bmatrix}0.022\\ 0.0005\end{bmatrix}\text{GPa} = \begin{bmatrix}22\\ 0.5\end{bmatrix}\text{MPa}$$

Note that due to the small value of e_1, its contribution to σ_1 is about 10% and it has little contribution to σ_b. It can be seen that the contribution of compaction to the axial load is quite significant. Note also that the above calculations are only approximate due to the application of linear assumption to a situation of large deformation.

The above example shows a simplified estimate for the compression stress σ_b due to existence of the compressive strain e_b. The compressive strain in the example is a function of the reduction in geometry of the die from the entrance to the point of interest. This, in turn, is related to the change in volume fraction of the fibers at the different positions. There are other parameters that also have an effect on the relation between the fiber volume fraction and the compressive stress as discussed below.

4.1.3. Parameters Affecting the Compression Stress

Figure 6.5 shows the relation between fiber volume fraction and compressive stress. It can be seen that even though the shape of the curve between compression stress and volume fraction is similar, Different types of fiber forms show different curves but the shapes of the curves are similar, the actual values of the stresses depend on the type of fibers. Apart from these, there are many parameters that affect the compressibility of the fiber bundle, one of which is the fiber volume fraction, as discussed previously. In addition, other parameters affect compressibility as well, such as the type of fiber material, fiber architecture (mat or weave pattern), combination of different types of fiber architecture (mat and roving alternating in different sequences), rate of loading, loading and unloading, repeated loading, and dry or lubricated fiber beds. The fact that the compressibility depends on so many factors makes it difficult to obtain good models that fit all systems. Experimental determination of each different system is therefore essential.

Shrinkage of resin has an important effect on the pull force. Polyesters and vinyl esters shrink about 8% while epoxies shrink about 5%. As such, it would be easier to pultrude polyester of vinyl ester composites than epoxy composites. The amount of fiber loading has an important effect on the pulling force, as shown in the above example.

4.2. Pull Rate

The pull rate depends on the time required for the resin to impregnate

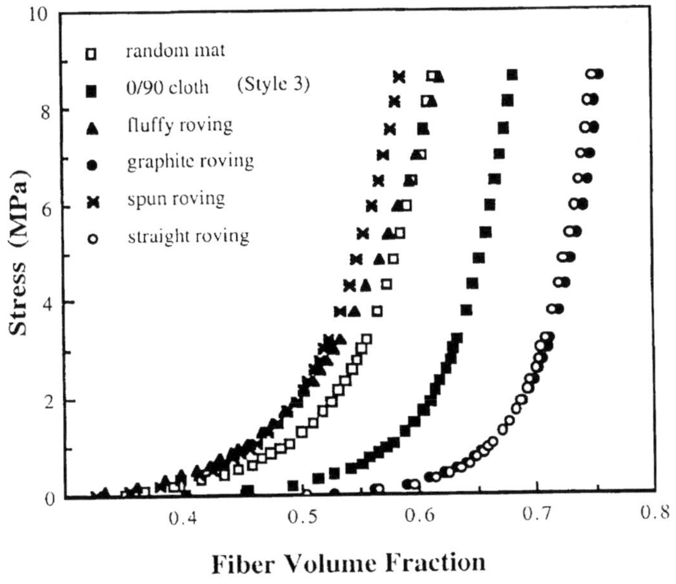

FIGURE 6.5 Compression stress as a function of fiber volume fraction (reproduced from "Pultrusion of composites," by J. P. Fanucci, S. Nolet, and S. McCarthy, in *Advanced Composites Manufacturing* by T. G. Gutowski, 1997, with permission from John Wiley and Sons).

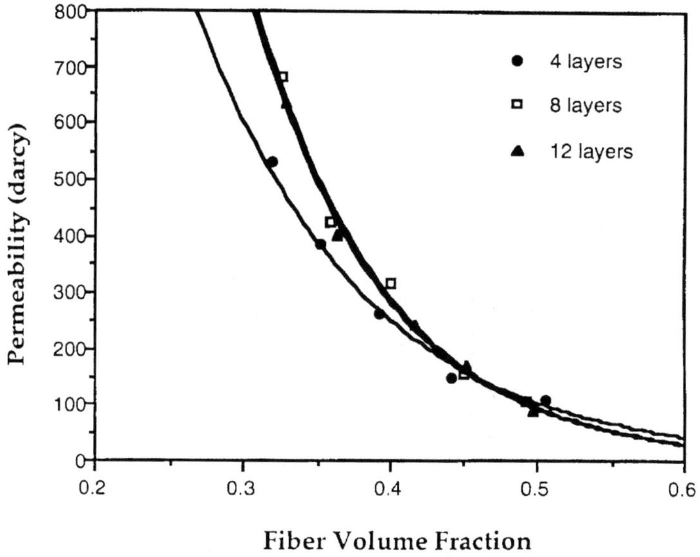

FIGURE 6.6 Effect of number of layers on permeability of 0/90 cloth (reproduced from "Pultrusion of composites," by J. P. Fanucci, S. Nolet, and S. McCarthy, in *Advanced Composites Manufacturing* by T. G. Gutowski, 1997,with permission from John Wiley and Sons).

the fiber tows and also for the resin to cure once complete wetting has taken place. The pultrusion die may consist of two sections along the direction of pulling, the first section is required for wetting to take place and the second section for curing to take place. The pull rate is related to the lengths of these sections. The length of the first section may be estimated using Darcy's law and the permeability of the fiber tows. The length of the second section depends on the reactivity of the resin and the temperature profile along this length.

4.3. Permeability

The permeability of the fiber tows depends upon the type of materials, the number of layers, and the volume fraction of the fibers. Figure 6.6 shows the effect of the number of layers of cloth on permeability.

5. SUMMARY

The state of development of pultrusion as a process for composite manufacturing is still in the experimental stage. This is due to the integration of many steps into the process which makes it complex. The permeability of the resin into the fiber bed varies greatly with the fiber bundles and also varies with the degree of compression of the fiber bed. The rate of reaction of the resin is rapid due to the fast production rate. Pultrusion can produce parts with low cost using low-cost materials such as glass and polyester.

6. REFERENCE

1. Meyer R. W., *Handbook of Pultrusion Technology,* Chapman and Hall, 1985.

CHAPTER 7

Liquid Composite Molding

1. INTRODUCTION

The last three chapters have presented a few processes for the manufacturing of composite structures. These have advantages and disadvantages.

The hand-lay-up on open molds provides flexibility and versatility in terms of different configurations; However it does not provide good quality of the part due to the lack of control of compaction and the entrapment of air during the laying-up process. In addition, this process suffers from the evaporation of styrene into the atmosphere, which is an environmental concern.

For the autoclave molding process, even though it also uses laying-up either by hand or by tape laying machine, the environmental concern is not critical because prepreg tapes are used and the evaporation of volatiles is not serious. The quality of the parts is very good due to the fact that the impregnation of the fibers is done off-line. The use of vacuum, pressure and temperature control also gives parts of good quality. However the autoclave molding process has disadvantages as follows:

- Since prepregs have to be used, the cost is high compared to cases where dry fibers are used.
- The prepregs have a shelf life, which imposes time constraint on their usage. This also can produce waste if the prepregs are not used during their shelf lives.
- Since laying up is required, the component cannot have fiber orientations other than in-plane. Having fibers along the thickness

direction of the part can improve properties such as interlaminar strength and toughness.
- The process requires an autoclave, which can be a substantial investment. The autoclave needs to be heated to a certain temperature and sometimes this can be costly as in the case where a large autoclave is heated to cure a small composite part.
- For parts with very large dimensions, such as those of a boat or a wind turbine blade, the use of an autoclave is economically impractical.

The filament winding and pultrusion processes are geared towards parts of special shapes such as those having surfaces of revolutions, or those having constant cross section along their length.

Liquid composite molding (LCM) is a process that may respond to the concerns mentioned above. The main steps of the process are shown schematically in Figure 7.1 and discussed below.

1. *Preforming:* During this step, dry fibers are packaged into a preform having the configuration of the part. The starting materials can be tows, random mats, or woven fabrics. The finished preform is usually woven, compression molded, braided or knitted together. Small amounts of adhesive or small-diameter stitches are usually used to hold the preform in shape.
2. *Tool:* After the preform is made, it is placed inside a tool (mold) for further processing. Usually the mold has two halves. Both of these can be made out of stiff metals (such as the case of SRIM, RTM, VARTM or RFIM) or one-half of the mold can be made out of stiff metal and the other half made out of a flexible membrane (such as the case of SCRIMP or its variations). The surface of the final part

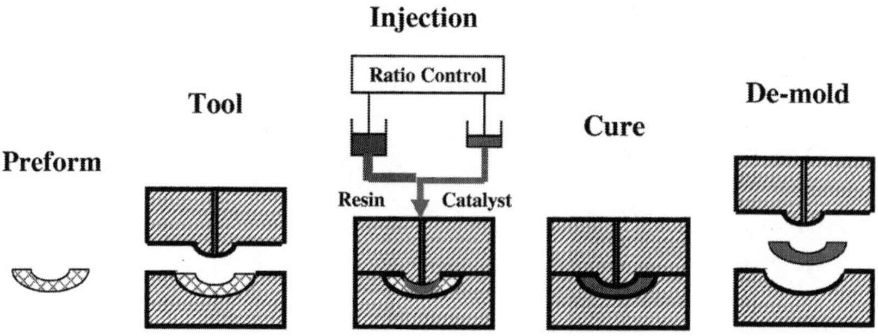

FIGURE 7.1 Schematic of the LCM process.

depends on the quality of the surface of the mold. Also, high pressure can be applied when both mold halves are made of stiff metals. The amount of voids that may be present in the final product depends on the ability of the resin to penetrate into small interstices between the fibers, and this may require high pressure. The type of tool used therefore depends on the required quality of the final part.

3. *Resin infusion:* After the preform is placed inside the mold and the two halves of the mold are closed, resin is infused into the mold. The objective of the infusion is to wet the fibers and to fill up any cavity within the preform. The infusion can be in the form of injection where high pressure [several hundred psi (tens of MPa)] for the case of SRIM, or moderately high pressure (around 100 psi or 6.89 MPa), for the case of RTM, is used. It can also be simply suction created by vacuum (such as the case of VARTM or SCRIMP). The duration of time for the infusion of resin depends on the size of the part and on the reactivity of the resin system. For resin with fast reactivity, such as cyanate for SRIM, the infusion takes place within a matter of seconds; whereas for slower reaction systems such as epoxies for RTM, the infusion time can be on the order of minutes or hours.
4. *Curing:* After the resin has been infused completely into the cavity of the fiber preform, curing takes place. Normally the resin already contains curing agents and catalysts for curing. It is important that the resin does not gel during the infusion process. If the resin gels before the preform is infused, short shots are obtained. Curing can be accelerated by heating.
5. *Demolding:* The part is demolded and removed from the mold.

The advantages of LCM are as follows:

1. The preforms are made using dry fibers and they do not have to contain the partially cured resin as in the case of prepregs (preforms may contain binders, which are small amounts of resin used to hold the shape of the preforms together). Because of this, fibers with different orientations can be built into the preforms. Composites made from the preforms may have reinforcements along the thickness direction in addition to those in-plane. Different techniques such as weaving, braiding, stitching, and knitting can be used to make the preforms.
2. The dry preforms do not have the constraint of shelf life.
3. The process is done in a closed mold. For manufacturing involving

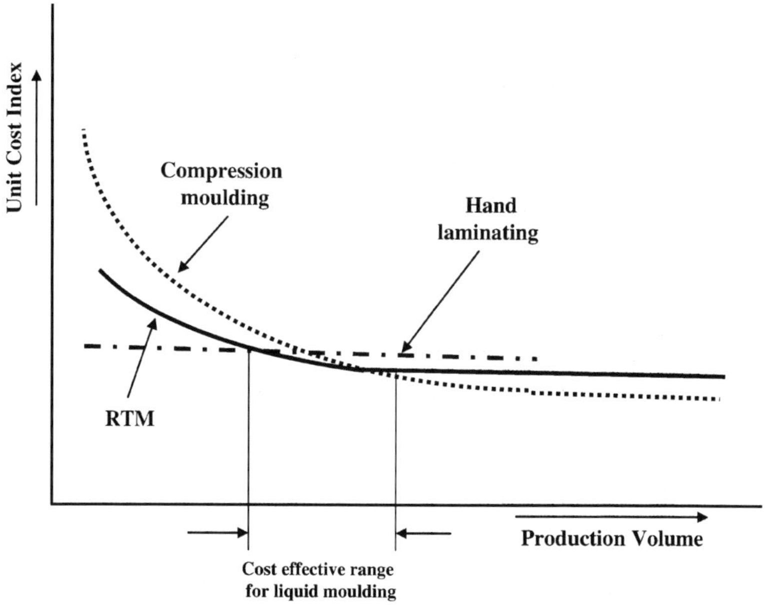

FIGURE 7.2 Cost versus production volume of different manufacturing processes.

polyester and vinyl ester resins, the issue of styrene evaporation into the atmosphere is not a great concern.

4. The cost-effective range for LCM is in the middle range in the production volume. Figure 7.2 shows the relation between production volume and the unit cost index for a few processes. LCM can be more cost-effective compared with the autoclave molding process when the production volume is on the order of 20,000–60,000 units per year [1].
5. The molds required for LCM are generally considered to be lightweight and low cost compared with conventional compression molding and metal forming, resulting in a lower investment to enter production.

Initially liquid composite molding was developed for low-cost applications derived from the injection molding of regular plastic components. Due to its low cost, relatively fast production rate and its ability to provide closed mold conditions that help to address the problem of styrene (in open mold process), LCM has found acceptance for the manufacturing of composites for automotive applications.

The disadvantages of LCM are as follows:

1. Preforms need to be held together by binders. The presence of binders may interfere with the flow of resin to wet the fibers. Binders also need to be dissolved in the resin to avoid the bundling of fibers, which may affect the resulting mechanical properties.
2. Preforms need to fit well into the tool. For the resin transfer molding (RTM) process, if the preforms do not fit well into the tool such that there is looseness at the peripheries of the preform, liquid resin can run quickly along these easy paths resulting in resin rich areas in the final part.
3. The permeability of the preform depends on many factors, such as the volume fraction of fibers, the compression pressure on the preform, the type of fiber form used, and the stacking sequence of the fibers. The variability of the permeability of the fiber preforms makes it difficult to predict the speed of flow of the liquid resin in them. This can result in lack of wetting, voids, and low mechanical properties such as interlaminar shear strength.
4. The quality of the part can be affected by the presence of voids, dry spots or resin rich areas.

Depending on the fiber volume fraction and the end-use applications, there are many variants of the LCM process as follows:

- *Injection molding (IM):* This is a pure plastic injection process where there are no fibers involved, which has been used to make injection molded parts for a long time. The resin is mainly engineering thermoplastics such as polypropylene, polystyrene, and polymethylmethacrylate (PMMA). Sometimes short fibers (such as short glass or carbon fibers) can be incorporated into the thermoplastics to make reinforced plastic components. In this case, the fibers are mixed with the resin and injected together, rather than in the form of fiber preform.
- *Structural reaction injection molding (SRIM):* This is similar to IM above except that in this case a fiber preform is placed inside the mold cavity before injection. Figure 7.3 shows a schematic for this process. Due to the high rate of reaction, the pressure in the mold is usually high and the duration of the reaction is on the order of seconds.
- *Resin transfer molding (RTM):* This is similar to the SRIM process except that the duration of the injection step lasts on the order of minutes and the pressure inside the mold is in less than

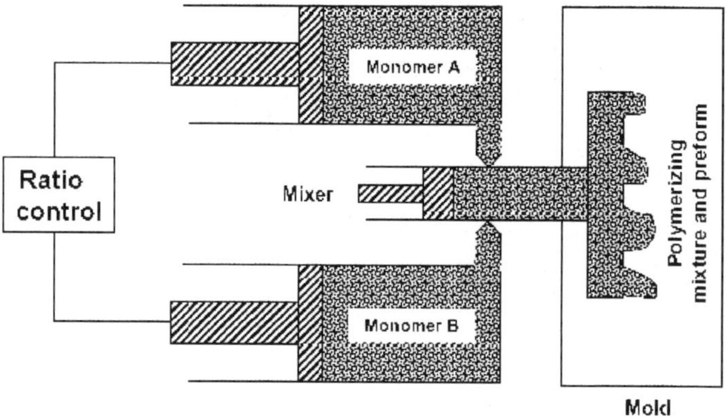

FIGURE 7.3 SRIM process.

100 psi (680 kPa). Figure 7.4 shows an RTM mold for automotive parts.

- *Vacuum-assisted resin transfer molding (VARTM):* This is similar to RTM except that rather than using pressure, vacuum is used. Because of this, the pressure differential is small. The advantage here is that a rigid mold is used only on one side of the part where on the other side a flexible bag can be used. This can result in significant cost savings. The disadvantage is that due to the low

FIGURE 7.4 An RTM mold for a curved piece.

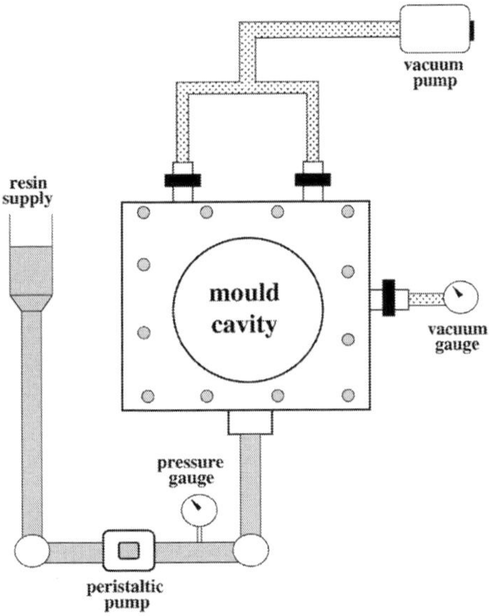

FIGURE 7.5(a) Vacuum-assisted molding arrangement.

pressure, more voids may appear in the part. Figure 7.5 shows schematics of the VARTM.

- *Seaman composite resin infusion molding process (SCRIMP):* This process is similar to VARTM in which only vacuum is used to provide the pressure gradient. In the case of RTM, both mold surfaces are hard, meaning both mold surfaces are made of some sort of metal. In the case of SCRIMP, only one mold surface is hard, the other mold surface is a flexible membrane that is used to contain the vacuum. Figure 7.6 shows a schematic of the process. In SCRIMP, the liquid resin flows in between the flexible membrane and the fiber preform. This type of flow is rapid since

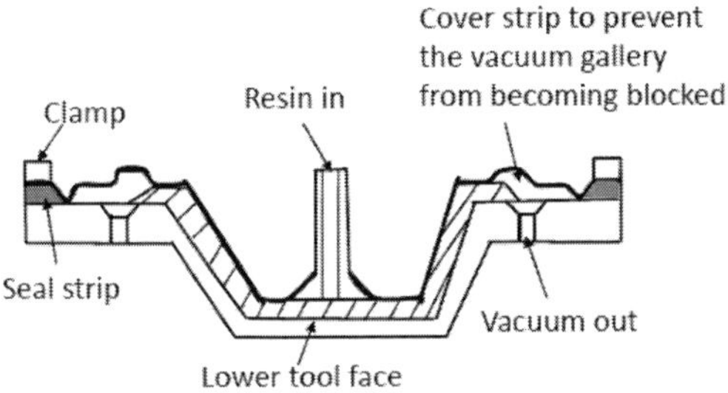

FIGURE 7.5(b) Cross section of a VARTM set up.

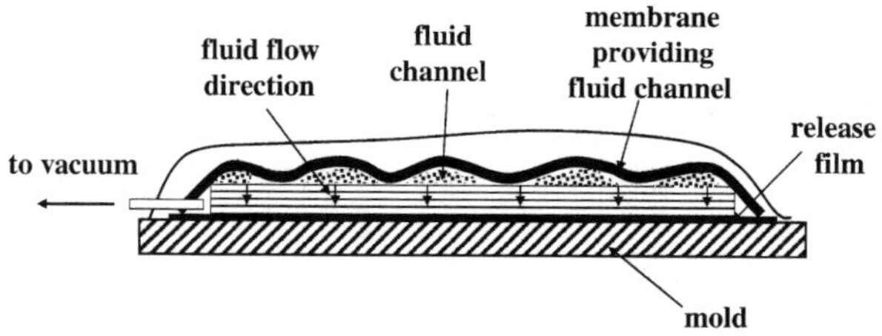

FIGURE 7.6 Schematic of the SCRIMP process.

the resin does not have to flow through the dense fiber preform along the planar dimensions of the part. To wet the fibers, resin only needs to flow through the thickness of the part. The distance *dx* in Darcy's law is therefore small and one does not need to have high pressure difference (*dp*) to get the resin to flow through the fiber beds. The advantage of this process is that it allows the ability to manufacture very large components on the order of several tens of meters (such as boat hulls or large turbine blades). One disadvantage of this method is that good surface finish is only provided on one side (the mold side). The other disadvantage is that the percent void content may be high. For critical applications such as aerospace, the amount of void content needs to be very small.

- *Resin film infusion molding (RFIM):* In this process, instead of injecting resin into the mold, thin films of resin are placed at the bottom of the fiber beds or between different layers of the dry preform. Upon heating and application of pressure, the resin film melts and the liquid resin permeates into the dry fiber preform. Figure 7.7 shows a schematic of the process.

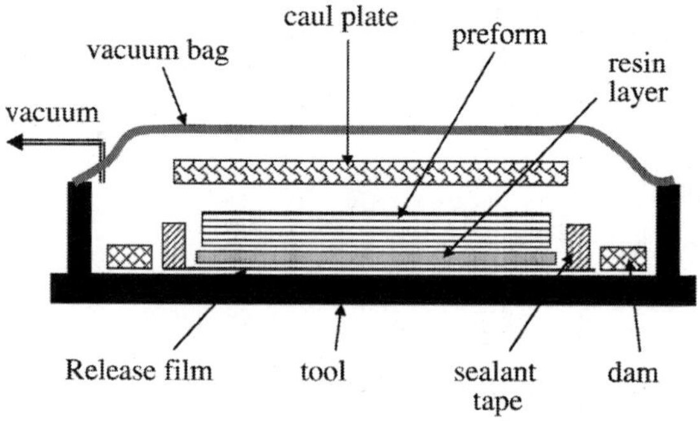

FIGURE 7.7 Schematic of the RFIM process.

2. MATERIALS

2.1. Fibers

Fibers used for LCM are usually glass or carbon. Discussion on fibers was presented in Chapter 3. In addition to the fiber forms presented in Chapter 3, there are other forms of fibers that are specifically applicable to the LCM process.

2.1.1. Flow Enhancement Fabrics

The macroscopic permeability (permeability referring to the fabric as a whole rather than the individual filaments) can be increased by creating effective flow channels between fiber bundles. This can be achieved by fiber clustering, which may still allow high volume fraction to be attained but with a less uniform fiber distribution. Commercially available flow enhancement fabrics are said to offer a number of advantages over the aligned fabrics, in particular, reduced injection times, which may make possible the production of relatively large parts at high volume fractions. The main disadvantage of these materials is the potential reduction in mechanical properties caused by less uniformity in the fiber distribution.

For VARTM or SCRIMP, a layer of impermeable plastic with flow channels can be placed on top of the fiber preform to facilitate fast penetration of the liquid resin. Flow of the resin will only need to go through the thickness of the preform (Figures 7.5 and 7.6).

2.1.2. Surface Veils

Surface veil is a random reinforcement with low superficial density and is produced from a fine (low tex) glass fiber. This material is used in LCM to provide a high quality surface finish by eliminating fiber strike-through and creating a resin rich surface layer, or alternatively where chemical resistance is required (where a C-glass tissue may be used). The use of a surface veil may eliminate the need for a gel coat. A number of materials are commercially available based on either chopped or continuous filaments held together with either a polyester or PVA binder and having superficial densities in the range of 30–100 g/cm^2.

2.1.3. Binder

Binder is applied to the fibers during the preform manufacturing stage to provide cohesion to the fiber architecture during subsequent handling and processing operations. Binding can be achieved mechanically by needling or stitching with a light yarn or roving, but it is more usual to use a chemical adhesive binder. This may be either a thermoplastic or thermoset in the form of a powder, an emulsion or a solution. For systems using polyester of vinyl ester, binders may be categorized by their solubility in styrene. A dissolution time of less than 60 seconds corresponds to high solubility, between 60 and 200 seconds indicates medium solubility, and over 200 seconds represents low solubility. Relatively low solubility binders result in improved flow characteristics at the expense of prolonged fiber wet-out times. One potential consequence of binder dissolution is a change in resin viscosity. It has been suggested [2] that the viscosity of a vinyl ester resin may be doubled by the addition of 5% by mass of thermoplastic polyester binder.

2.2. Preforms

Preform is an assembly of fibers having the configuration of the part. Once the preform is wetted by the liquid resin and after the resin is cured, a composite part is obtained. Figure 7.8 shows an example of a preform.

There are several good reasons to preform the reinforcement before loading it into the mold.

- Preforms speed up the process and free the mold from everything except loading, injection, in-mold cure, and demolding.

FIGURE 7.8 Knitted glass fiber preform for a stiffened T joint (courtesy of Preform Technologies Ltd.).

- Preforms improve quality and reduce part-to-part variations. For fast cycle times, the ideal is to make the preform so stiff that it becomes self-locating in the production mold. In other cases, when the mechanical properties are of paramount importance, one often wants to minimize the preform binder since the mechanical properties can be adversely affected by the binder.

In general, a good preform is required to be inexpensive to make and stiff enough to be stacked and handled before injection. The fibers must stay in the direction in which they have been placed both during handling and injection. To achieve all these goals it is common to apply some form of preforming agents (binders). Both thermoplastic and thermosetting powders are commonly used for this purpose.

The compaction behavior of the preform differs a lot depending on the performing method and the type of reinforcement that has been used. An important observation is that the fiber volume fraction at zero compaction pressure can differ significantly between different fabrics. A lower nominal fiber volume fraction can result in movement of the reinforcement during filling and incomplete impregnation. Another typical feature of the compaction behavior is that all fabrics behave like nonlinear (stiffening) springs and that the possible increase in fiber volume fraction from the value at rest is limited.

In practice, most molds are flexible and will deform when the fiber volume fraction becomes too high (which corresponds to a high compaction pressure). The forces associated with the compaction can be so high that the mold surface becomes deformed or even that the entire mold breaks (if the clamping force is powerful enough). High compaction pressure is a particularly difficult problem when the thickness of the laminate in a part varies in different positions (stepping of the thickness). In this case, even a small error in the placement of the fabric with a corresponding increase in local fiber volume fraction can lead to a dramatic increase in compaction pressure. Both thermosetting and thermoplastic powders are commonly used to stiffen the preform. Ideally, the preforming agent should not decrease the permeability, the wettability, or the mechanical properties of the finished part, but it should still stiffen the perform so that it can be handled. In practice, however, this compromise is difficult to achieve. For example, the mechanical properties can be significantly reduced by the preforming operation, but they can also be close to the value without preform binder with a judicious choice of preforming agent. The raw material suppliers can usually recommend suitable preforming agents for a given matrix system or provide pretreated reinforcement with binder.

It should be noted that the permeability of the resin in the fiber preform depends to a great extent on the fiber volume fraction. Since the compaction pressure has great influence on the fiber volume fraction, this pressure therefore has important influence on the flow of the resin through the preforms.

2.2.1. Preforming Methods

The preforming methods can roughly be classified into five basic types: (1) cut and paste, (2) spray-up of chopped fiber on preformed models, (3) thermoforming, (4) weft knitting, and (5) braiding.

2.2.1.1. Cut and Paste

In this technique, sheets of fabrics are cut to simple shapes and these pieces are fit together with an adhesive or by stitching to make up the configuration of the part. It should be remembered that an adhesive can adversely affect the mechanical properties.

2.2.1.2. Spray-up

In this technique, continuous rovings are chopped using a chopping device. These rovings are deposited on a perforated former having the shape of the perform. Binder solution is sprayed on the chopped fibers to provide adhesion. After spraying, hot air is allowed to circulate for about 1 minute so that the thermoplastic binder melts. After melting, the air stream is switched to cold and the preforming powder solidifies. Figure 7.9 shows an example of a preform made by spray-up.

2.2.1.3. Thermoforming

In thermoforming, the fiber bed along with binders is compression molded at the proper temperature. The formability of woven fabrics is limited and only moderately double-curved shapes have been formed commercially. Wrinkles and folds formed by draping of woven fabrics can to some extent be predicted with computer simulations. There is a possibility that the fiber orientation may change during forming.

2.2.1.4. Weft Knitting

There are two kinds of knitting operations: weft and warp. Both of these produce interlooped structures. These methods differ in that weft

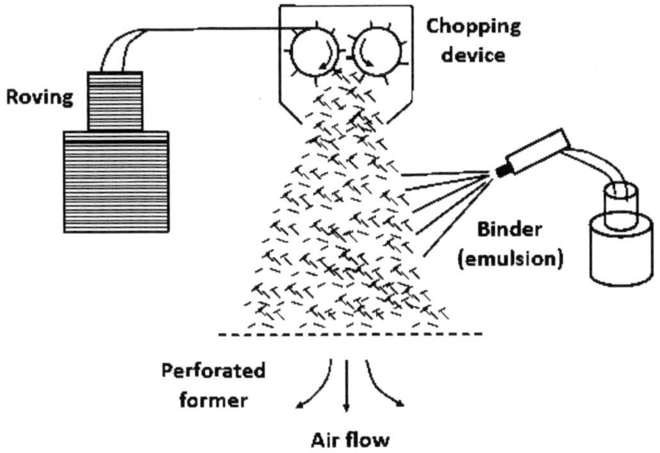

FIGURE 7.9 Preform by spray-up.

knits are formed in the weft or horizontal direction, whereas warp knits are formed in the warp or vertical direction. The weft-warp knits have the advantages of conformability and can be automated easily. The disadvantages are that high fiber volume fraction is difficult to achieve and the knit is anisotropic.

2.2.1.5. Braiding

Braiding is a good technique for forming tubular structures. Braiding is available in diameters up to about 300 mm and with different types of fibers. Figure 7.10 shows an example of a braided perform.

2.3. Matrix

Common requirements for resin systems for LCM processes are:

- Sufficiently low viscosity (about 500 cP) and long gel time to permit complete impregnation, mold filling and fiber wetting
- Appropriate curing characteristics to provide acceptable cycle times
- Adequate mechanical properties and physical characteristics to meet the performance requirements

RTM processes rely heavily on polyesters, vinyl esters and (for aerospace applications) epoxies, while SRM is almost exclusively based on polyurethanes.

2.3.1. *Polyester and Vinyl Ester*

In common with the hand-laminating industry, RTM has been dominated (in tonnage) by the use of polyester resins. Unsaturated polyesters are produced via a condensation reaction of organic acids (maleic and phthalic anhydride) with ethylene or propylene glycol to produce esters (Chapter 2). The styrene content of an unsaturated polyester is important in that it controls the resin viscosity and thereby the impregnation process. Increasing styrene content will decrease the viscosity but will increase the heat of reaction and peak exotherms at the expense of the final mechanical properties. Excessive styrene content may be detrimental to product quality since any residual monomer following curing or post-curing may continue to be lost in service with dimensional changes in the finished part.

2.3.1.1. Shrinkage Control Additives (Low Profile Additives or LPA)

Polyesters and vinyl esters exhibit a significant amount of shrinkage

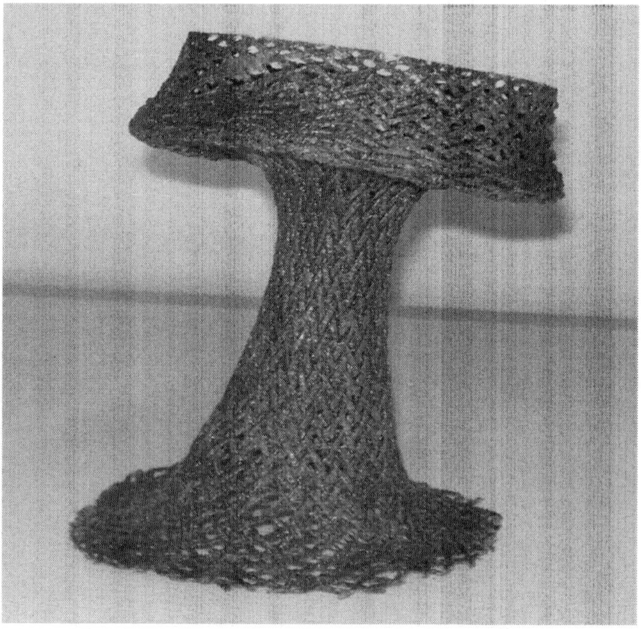

FIGURE 7.10 Preform made by braiding.

upon curing (about 8%). This shrinkage can produce undesirable effects such as poor surface finish, out-of-dimensions, residual stresses and cracks. Shrinkage control additives or Low Profile Additives (LPAs) have been added into the resins to reduce the shrinkage. LPAs in the form of thermoplastic additives, have been widely used in molding compounds to produce molded parts requiring smooth surface and dimensional stability. A variety of thermoplastics include: polyethylene, polystyrene, polymethyl methacrylate, and polyacetate. The precise mechanism whereby shrinkage control is achieved is the subject of debate. However, it is generally agreed that during the curing reaction the shrinkage control additive and cross-linked unsaturated polyester phases must separate. The micro-voidage that arises from this compensates for the bulk shrinkage and reduces the effects of any dimensional changes or surface defects.

Although the major and desired effect is the reduction of surface shrinkage for cosmetic reasons, thermoplastic additives can have a number of effects on the processing characteristics of the resin and on the properties of the final composite as follows:

- Modified resin viscosity
- Reduced heat of reaction due to the dilution of the reactive mass
- Modified reaction rate
- Reduced laminate strength and modulus

2.3.2. Epoxies

Discussion on epoxy resin was presented in Chapter 2. Typically epoxy resins cost about four times as much as general purpose polyester and two times as much as vinyl ester resins. Epoxies have major performance advantages over general purpose unsaturated polyesters including higher strength, modulus and fracture toughness. The good adhesion of epoxies to substrates generally leads to a stronger interface with fibers, which, in turn, determines the performance of the composite. Epoxies generally have shrinkage of about 5%. For RTM applications, the most common hardener is a low viscosity (cycloaliphatic) amine. Gel times in the range of 2 minutes to several hours are possible by correct matching of the hardener and mold temperature. Cure times are typically 6 times the gel time and a secondary post-cure is usually required. Anhydrides are used almost exclusively for elevated temperature curing and provide extended pot lives at room temperatures with several days being practical.

3. MOLD FILLING

The objectives of mold filling in LCM are to fill the mold completely, to wet the fibers well, to avoid dry spots and voids, and to avoid modifying the fiber orientation during the filling process. The discussion on mold filling will be divided into two parts. In the first part, ideal filling conditions are assumed where flow through porous medium is examined. In the second part, problems and issues arising from the practice of filling will be discussed.

3.1. Part I. Theoretical Considerations

The flow of resin through a fiber preform is usually assumed to be equivalent to that of an incompressible fluid through a porous medium. Therefore, the physics of the fill phase during liquid composite molding is based on in-plane incompressible mass conservation and uses Darcy's law as a momentum balance.

The equation of mass conversation for the fluid phase can be written as:

$$\nabla.\bar{u} = 0 \tag{7.1}$$

where $\bar{u}$ is the superficial fluid velocity vector (that is the velocity at which the fluid actually travels, rather than the observed or macroscopic velocity).

Darcy's law in three dimensions can be written as:

$$\bar{u} = -\frac{1}{u}[K]\nabla p \tag{7.2}$$

In which $[K]$ is the permeability tensor, taking the form of a $[3 \times 3]$ matrix, as:

$$[K] = \begin{bmatrix} K_{xx} & K_{xy} & K_{xz} \\ K_{yx} & K_{yy} & K_{yz} \\ K_{zx} & K_{zy} & K_{zz} \end{bmatrix}$$

If the flow is predominantly one-dimensional, the above two equations can be significantly simplified. This is shown for the two cases of rectilinear and radial flows below.

Rectilinear flow. If the flow is rectilinear (such that the fluid velocities in both the *y* and *z* directions are zero), then Darcy's law is reduced to the following equation:

$$u_x = \frac{Q_x}{A} = -\frac{K_{xx}}{\mu}\frac{dp}{dx} \tag{7.3}$$

where *A* is the cross-sectional area of the cavity. Figure 7.11 shows a schematic of rectilinear flow. The mass conservation equation is reduced to:

$$\frac{du_x}{dx} = 0 \tag{7.4}$$

Combination of the above two equations gives:

$$\frac{d}{dx}\left[\frac{K_{xx}}{\mu}\frac{dp}{dx}\right] = 0 \tag{7.5}$$

The Case of Constant Pressure at the Injection Gate

If the pressure at the injection gate is constant, and if the permeability and viscosity remain constant throughout the mold, then this equation implies a linear pressure distribution between the injection gate and the flow front. For example, if the resin pressure at the injection gate is P_o and the pressure at the flow front is 0 gauge, then the pressure distribution can be written as:

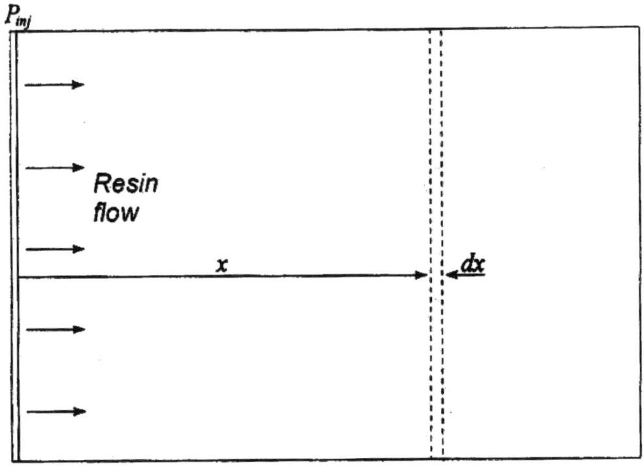

FIGURE 7.11 Schematic of rectilinear flow.

$$P = P_o\left[1 - \frac{x}{x_{ff}}\right] \tag{7.6}$$

where x_{ff} is the position of the flow front. The resulting pressure gradient can be substituted into Darcy's law to obtain an expression for the macroscopic flow velocity. Note that the superficial flow velocity is defined by Equation (7.3) where the macroscopic flow velocity is defined as:

$$v_x = \frac{Q}{A_{\text{cavity}}} = \frac{Q}{A\phi}$$

$$v_x = \frac{u_x}{\phi} = \frac{K_{xx}P_o}{\phi\mu x_{ff}} \tag{7.7}$$

where,

φ = the porosity and is equal to $(1 - V_f)$
v_x = the macroscopic flow velocity
u_x = the superficial flow velocity

It can be seen that the macroscopic flow velocity is independent of the distance from the injection gate. Equation (7.7) can be expressed as:

$$\frac{dx}{dt} = \frac{K_{xx}P_o}{\phi\mu x} \tag{7.8}$$

which can be written as:

$$xdx = \frac{K_{xx}P_o}{\phi\mu}dt \tag{7.9}$$

If the injection pressure is constant, integrating this equation from 0 to x_{ff} and from 0 to t_{ff} we have:

$$t_{ff} = \frac{\phi\mu}{2K_{xx}P_o}x_{ff}^2 \tag{7.10}$$

By substituting the mold length x_{ff}, this expression can be used to calculate the maximum fill time for the rectilinear flow under constant injection pressure. The above equation can also be arranged to determine the permeability:

$$K_{xx} = \frac{\phi\mu}{2KP_o t_{ff}}x_{ff}^2 \tag{7.11}$$

where the position of the flow front x_{ff} is determined at the corresponding time t_{ff}.

The Case of Constant Flow Rate at the Injection Gate

Alternatively if the injection flow rate is held constant, then it can easily be shown that:

$$t_{ff} = \frac{\phi A x_{ff}}{Q_o} \tag{7.12}$$

where Q_o is the constant injection flow rate.

It can be seen that now the fill time is directly proportional to the distance from the injection gate, and is independent of the resin viscosity and reinforcement permeability.

The pressure at the injection gate can be found by using Equation (7.7) as:

$$P_o = \frac{u_x \mu x_{ff}}{K_{xx}} = \frac{Q_o \mu x_{ff}}{A K_{xx}} \tag{7.13}$$

Equation (7.13) suggests that, for constant flow rate injection, the pressure at the injection port will increase as the flow front progresses.

Radial flow: If the resin is injected at the center of the mold from a point source, then flow will proceed radially until the resin reaches the mold wall. Figure 7.12 shows the schematic of radial flow. Darcy's law can be applied in radial coordinates:

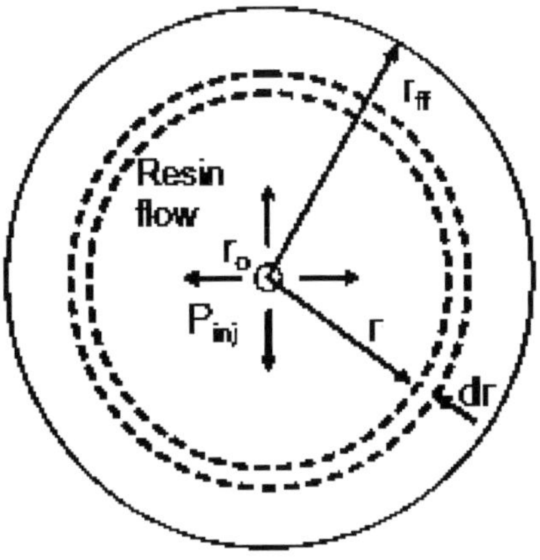

FIGURE 7.12 Schematic of radial flow.

$$Q_r = -\frac{K_r A}{\mu}\frac{dP}{dr} = -\frac{K_r}{\mu}2\pi rh\frac{dP}{dr} \tag{7.14}$$

where h is the thickness of the preform. Note that the volumetric flow rate is used as this allows the change in cross-sectional area of the flow front to be considered.

Assuming that the reinforcement permeability and resin viscosity remain constant throughout the mold, Rudd et al. [3] obtained the following equation for the pressure.

$$P = P_o \frac{\ln\left[\dfrac{r}{r_{ff}}\right]}{\ln\left[\dfrac{r_o}{r_{ff}}\right]} \tag{7.15}$$

where r_o is the radius of the injection port. This equation shows that the pressure decays rapidly as the distance from the injection gate increases. Equation (7.15) can be substituted into Equation (7.14) to obtain the macroscopic fluid velocity in the radial direction as:

$$v_r = \frac{Q_r}{A\phi} = \frac{K_r P_o}{\phi\mu r \ln\dfrac{r_{ff}}{r_o}} \tag{7.16}$$

If the injection pressure is held constant, then the time required to fill a region of radius r_{ff} is:

$$t_{ff} = \frac{\phi\mu}{2K_r P_o}\left[r_{ff}^2 \ln\frac{r_{ff}}{r_o} - \frac{1}{2}(r_{ff}^2 - r_o^2)\right] \tag{7.17}$$

For constant flow rate injection, the fill time is given as:

$$t_{ff} = \frac{\phi h\pi(r_{ff}^2 - r_o^2)}{Q_o} \tag{7.18}$$

and the resulting pressure at the injection gate is given as [from Equation 7.16)]:

$$P_o = \frac{Q_o\mu}{2\pi hK_r}\ln\left[\frac{r_{ff}}{r_o}\right] \tag{7.19}$$

3.1.1. Coefficient of Permeability

The coefficient of proportionality k (k_{xx} or k_r) is called the permeability of the reinforcement. According to theory, k is only dependent on the geometry between the fibers in the reinforcement (the pore space). Several models for the dependence of k on the fiber volume fraction have been proposed. The most cited is the Kozeny-Carman model, which predicts the quadratic dependence on the fiber radius r in addition to the dependence on V_f.

$$K = \frac{r^2}{4k_o}\frac{(1-V_f)^3}{V_f^2} \tag{7.20}$$

The constant k_o is called the Kozeny constant and it attains a value of 0.7 for well-ordered reinforcements with uniformly distributed fibers (e.g. unidirectional prepreg). The effective values for the Kozeny constant for angle ply laminates at $\pm\alpha$ (Figure 7.13) are shown in Table 7.1.

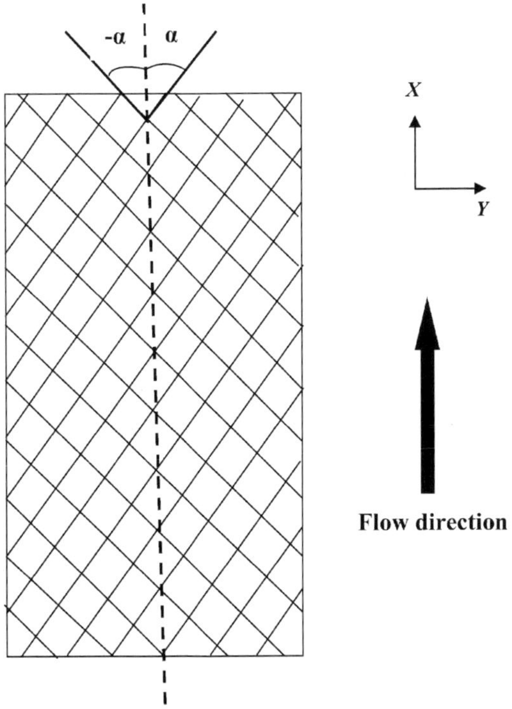

FIGURE 7.13 Flow in angle-ply laminate.

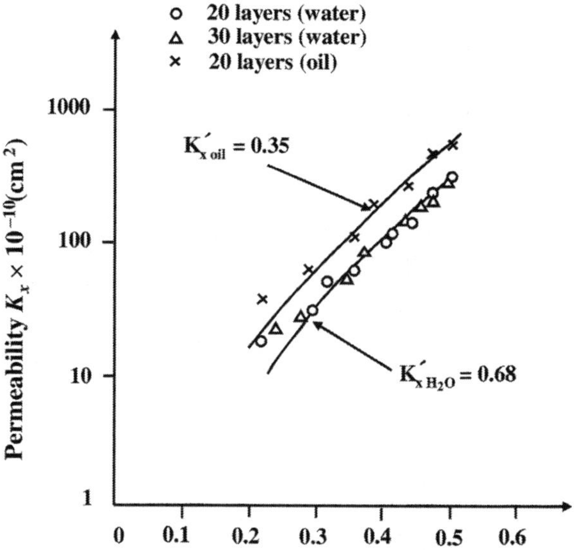

FIGURE 7.14 Axial permeabilities for a uniaxially aligned graphite fiber bed [5].

For flow along the fiber direction in unidirectionally aligned fiber beds, Lam and Kardos [5] show that the Kozeny-Carman equation predicts the permeability well with a Kozeny constant of $k_o = 0.68$ for water and $k_o = 0.35$ for silicone oil as permeants for the liquid volume fraction range of 0.25–0.5 (Figure 7.14).

For flow transverse to the fiber direction, using $k_o = 11$, Lam and Kardos [5] found that the data points fit well to the Kozeny-Carman equation for liquid volume fraction range of 0.25–0.5, as shown in Figure 7.15.

Lam and Kardos [5] found that for unidrectionally aligned graphite fiber reinforced resin prepregs, the ratio of the transverse to the axial fiber bed permeabilities is $K_z/K_x = 1/19$ whereas Gutowski et al. [6] found a similar ratio of 0.7/17.9.

TABLE 7.1 In-plane Kozeny Constants for ±α Preform [4].

α	$k_{ox}(0°)$	$k_{oy}(90°)$
0	0.68	11.0
15	1.18	10.1
30	1.49	6.65
45	2.70	2.70

3.1.2. Effect of Off-Axis Fiber Orientations on Axial Permeability

Following the suggestion of Scheidegger and Marcus, the anisotropic axial permeability K_x can be represented by:

$$\frac{1}{K_x} = \frac{\cos^2 \alpha}{K_{x\,uni}} + \frac{\sin^2 \alpha}{K_{x\,90}} \tag{7.21}$$

where $K_{x\,uni}$ is the axial permeability for a unidirectional fiber bed, $K_{x\,90}$ is the axial permeability for a 0–90 bed of fibers, and α is the angle between the fibers in successive plies. Equation (7.21) can be written in terms of the Kozeny constants as:

$$k_x = k_{x\,uni} \cos^2 \alpha + k_{x\,90} \sin^2 \alpha \tag{7.22}$$

Figure 7.16 shows a comparison between experimental results and those determined from Equation (7.21).

For commonly used fabrics in LCM (e.g., with continuous strand mat or weaves), $r^2/4k_o$ should be seen as an adjustable model parameter with only weak coupling to the fiber or fiber bundle diameter.

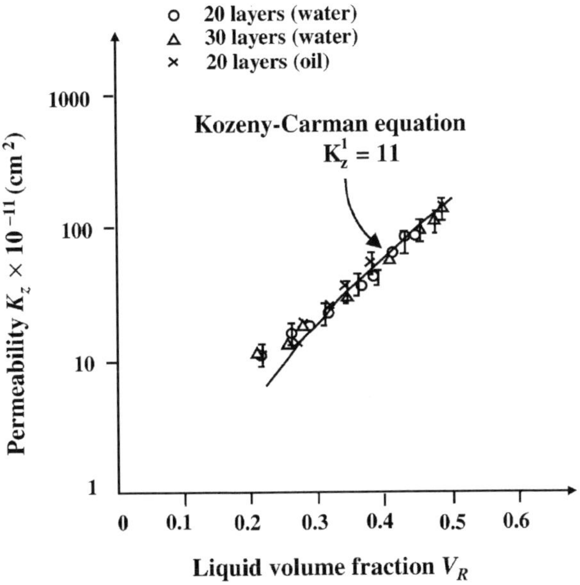

FIGURE 7.15 Transverse permeabilities during consolidation for a uniaxially aligned graphite fiber bed [5].

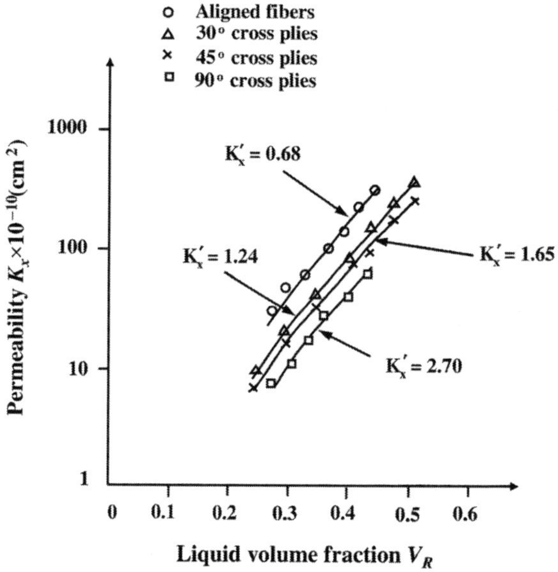

FIGURE 7.16 Axial permeability during consolidation as a function of ply orientation for a water permeant [5].

Continuous strand mats are approximately isotropic and have almost the same permeability in all directions (in the plane of the fabric). Many other fabrics, however, are strongly anisotropic and have different permeability in different directions.

The best way to use the Kozeny-Carman model is to use it as an interpolation formula for intermediate volume fractions between known values. Extrapolation should be done with extreme caution because the models are developed for idealized reinforcements. Table 7.2 shows typical values for the permeability of different types of reinforcement.

TABLE 7.2 Typical Permeability Data for a Few Reinforcement Materials [7].

Type of Material	Fiber Volume Fraction	Permeability (m^2)
Continuous glass strand mat	0.25	1×10^{-9}
Unidirectional glass (along fiber direction)	0.59	7.1×10^{-11}
Unidirectional glass mat (transverse to the fiber direction)	0.59	1.2×10^{-11}

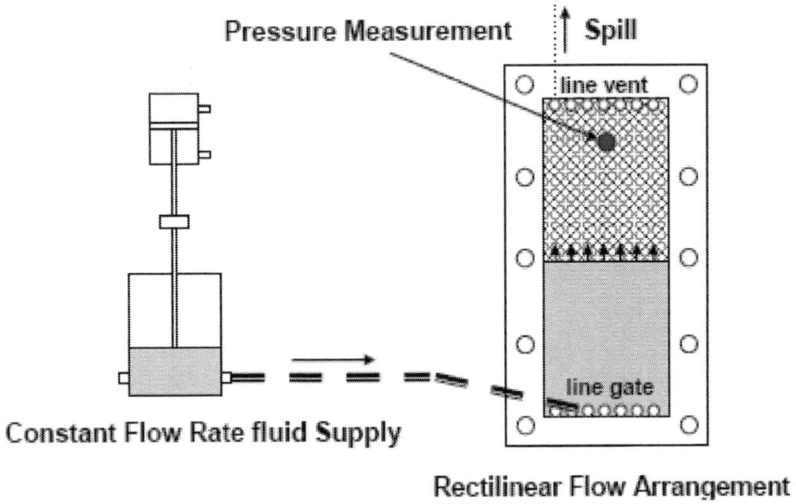

FIGURE 7.17 An experimental setup to determine the permeability of fiber preforms using rectilinear flow.

The permeability can be determined experimentally in several different ways (e.g., in a radial flow or unidirectional flow experiment). The experiments can also be done with either an advancing flow front (wetting flow or unsaturated) or a fully saturated reinforcement under steady-state conditions.

A convenient way to estimate the permeability is to use a rectangular mold where the resin is injected from one of the sides unidirectional to the opposite side. The other two sides are sealed tightly against the reinforcement so that the tow front becomes a straight line. Experimental setups to determine the permeability are shown in Figures 7.17 and 7.18.

One of the major sources of errors in permeability measurements is mold deflection, and it is a particular nuisance in the radial flow method because the smallest in-plane dimension of the mold (which governs the mold deflection) has to be larger than that of the unidirectional flow method. The major difficulty with the unidirectional flow method is preventing leakage at the edges.

3.1.3. Injection Strategies

The mold filling time and the quality of the part are affected by the mold-filling strategy (the way in which the resin is introduced and air is vented out of the mold). The mold filling strategies can be divided into three main types:

1. *Point injection.* The resin is introduced through a port in the center of the part, the resin flows essentially radially into the reinforcement, and air is vented at the periphery of the part. Figures 7.12 and 7.18 show the schematic of the point injection.
2. *Edge injection.* This is accomplished by injection through an inlet at one edge of the part. The flow is more or less unidirectional over the part, and air is vented at the opposite side. Figures 7.11 and 7.17 show the schematic of edge injection.
3. *Peripheral injection.* The resin is introduced in a resin distribution channel around the periphery of the part. The flow is radially inward and air is vented at the center of the part. Figure 7.19 shows a schematic of the peripheral injection.

The mold filling time differs considerably between the different strategies, with peripheral injection being much faster than the other two. Depending on other problems that may occur, however, all three alternatives are commonly used. The three basic strategies can also be combined, as in multipoint injection, to obtain faster filling or better impregnation. The position of the inlet(s) and outlet(s) is crucial with all three strategies because dry spots or areas of high void content will result if the gates are improperly positioned.

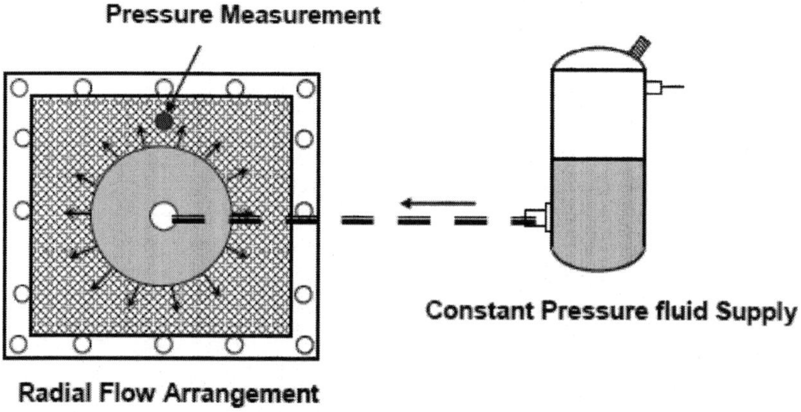

FIGURE 7.18 Experimental setup to determine the permeability of fiber preforms using radial flow.

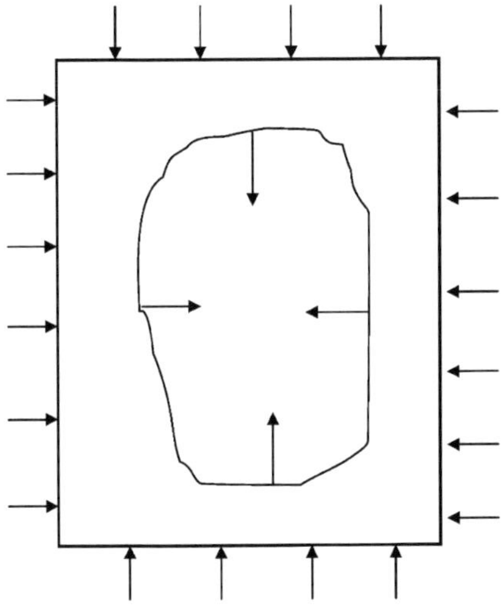

FIGURE 7.19 Peripheral injection.

3.1.3.1. Estimation for the Required Filling Time

The fill time can be estimated using the equations derived earlier. This is illustrated in the following example.

Example 7.1: Estimation of Fill Time for a Large Composite Part

It is desired to fill a plate made of glass fiber perform using epoxy resin. The dimension of the plate is 1 m × 1 m. The injection pressure at the gate is kept constant at 500 kPa. The viscosity of the resin is 800 cP. The permeability of the fiber preform is 1×10^{-10} m^2. Fiber volume fraction is 0.50. Determine the estimated time to fill the mold using the different techniques as follows:

a. Point injection with a 3 cm diameter of the injection port. Assume that the equation is valid over the whole domain of the plate.
b. Edge injection
c. Peripheral injection

Solution

a. Point injection

The time required to fill the space is given by Equation (7.17) previously as:

$$t_{ff} = \frac{\phi\mu}{2K_r P_o}\left[r_{ff}^2 \ln\frac{r_{ff}}{r_o} - \frac{1}{2}(r_{ff}^2 - r_o^2)\right] \quad (1)$$

The equation is valid for the region of radius r_{ff}. However the statement of the problem assumes that the equation is valid over the whole domain of the plate. Since the plate is a square 1 m × 1 m, the effective radius r_{ff} can be taken to be half the square root of (1 + 1) = (0.5)(1.414) m = 0.707 m. The injection port has a diameter of 3 cm giving r_o = 1.5 cm. Substituting the values as given in the problem statement and above information into Equation (1) yields:

$$t_{ff} = \frac{(0.5)(800\times10^{-3}\ \text{Pa - sec})}{(2)(10^{-10}\ \text{m}^2)(500\times10^3\ \text{Pa})}\left[(0707\ \text{m})^2 \ln\frac{0.707}{0.015} - \frac{1}{2}(0.707^2 - 0.015^2)\ \text{m}^2\right]$$

$$t_{ff} = 6{,}700\ \text{sec} = 1.86\ \text{hrs}$$

b. Edge injection

The characteristic dimension in this case is the length of the plate, or L = 1m. Equation (7.10) gives:

$$t_{ff} = \frac{\phi\mu}{2K_{xx}P_o}x_{ff}^2 = \frac{(0.5)(800\times10^{-3}\ \text{P.sec})}{(2)(10^{-10}\ \text{m}^2)(500\times10^3\ \text{Pa})}(1\ \text{m})^2$$

$$= 400\times10^1\ \text{sec} = 4000\ \text{sec} = 1.11\ \text{hrs}$$

c. Peripheral injection

For the peripheral injection, the distance is equal to half that of the edge injection. The time would be 1/4 that of the edge injection for the theoretical method.

For theoretical method, t_{ff} = 1000 secs.

The fill time for a given injection strategy can be reduced by:

- Decreasing the viscosity (raise temperature or change resin)
- Increasing the pressure (beware of fiber washing and mold deflection)
- Changing the reinforcement

The most dramatic change in fill time, however, can be achieved by a reduction of the flow length through:

- Additional inlets
- Resin distribution
- Other changes in the injection strategy

The simplest way to estimate the fill time more carefully is to guess the flow path during filling. The longest flow distance is then estimated from the guess and the fill time can be computed from the formula for unidirectional filling. This method is surprisingly powerful, at least in cases where it is easy to guess the flow path. A useful method to guess the path is to try to imagine how a heat wave from a sudden temperature rise at the inlet would propagate through the part. In the preceding example and point injection the flow front will develop as a circular front, starting at the inlet, until it meets the closest side. From then on, the front will tend to move unidirectionally in both directions toward the far side (if there is no leakage at the sides). A reasonable estimate of the fill time is somewhere between the time to fill radially and the time to fill unidirectionally to the far side (flow distance 1.5 m). However, the estimate of the fill time shows that a more detailed study of injection strategy and process parameters is necessary because it is difficult to reach an acceptable production economy with a fill time of about 1 hour. An obvious action would be to choose a more efficient resin distribution method than a one-point inlet (e.g. peripheral injection or a multipoint injection). In addition, it would be worthwhile to try to lower the viscosity and increase the pressure.

There are several phenomena that occur in practice that invalidate the assumptions that lead to the fill time formula. Examples of such phenomena are:

- Fiber washing
- "Race tracking" at edges or on top of the reinforcement
- Significant mold deflection
- Significant cure during injection
- Significant pressure drop in resin distribution channels
- Non-Newtonian behavior of the resin
- Binder dissolution in the resin (increases the viscosity)
- Preform variations

The preceding formulas are useful for rough estimates but can sometimes yield significantly shorter fill times than in reality. More accurate predictions can be obtained through computer simulations based on Equations (7.2) and (7.3), and this is the recommended method when sufficient time and resources are available.

One problem with computer simulations is to obtain realistic values for the material parameters, particularly the permeability. As presented before, permeability depends on many parameters and it is difficult to have an accurate prediction for this property at all locations along the flow path at different temperatures and wetting conditions. These depend on the preforming step (fiber orientation) and to some extent on the loading step (improper location that leads to locally high- or low-fiber volume fraction). However considerable progress has been made on this problem and it seems likely that the accuracy of simulations will increase further as more progress is made.

3.2. Part II. Problems and Issues Related to Mold Filling

The discussion in part was based on the assumption that the flow behaves normally and that there are no deviations from the ideal situation. However, in reality, there are many issues and problems associated with mold filling. These include: different types of permeability, race tracking, fiber washing, occurrence of voids, limited fiber wetting time, and dry spots.

3.2.1. Different Types of Permeability of the Fiber Perform

Permeability can be defined as the compliance to the flow, i.e., the ability of the preform to let the flow go through it. This definition is simple but permeability of the fiber preform is probably one of the most challenging aspects in LCM. Permeability is influenced by the physical characteristics of the reinforcement; type of fabrics used; and thickness of the preforms, including pore size, roughness, tortuosity and channel lengths. These factors, in turn, are likely to be influenced by the compaction pressure, fiber volume fraction, orientation of the fibers, fiber architecture, history of compression of the preform, temperature, wetness or dryness of preforms, part thickness, and stacking sequence.

When a liquid is forced to flow through a simple layer of fiber perform with fibers all oriented along one direction, two types of flow front are observed: macro flow front and micro flow front, as shown in Figure 7.20. The reason for this is because, inside a fiber bed, there are two types of flow channels. One is the large channel that arises from the gaps between the tows. The other is the small channels that arise from the space between fibers within a tow. When the flow is dominated by the applied pressure gradient rather than capillary effects, the resin proceeds faster outside the fiber bundle than within and creates voids as the faster flowing resin enters the fiber bundle. Unless the voids can be flushed from the fiber bundles they are retained and the final void content depends on the cavity pressure. At low flow rates, the flow front is able to progress more rapidly within the fiber bundle than outside it. At moderate flow rates the capillary and viscous flow rates are approximately equal which results in more or less simultaneous impregnation of the small and large gaps between the fibers. At high flow rates the viscous forces dominate and only the large capillaries become infiltrated.

The permeability also depends on whether the fibers are dry or wet. In the case of advancing flow front (unsaturated permeability), the permeability is different from the case of saturated flow when the fibers are already wet (saturated permeability).

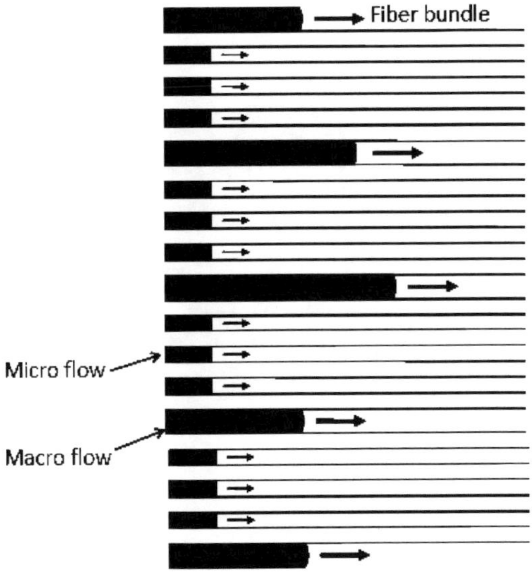

FIGURE 7.20 Macro flow front and micro flow front.

3.2.2. Race Tracking

In LCM, particularly RTM, where two solid mold surfaces on both sides of the preforms are used, it is essential that there are no large gaps between the preform and the mold wall. The existence of the large gaps will provide easy paths for the resin to flow. One example of areas of large gaps is shown in Figure 7.21. In this case, the preform has a tendency to follow the "inner lane" around corners so that the fiber content

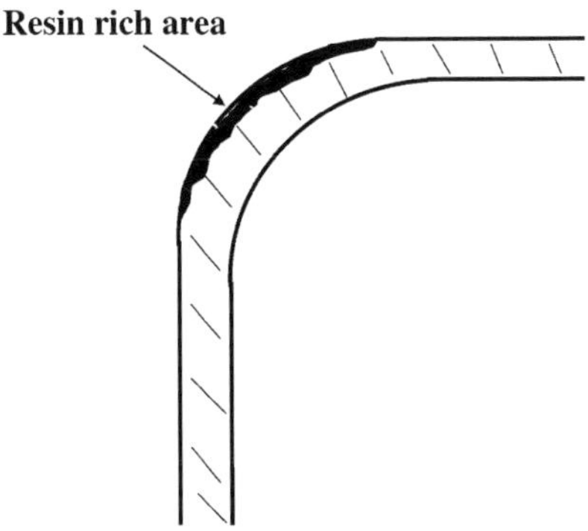

FIGURE 7.21 Race tracking due to a large gap between preform and mold wall.

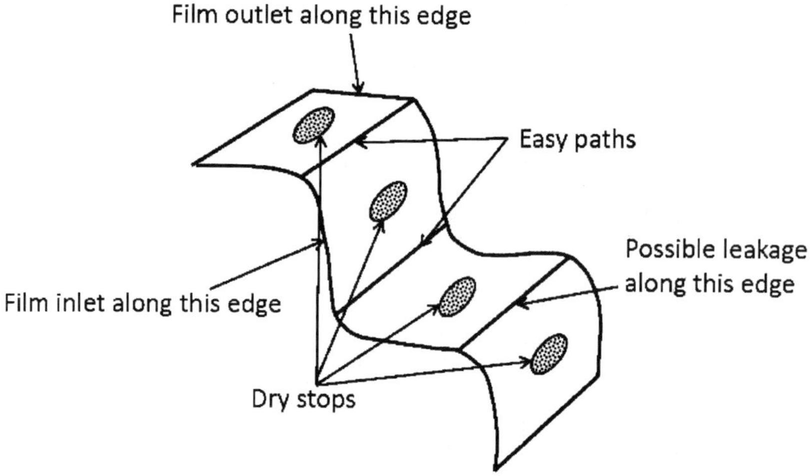

FIGURE 7.22 Race tracking and dry spots.

becomes low or negligible at the outer lane. One way of reducing the problem is to "compensate" the preform tool and modify the geometry so that the preform fills the real mold well at corners. The problem with varying fiber content at the comers is less pronounced for high fiber volume fractions.

Another example of race tracking is shown in Figure 7.22. In this case, the regions of large gaps occur at the bend lines. The flow will follow these gaps along the edge and along the bend lines. The flow then follows along radial directions. If air pockets are not completely removed, the flow cannot enter into the central part and these become dry regions. The dry region will decrease in size during the rest of the injection due to the increasing pressure around it, but a permanent dry spot will usually result if the enclosed region is large enough. The best solution to address this problem is to adjust the preform by increasing the volume fraction at a race track or by compensating the preform geometry. It is also worth considering a change in the injection strategy that makes the formation of dry spots less likely.

3.2.3. Fiber Washing

The reinforcement can be displaced significantly by fluid forces if injection velocity is too high or equivalently if the injection pressure is too high compared with the friction forces between the mold and the reinforcement. The problem will be less pronounced at higher fiber volume fractions when the compaction pressure is higher. As a consequence, it is harder to hold the reinforcement in place. The optimum fiber volume fraction depends on the type of reinforcement used. One way of estimating suitable processing conditions is to measure the "bulk factor" of the fabric (i.e. the fiber volume fraction when the fabric is uncompressed). Fiber volume fractions below this level are completely unacceptable because they will result in fiber washing, flow on top and so on. On the other hand, too high of a fiber volume fraction will make it difficult to close the mold and also reduce the permeability so much that it will be difficult to fill the mold in an acceptable time. Fiber washing changes the orientation of the fiber and this has detrimental effects on the final properties. Fiber washing usually occurs at locations close to the inlet of the resin.

3.2.4. Occurrence of Voids

One of the fundamental problems to be addressed during the impregnation phase is the removal of air from the mold cavity. This is necessary to produce high quality components with low void content. Air is present both within and between the fiber bundles and the displacement of each is necessary for minimum voidage.

Low pressure processes such as vacuum impregnation are best operated with the mold inclined so that the flow can proceed vertically upwards. This ensures that the effective pressure gradient is controlled by the suction pressure and minimizes race tracking effects.

Due to the dual flow mechanisms discussed earlier, where low void contents are critical, it is important that care is taken to control the resin flow rate such that it proceeds with a speed that is comparable with air removal from within the fiber bundles. This may need to be determined empirically since the micro-scale flow depends upon the resin properties and fiber architecture. Flow front velocities between 0.1 m/min and 0.6 m/min were found to be successful for a glass/polyester system, where typical resin viscosities of 300 cP are common. For vacuum driven processes, it is also important to avoid any air ingress following the mold filling phase which can be done conveniently by maintaining a positive pressure in the mold between mold filling and resin gel.

3.2.4.1. High Void Content at Outlet

Voids are usually formed at the flow front during mold filling. The voids move with the resin, but there will always be a region close to the flow front where the void content is higher than it is in the rest of the part.

To solve the problem, one can either use vacuum assistance during filling or allow a longer time for resin flow (resin flushing) after complete mold filling. Optimization of the process parameters, particularly injection pressure and temperature, can also reduce the problem.

3.2.4.2. High Void Content and Vacuum

Vacuum assistance during mold filling will usually reduce the void content significantly. The void content will be approximately proportional to the absolute pressure in the air inside the mold during mold filling. The mold, the sealings, and all gates, however, must be vacuum tight for this to be true. Even small leaks may be sufficient to give a very high void content. The sensitivity to leaks depends on the injection strategy. Point injection is usually the most robust method and peripheral injection the most sensitive method (air is "sucked" into the molding). Another cause for this problem can be the presence of volatile components in the resin. This can be tested by placing a beaker with resin inside a transparent container that is evacuated to the desired vacuum level. It is normal for some dissolved gas to come out of solution, but if gas bubbles continue to form after a long time they are most likely the result of evaporation of volatile components. For some resins this effect can be so strong that the resin "boils over." The best solution in this case is to change to a resin that can be processed at the desired vacuum level. An alternate solution is to reduce the vacuum level until the problem disappears.

3.2.5. Fiber Wetting

One of the limitations of the liquid molding processes compared to those based on preimpregnated materials is the relatively short length of time that elapses between the macroscopic impregnation of the preform and the rapid viscosity rise that accompanies the curing reaction. One of the consequences of this is the limited time available for the wetting of the individual fibers and development of the fiber-matrix interface. High speed processes such as SRIM involve such rapid gel of the resin system that the time available for fiber wetting may be limited to a few seconds. Conventional RTM almost certainly provides a greater window for wetting over the majority of the area of the mold but in a limiting case, i.e. adjacent to the vent, the wetting time may be limited to 1 minute or less in extreme cases. Because of the limited time available for wetting and bond formation the compatibility of the fiber surface (imparted by the sizing) and the resin system is of critical importance.

3.2.5.1. Effect of Fluid Surface Tension

It was presented in Chapters 1 and 2 that surface tension of the fluid plays an important role for the wetting of fibers. It is essential that the surface tension of the fluid resin be less than that of the fibers so that wetting can occur. The value of the fluid surface tension may also have an effect on the speed that wetting takes place. This is because the capillary action of the fluid in small channels has influence on the micro flow. By using different permeants with a range of viscosities and surface tensions and by changing the flow rate, the effect of the capillary number on void formation and retention was studied. The capillary number C_a is defined as:

$$C_a = \frac{\mu v}{\gamma} \tag{7.23}$$

where,

μ = the Newtonian viscosity.
v = the interstitial velocity.
γ = the fluid surface tension.

It was established [8] that a critical value of capillary number exists. Figure 7.23 shows the void content versus capillary number for a number of glass fiber preforms and oil. It shows a critical capillary number of 0.0025, below which the void content increases exponentially with decreasing capillary number. Above the critical value the void content was found to be negligible. Equation (7.23) shows that for a certain fiber network, the capillary number is inversely proportional to the surface tension of the fluid. The existence of a critical capillary number therefore means that there is also a critical surface tension of the fluid above which a large number of voids may occur.

A large body of work in this field has been collated by Lundstrom [9], which concludes that the architecture of the preform influences both the formation and the transportation of voids and that the following steps should be taken to minimize their occurrence in RTM laminates:

- Resin degassing
- Vacuum assistance during impregnation
- Positive pressure following mold fill and during heating and curing
- Purging the cavity with an excess of resin following first fill

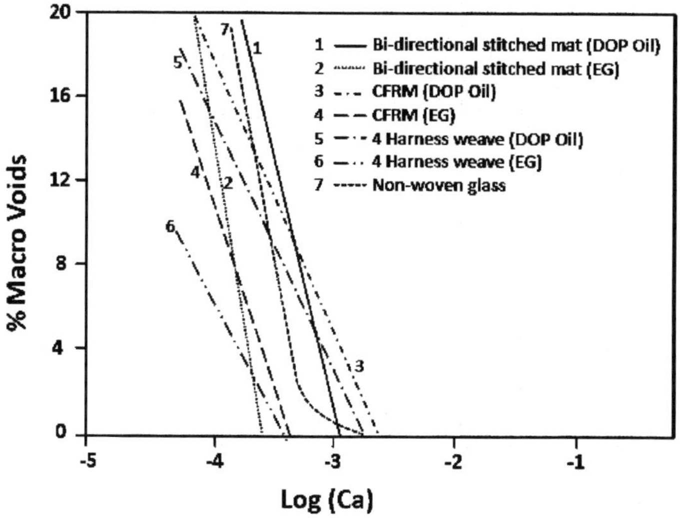

FIGURE 7.23 Void content versus capillary number (reproduced from Patel N., Rohatgi V. and Lee L.J. "Influence of processing and material variables on resin/fibre interface in liquid composite moulding," *Polymer Composites,* April 1993, Vol. 14, No. 2, pp. 161–172, with permission).

3.2.6. Dry Spots

Large spots with unimpregnated reinforcement can occur, even without race tracking, due to improper position of the injection or ventilation gates. The severity of the problem can vary. A longer time for resin flow out of the outlet gate (resin flushing) may be sufficient to solve the problem of the dry spot close to an outlet. In other cases, it may be necessary to move the outlets in the mold. If the mold is made from composite material (e.g. mass cast or laminated), then this may require making a completely new mold. With steel molds, new holes can be drilled and old ones can be filled if necessary.

3.3. Maximum Mold Filling Time

The mold filling time depends both on the permeability of the reinforcement and on the viscosity of the resin. As a rule of thumb, the catalyst system of the resin and the processing temperature can be chosen so that the gel time is about three times longer than the fill time. The time when the viscosity increases so much that no flow can occur is sometimes called the noninjection point or NIP time. The NIP time is related to the gel time of the resin, but it occurs considerably earlier because a moderate increase in viscosity (compared with gelation) will already make further flow impossible.

Most resin systems that have viscosity below 1 Pa·s (1000 cP) can be resin transfer molded. Even higher viscosity can be accepted, but the price for this is usually a very long injection and cure time. High viscosity systems can often be preheated before injection so that the viscosity is reduced sufficiently. A fairly low temperature increase can already be sufficient to reduce the viscosity to the recommended level because the viscosity dependence on temperature is exponential.

4. IN-MOLD CURE

4.1. Fundamentals

The resin cure must proceed in such a way that the curing reaction is slow or inhibited in a time period that is dictated by the mold fill time plus a safety factor. Otherwise, the increase in viscosity will reduce the resin flow rate and prevent a successful mold fill. On completion of the mold filling, the rate of cure should ideally accelerate and reach a complete cure in a short time period. There are limitations, however, on how fast the curing can proceed set by the resin itself, and by heat transfer rates to and from the composite part. An ideal resin processed in optimized conditions should have:

- A suitable and low resin viscosity during mold filling
- A short cure time from completed mold filling to demolding
- No material defects
- No shrinkage

4.2. Optimization of Cure

The most common objective for cure optimization is to minimize cycle time. Other factors to be considered are:

- Residual stresses
- Warping
- Void formation
- Surface quality
- Special knowledge of design work for composites

The parameters for cure optimization are: variations of the resin cure system, temperature control during cure, and time. Although the parameters cited here are limited to three there is a large number of degrees of freedom within these high-level parameters. First, it is necessary to identify the primary function in the system we aim to control. From a theoretical point of view, the key function is the reaction rate.

Figure 7.24 is a schematic illustration of the dependence of gel time and complete cure time on resin reactivity. Resin reactivity can be changed by altering the temperature or changing the resin formulation. In this figure, the process window is formed by the common part of the rectangle formed by dashed lines and the lines indicating the time to gelation and complete cure. The rectangle is formed by the horizontal lines for fill time and required maximum cycle time and the vertical lines by the requirement for complete cure and no thermal degradation.

Several choices exist to optimize cure. The most straightforward and obvious are:

- Use of existing knowledge from similar material combinations
- Use of guidelines from resin manufacturers
- Small-scale tests to guide in the choice of parameters
- Fundamentally based calculations
- Full scale tests in the production mold

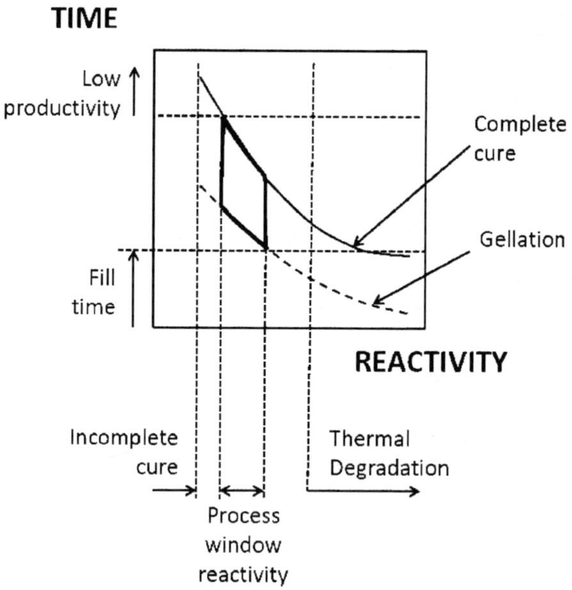

FIGURE 7.24 Processing window.

The use of simulation tools (e.g. computer programs) to calculate curing conditions is an area of great interest and is increasingly used to optimize cure for several processes, including LCM. However there are several obstacles to this advanced route, such as the need of:

- Skill to use computers
- A computer and a suitable program
- Accurate material properties for all constituents
- Accurate kinetic model for the resin

In addition to the above, experience and a good knowledge of processing is needed to evaluate the results from the simulation. The key problem today is often the last point (i.e., to obtain a sufficiently accurate kinetic model). Because these models are different for all resins and also different when the curing system is changed, the option is often only one: to obtain the model by one's own measurement. Hence, skill, knowledge, instrumental capacity, and quite a lot of experimental work are needed just to obtain a model. In addition, fiber sizing can influence the cure significantly, adding one more factor to take into account.

4.3, Cure Problems

Cure problems can be divided into two types: problems that give material defects and problems that make the process inefficient. Some examples of common problems of the first type are discussed below.

4.3.1. Delaminations

Delaminations can occur during cure as a result of high internal stresses. Those stresses develop due to resin shrinkage and thermal volume changes. The level of stresses depends on several material properties, such as Young's modulus, Poisson ratio, and thermal coefficient of expansion for both the resin and fibers. The level of stresses also depends on several conditions, such as fiber orientations, fiber volume fraction, and part geometry. The strength of the matrix plays a primary role for delaminations to occur, but the solution to delamination problems is usually not to increase matrix strength. In addition, the matrix material is changing during the curing process, and all matrix properties vary with the degree of cure. At low degree of cure, therefore, both modulus and strength are low regardless of matrix type.

Important parameters that affect delamination are:

- Thick composite parts

- Geometrically complex parts (varying thickness and closed parts)
- Resin with high shrinkage
- High temperature variation during the cure process

Possible actions to solve delamination problems are:

- Reducing temperature and, hence, prolonging cure
- Controlling cure propagation in thick parts by curing from inside to outside (hollow parts) or curing sequentially along the part, which can be done by partial heating of the mold
- Changing the resin system to one with less shrinkage or using additives that reduce shrinkage (this will create micro voids)
- Increasing pressure during cure

4.3.2. Surface Finish

The surface finish can be good in LCM but is usually good on one side of the laminate. A prerequisite for a good surface finish is to have a high gloss mold surface finish. The mold material and the release agent can also influence the surface finish indirectly. The primary cause of rough surface is resin shrinkage, provided the mold surface has a high gloss. Hence, if a good surface is required some action must be taken to reduce the resin shrinkage.

Possible actions to obtain a good surface finish are:

- Using high temperature on the mold side, where high surface finish is required, combined with internal mold pressure during cure
- Using LPA (low profile additive) resin systems

4.3.3. Porous Areas

Porous areas in LCM are often caused by mold filling problems. However, other causes also exist (e.g., high temperatures during cure may evaporate resin monomers or dissolved gas that form voids). Actions to be taken to optimize cure for minimum void content are:

- Degassing of resin before injection
- Reducing temperatures during cure

4.3.4. Incomplete Cure

The reactions are seldom allowed to go to completion in the mold. The usual practice is to let the part cure until it is sufficiently stiff to de-mold and then to perform a post cure either in an oven or at room temperature. If the degree of cure is too low at de-molding, then a number of problems can occur:

- Inability to de-mold without damaging the part
- Permanent deformation of the part

Solutions to these problems are to increase the cure temperature, to let the part cure for a longer time in the mold or to change the resin formulation. It is a good idea to make a serious effort to optimize the cure cycle and resin formulation if incomplete cure should occur.

5. REFERENCES

1. Verrey J. et al. "Manufacturing cost comparison of thermoplastic and thermoset RTM for an automotive floor pan," *Composites. Part A,* 2006, Vol. 37, pp. 9–22.
2. Chen J., Backes D and Jayarman K. "Dynamics of binder displacement in liquid molding," *Polymer Composites,* 17, 1996, pp. 23–33.
3. Rudd C. D. et al. *Liquid Moulding Technologies,* Woodhead Publishing Ltd., Cambridge, England, 1997.
4. Advani S. G., Bruschke M. V. and Parnas R. S. "Resin transfer molding flow phenomena in polymeric composites," Chapter 2 in *Flow & Rheology in Polymer Composites Manufacturing,* S. G. Advani, ed., Elsevier Science BV, 1994.
5. Lam R. C., and Kardos J. L. "The permeability and compressibility of aligned and cross plied carbon fiber beds during processing of composites," *Polymer Engineering and Science,* July 1991, Vol. 31, No. 14, pp. 1064–70.
6. Gutowski T. G. et al. *J. Composite Materials,* 1987, 21, p. 650.
7. Gebart R. and Strombeck L. A. "Principles of liquid composite molding," in *Processing of Composites,* R. S. Dave and A. C. Loos, eds., Hanser, 2000.
8. Patel N., Rohatgi V. and Lee L. J. "Influence of processing and material variables on resin/fibre interface in liquid composite moulding", *Polymer Composites,* April 1993, Vol. 14, No. 2, pp.161–172.
9. Lundstrom T. S. "Formation and void transport in manufacturing of polymer composites," Doctoral thesis, 1996, Lulea University of Technology.

6. HOMEWORK

1. It is desired to fill a plate made of glass fiber preform using epoxy resin. The dimension of the plate is 1 m × 1 m. The injection pres-

sure at the gate is kept constant at 500 kPa. The viscosity of the resin is 800 cP. The permeability of the fiber perform is 1×10^{-10} m^2. Fiber volume fraction is 0.50. Determine the estimated time to fill the mold using the different techniques as follows:

a. Point injection with a 3 cm diameter of the injection port. Assume that the equation is valid over the whole domain of the plate.

b. Edge injection

c. Peripheral injection

CHAPTER 8

Long Fiber Thermoplastic Matrix Composites

1. INTRODUCTION

The techniques of autoclave, hand laminating, filament winding, pultrusion and liquid composite molding have been developed mainly using thermoset matrix composites. This does not mean that these techniques cannot be used for thermoplastic composites. An autoclave can be used to process thermoplastic composites provided that the temperature and pressure are high enough. Filament winding (or rather fiber placement process) can also be used and pultruded thermoplastic composites parts have also been made. Liquid composite molding can also be used provided that the viscosity of the resin is small enough. Recent development of low viscosity thermoplastic composites [1] facilitates this process.

The reason for a special chapter dedicated to discussion of thermoplastic composites is due to the high viscosity of the thermoplastic resin. This high viscosity demands the use of high temperature and high pressure to enable the resin to flow to the surface of the fibers, to wet and bond with them. This significance can be illustrated via the use of Darcy's law as:

$$u = \frac{K}{\mu}\frac{dp}{dx} \tag{8.1}$$

For the same fiber network, the permeability is the same, whether thermoset matrix or thermoplastic matrix is used. However, the viscosity of thermoset resin is about 500 cP at 20°C (Table 2.1, Chapter 2), whereas that of thermoplastic resin such as PEEK is 1,000,000 cp at 400°C. If the distance of flow dx is the same in both cases, then the pressure required for processing PEEK at 400°C would be 2000 times more than that of the case of epoxy at 20°C. If a pressure of 689 kPa is required for the processing of epoxy at 20°C, then it would take a pressure of 1378 MPa to process PEEK at 400°C for the same processing time. This is not practical.

In the face of the above difficulties, it is desirable to find ways to manufacture thermoplastic matrix composites because they can offer many advantages, such as the following.

a. Thermoplastic resin does not have the constraint of shelf life. To put it another way, thermoplastic composite preforms have infinite shelf life. This is because thermoplastic matrix solidifies upon cooling and there is no chemical reaction taking place to change its liquid state to solid state.
b. Thermoplastic resins are more ductile than thermoset resins. Composites made of thermoplastic matrix composites therefore can have larger fracture toughness than those made of thermoset matrix composites. The strain energy release rate for carbon/epoxy is about 100 J/m^2 while that of carbon/PEEK is about 1000 J/m^2. This means that the thermoplastic matrix resin has large plastic deformation whereas thermoset matrix material does not. This is very important for structures that need to resist impact and fracture load.
c. Thermoplastic resins can be recycled, i.e., they can be reheated to take a different form. This is also due to the same reason indicated above in which the material can be heated up to change the solid state to the melt state. At temperature close to the melt point, different shapes can be formed to make a new product. This behavior can also be used to heal the defects that may exist in the structure.
d. The processing time can be fast. The heating and cooling of a thermoplastic composite material can take place within the order of minutes. This is in comparison with the order of hours for thermoset matrix material. The reason for the long processing time for the thermoset matrix resin is because of the time required for all the chemical bonds to take place. In the case of thermoplastic resin, cooling will solidify the material and this can take place very quickly.
e. Thermoplastic matrix resin is weldable. This means that solid parts made of thermoplastic matrix composites may be welded together. For thermoset resins, because of the chemical bonding required for solidification, once the bond is formed, it cannot be used again to form another bond. The weldability of the thermoplastic resin en-

ables thermoplastic matrix composites to exhibit some degree of healing upon heating.

For the above reasons, special techniques need to be developed for thermoplastic composites manufacturing. The main focus of different strategies to handle thermoplastic matrix composites is to enhance the availability of the resin to the fiber surface. This can be seen in various approaches such as the formation of tapes, the use of fiber commingling, coating fibers with thermoplastic powder, and using fabrics with resin film sandwich. These techniques will be discussed in the subsequent sections of this chapter.

2. MATERIALS

As the name implies, thermoplastic matrix composites are materials made by the combination of fiber materials and thermoplastic matrix materials. Fibers are usually made of carbon, glass or Kevlar. Matrix materials can be made of engineering thermoplastic resins such as nylon, polypropylene or high performance thermoplastics such as polyetheretherketone (PEEK), polyetherketoneketone(PEKK), polyethermide (PEI), polyphenylenesulfide (PPS) or polyethersulfone (PES). Table 8.1 (repeated from Table 2.6) shows the common high performance thermoplastic matrices along with their properties. A more comprehensive list of thermoplastic matrix that can be used to make composites can be found in Reference [2].

Among the semi-crystalline thermoplastic resins, PPS is the least expensive but has low fracture toughness. Both PEEK and PEKK have higher fracture toughness but have higher processing temperatures than PPS. PEEK has been used more than PEKK and has a larger database. PEKK has a lower processing temperature than PEEK and yet PEKK has a higher T_g and lower cost.

Cogswell [3] gave comprehensive information about the physical properties of PEEK. Table 8.2 shows the values.

Due to the high viscosity problem, the manufacturing strategies for thermoplastic matrix composites are different from those of thermosets. The objective of these strategies is mainly to compensate for this large viscosity and attempts to get the resin to the surface of the fibers and to wet the fibers.

The strategies of manufacturing using thermoplastic composites can be considered to consist of two stages. In the first stage, the preliminary material form is done, and in the final phase, the final product is made. At

TABLE 8.1 Commonly Used high Performance Thermoplastic Matrices (courtesy of CYTEC Engineered Materials).

	PEI	PPS	PEEK	PEKK (DS)
Morphology	Amorphous	Semi-crystalline	Semi-crystalline	Semi-crystalline
T_g(°C)	217	90	143	156
Process Temp (°C)	330	325	390	340
Comments: Pros	✓High T_g ✓Moderate processing temperature	✓Excellent environmental resistance ✓Moderate processing temperature	✓Extensive database ✓Excellent environmental resistance ✓High toughness	✓Excellent environmental resistance ✓High toughness ✓Lower process temperature than PEEK ✓Bonding and painting
Comments: Cons	✓Environmental resistance	✓Low T_g ✓Low toughness ✓Low paint adhesion	✓High process temperature ✓High cost	✓Limited database in composite form

TABLE 8.2 Physical Properties of PEEK [3].

	Resin	Composite	Resin in Composite
Cooling: crystallization temperature (°C)	300	294	
Cooling: latent heat (kJ/kg)	49	14	43
Heating: melting temperature (°C)	343	342	
Heating: Latent heat (kJ/kg)	44	12	39
Heat content at 400°C relative to 20°C (kJ/kg)		559	
Coefficient of thermal diffusivity across the fiber direction (cm^2/s)		3×10^{-3}	
Coefficient of thermal diffusivity along the fiber direction (cm^2/s)		20×10^{-3}	
Thermal expansion along fiber direction (10^{-6}/°C)		0.5 (23–143°C), 1.0 (143–343°C)	
Thermal expansion across fiber direction (10^{-6}/°C)		30 (23–143°C), 5 (143–343°C)	
Thermal expansion—quasi isotropic (10^{-6}/°C)		29 (23–143°C)	

the end of the preliminary phase, a preliminary combination of fiber and matrix is made. This combination can be in the form of a tape, fibers with clinging powder, reinforcing fibers commingled with filaments made from thermoplastic matrix, or fabric/film sandwich. Figure 8.1 shows the different preliminary material combinations (PMCs).

In the second (and final) stage, the preliminary material combinations (PMCs) are transformed into the final composite product. This transformation is usually done using either compression molding or by fiber placement process. Figure 8.2 shows the two processes as applicable to different material combinations.

3. PRELIMINARY MATERIAL COMBINATIONS (PMCs)

In the PMC, it is essential for the matrix to be in the vicinity of the fibers. This is important to reduce the time and pressure required to get the matrix to get to the fibers during the final product fabrication. Some of the approaches used will be discussed below.

3.1. Tape

A tape consists of unidirectional filaments bonded together by the ma-

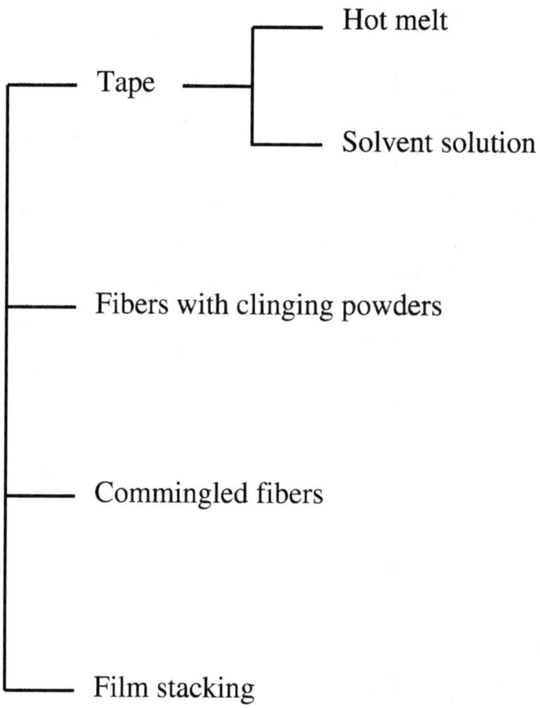

FIGURE 8.1 Preliminary material combinations (PMCs).

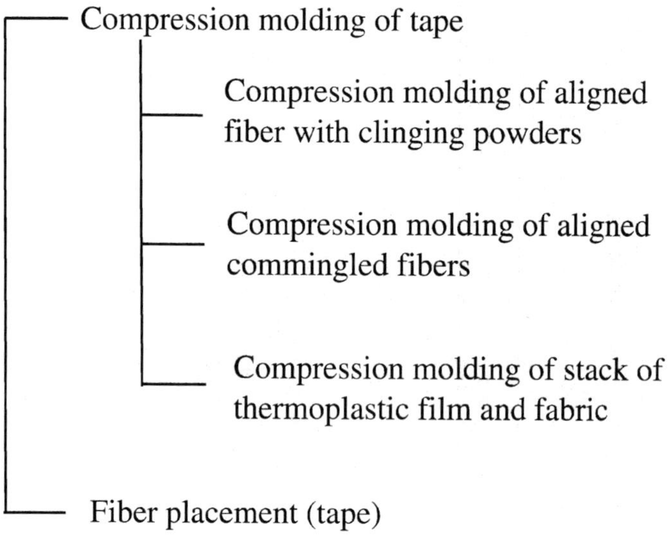

FIGURE 8.2 Processing the simple form into the final part.

trix resin. Resin needs to exist in liquid form to wet the fibers. This can be done by heating and melting the resin and running the fibers through the bath of liquid resin. Hot melt processes are probably most common. In the hot melt process, the matrix is heated until melting. Its viscosity should become low enough such that flow to the surface of the fiber is possible and wetting can take place. Figure 8.3 shows a schematic of the fibers running through a melt of resin. This can also be done by dissolving the matrix in a solvent and running the fibers through a bath of the solution. Solution processes are well established for thermosetting prepolymers. This process is used by Dupont to produce prepreg of Avimid K-III, a thermoplastic polyimide. The prepreg contains a substantial amount of residual solvent and must be cured. Therefore the production of Avimid K-III composite structures must be conducted in a manner similar to thermosetting composites. The complication in this technique is that solvent needs to be subsequently evaporated, which may give rise to voids and residual solvents.

Unidirectional tape is the most common form of thermoplastic ply. By convention, the tape is 0.127–0.152 mm (5–6 mils) and 7.62–30.48 cm (3–12 in) wide. To produce a 30.48 cm wide tape requires approximately 24 tows with 12,000 filaments each of 8 μm diameter to the combining process. Conversely, wide tape can be filament wound from a single tow using a large diameter mandrel. This latter approach is convenient for experimental ply production but may not be appropriate for low cost fabrication. Conversely, filament winding towpreg directly to produce a consolidated structure is potentially low in cost. This is due to process integration. Fabric plies are difficult to produce from thermoplastic towpreg due to the stiffness of most towpregs. Consolidated towpreg, typically from slit tape, can be braided into two dimensional fabrics, but

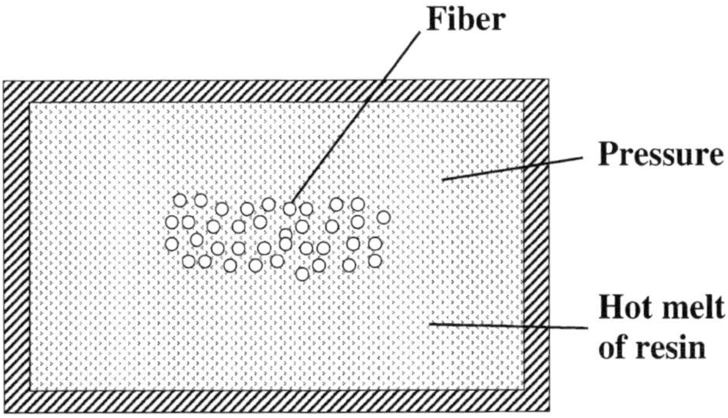

FIGURE 8.3 Impregnation of fibers by running fibers through a bath of melted resin.

FIGURE 8.4 Photo of a roll of unidirectional tape made of carbon/PEKK.

three dimensional fabrics are difficult to produce. Figure 8.4 shows a photograph of a roll of tape.

The speed of production of the hot melt process (meters per minute) depends on the viscosity of the melt, the thickness of the prepregs to be made and the applied pressure. Darcy's law can be used to estimate the rate of production as illustrated in the example below.

Example 8.1

A hot melt process is used to produce prepregs for carbon/PEEK 0.2 mm thick. The temperature of the process is 380°C giving rise to the viscosity of the resin of 1000 Pa(sec. A pressure of 1 MPa is applied to induce the flow across the thickness of the prepreg. Determine the maximum rate of production, if the length of the die is 50 cm and the permeability of the fiber preform is assumed to be 10^{-12} m^2.

Solution

The fastest rate of production occurs when the resin has sufficient time to flow across the thickness of the layer. Using Darcy's law, one has:

$$u = -\frac{K}{\mu}\frac{\Delta p}{\Delta x}$$

where,

u = the flow velocity across the thickness of the prepreg
K = the permeability of the fiber preform
Δp = the pressure gradient across the thickness of the prepreg
Δx = thickness of the laminate

Substituting in the values yields:

$$u = \frac{10^{-12}\ \text{m}^2}{1000\ \text{Pa - sec}} \frac{1\ \text{MPa}}{0.2\ \text{mm}} = 5 \times 10^{-3}\ \text{mm / sec}$$

Time required to traverse the thickness of the preform:

$$t = \frac{h}{u} = \frac{0.2\ \text{mm}}{5 \times 10^{-3}\ \text{mm / sec}} = 40\ \text{sec.}$$

For the length of the die of 50 cm, the maximum rate of production R would be:

$$R = \frac{L}{t} = \frac{50\ \text{cm}}{40\ \text{sec}} = 1.25\ \text{cm / sec} \qquad \text{or} \qquad 1.25\ \text{cm / sec}$$

3.2. Fibers with Clinging Powders

In the powder clinging process, the matrix powder is made to stick to the surface of the fibers. Figure 8.5 shows a schematic of the process. First, the dry fiber tow is fed from a creel to an air-conditioned spreader. The tow is spread to expose the fiber and grounded in order to pick up charge powder. By spreading a tow to expose virtually every fiber, it is easier to get the liquid resin to the surface of every fiber and it takes less pressure to force a polymer melt through a fiber bed. The fiber tow then enters into a heated chamber where matrix powder is electrified such that it carries an electrical charge, then it is fluidized. The powder is deposited on the band of fibers due to static electricity. At the next station of the

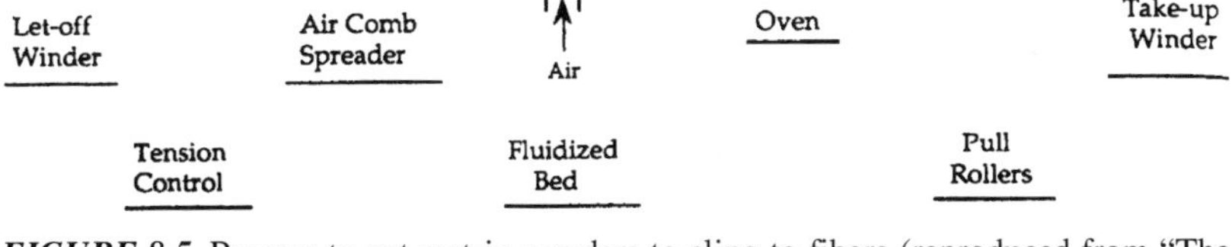

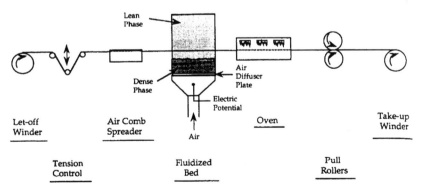

FIGURE 8.5 Process to get matrix powders to cling to fibers (reproduced from "The processing science of thermoplastic composites," by J. D. Muzzy and J. S. Colton, in *Advanced Composites Manufacturing,* T. G. Gutowski, ed., with permission from Wiley Interscience).

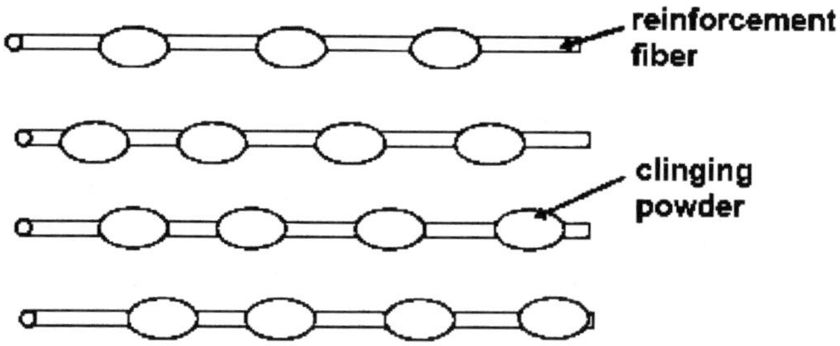

FIGURE 8.6 Fibers with clinging powders.

process, the material is heated. The resin melts, flows and then wets the fibers. After cooling, the powder is fused into the fibers. After passing through a fluidized bed, the tows enter a tunnel oven to melt the polymer onto the fiber. After cooling, the towpreg is wound onto a take-up roll.

There are advantages to the powder mixing process. By avoiding solvent or water in the combining operation, there is no need to remove volatiles. The extent of mixing between fiber and powder depends upon the extent to which the tow is spread. It is possible to spread the tow to expose virtually every fiber, thereby achieving good mixing. Spreading the tow and not collapsing it when the polymer is molten leads to a flexible tow that can be braided or woven. The coated tow can be heated and cooled rapidly to minimize polymer degradation. The tow is not exposed to high stress, which minimizes fiber damage. Since powder coating can be accomplished quickly and continuously, dry powder combining process is potentially inexpensive.

The production of towpregs using the electrostatic fluidized bed process has been demonstrated using numerous thermoplastics and thermosets as well as carbon, glass and aramid fibers. Good fiber wetting was obtained even when the particle size of the powder was substantially greater than the 8 μm fiber diameter. In a commercial scale version of this process a line speed of 40 cm/s has been achieved while attaining 40% vol polymer content. Figure 8.6 shows a schematic of the fibers with clinging powders.

3.2.1. Slurry and Foam

The above approaches appear to work particularly well for fine powders below 25 μm. Electrostatic cloud coating has worked successfully for powder well over 100 μm. Since many polymers are difficult to grind, the ability to accommodate large particles is a definite benefit. Slurries and foams are being explored as alternative combining methods.

3.3. Commingling/Bundling and Microbraiding

As long as the thermoplastic matrix can be spun into fiber form, it is possible to combine the thermoplastic matrix and reinforcing fibers by commingling (Figure 8.7). Wetting of the fiber in this case is deferred until the composite is consolidated. This approach leads to flexible preforms since the independence of the structural elements is maintained until the preform is consolidated. Because the matrix filament has not wetted out the reinforcing fiber until consolidation, more effort is required to complete consolidation compared to preconsolidated prepreg.

Another way of bundling the matrix filament together with the reinforcing fibers is to microbraid the matrix filaments with the reinforcing (Figure 8.8).

3.4. Fabric/Film Sandwich

Film stacking with dry fiber or commingled slit film and dry fiber are examples of combining processes where wetting is deferred until consol-

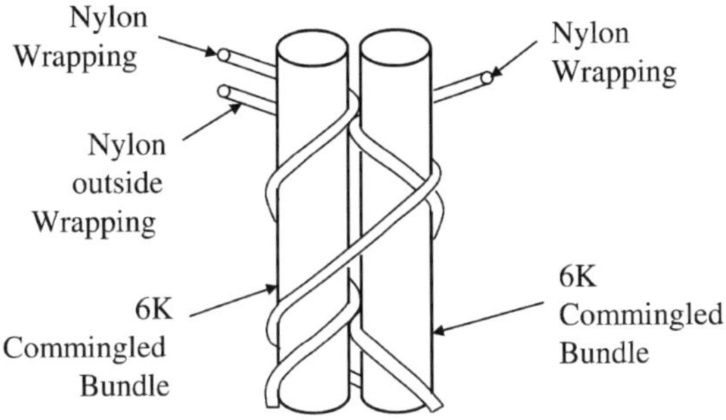

(a)

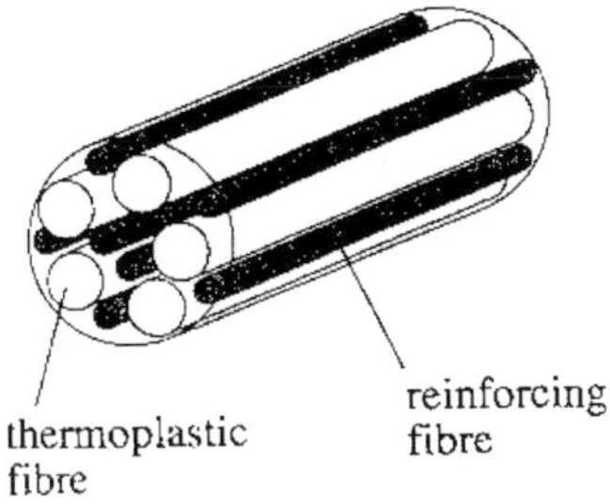

(b)

FIGURE 8.7 Commingling matrix filaments with reinforcing fibers.

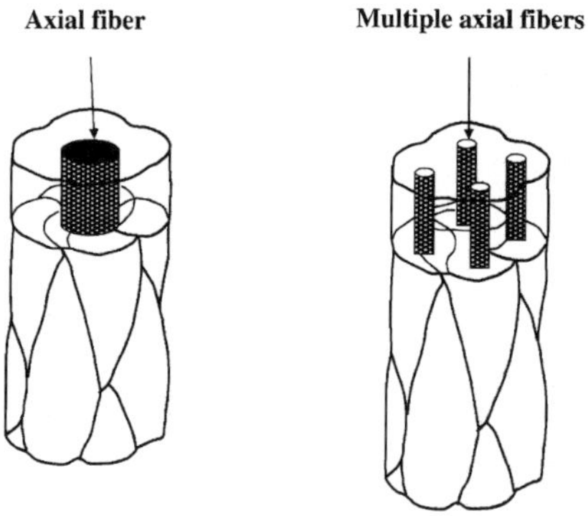

FIGURE 8.8 Microbraiding of matrix filaments with reinforcing fibers. The left shows one single reinforcement fiber tow at the center surrounded by matrix filaments. The right shows four reinforcement tows surrounded by matrix filaments (adapted from Reference [4], with permission from Canadian Association for Composite Structures and Materials).

idation (Figure 8.9). These options are less expensive than commingled fibers, but the degree of physical mixing is not as good as with commingled fibers.

3.5. Surface Polymerization

Surface polymerization of the polymer on the fiber consists of electro-polymerization on graphite fibers or vapor deposition polymeriza-

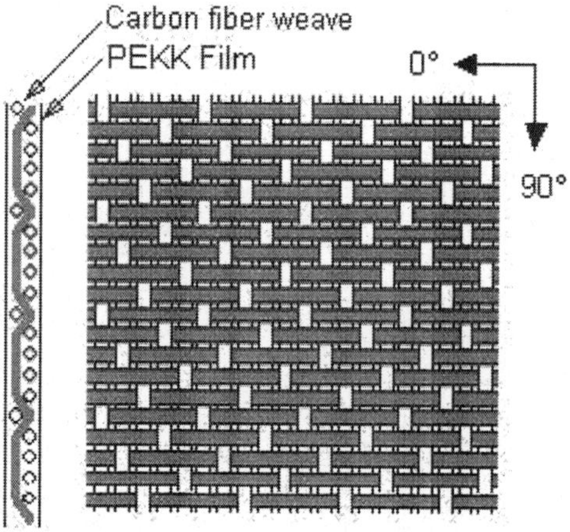

FIGURE 8.9 Fabric/ilm sandwich.

tion of paraxylene on graphite fibers, analogous to chemical vapor deposition. These approaches can produce individually coated fibers, hence the prepreg can have virtually the same flexibility as commingled fibers as well as complete fiber wetting.

4. FABRICATION OF THE FINAL PRODUCT

The transformation of the preliminary material combinations into the final thermoplastic composite product is mostly done using compression molding, and to a certain extent, fiber placement process.

4.1. Compression Molding

All of the PMCs (tape, tows with clinging powder, commingled tows, microbraided tows, tows with surface polymerization, or fabric/film sandwich) can be stacked up and compression molded to make the final composite product. The form of the PMC depends on the complexity of the configuration of the final product.

Thermoplastic tapes do not have sufficient drape and tack at room temperature to permit lay-up of complex shapes. Normally, flat pieces are made if tapes are to be used. Since cold thermoplastic tapes lack tack, they must be spot welded to the ply immediately below the ply being laid down. Either heat gun or ultrasonic gun is normally used for this task.

Tows with commingled matrix filaments, microbraided tows, tows with surface polymerization, or fabric/film sandwiches are more flexible than tapes and they can be arranged to make shapes other than flat plates. However, since the fibers in these configurations are not well constrained, alignment and tension need to be applied to the tows to assure proper fiber orientation control. For example, Figure 8.10 shows the ar-

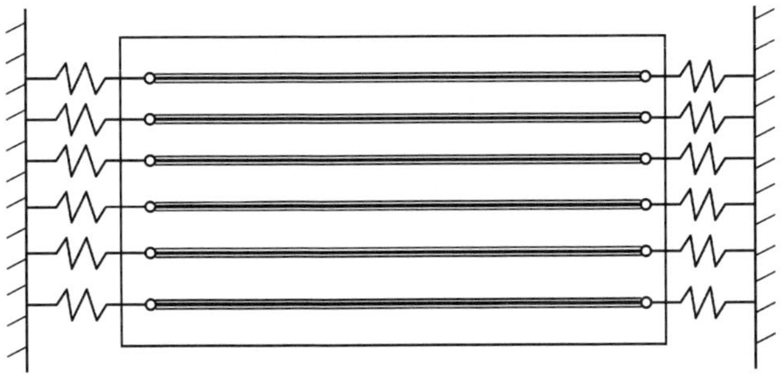

FIGURE 8.10 Springs are used to align and tension commingled fiber tows.

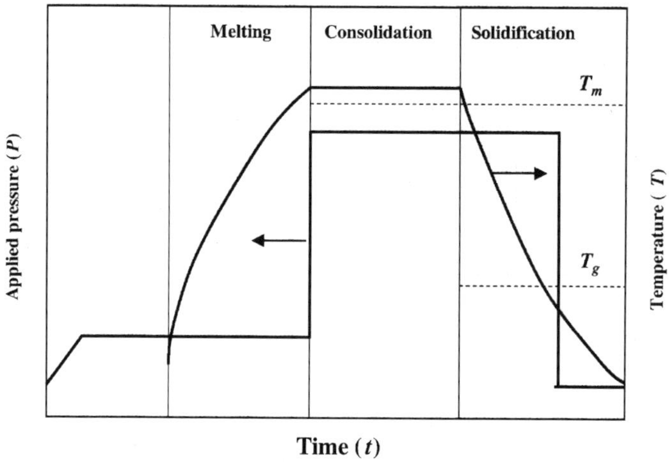

FIGURE 8.11 Heating and pressurizing cycles.

rangement where tension springs are used to tighten the commingled tows before being placed inside a compression molding press to make a flat plate.

Commingled tows, microbraided tows or tows with clinging powder can also be woven into fabrics, or braided into preforms before molding.

Subsequent to the preliminary assembly, the assembly of the dry fibers is placed in a compression molding machine for final molding. The molding parameters consist of application of heat and pressure for a certain amount of time. The general schedule for this application is shown in Figure 8.11, where the temperature needs to be increased before the pressure is applied. This allows the resin to be liquefied and to flow through the interstices to wet the fibers before the liquid resin is squeezed out for its minimization. The pressure is also maintained after the temperature is reduced from the maximum temperature. This prevents the fiber network from springing back before the resin solidifies.

There are two distinct approaches to making shapes from thermoplastic composite materials: preimpregnate the fibers with the resin and then mold into shape, or arrange the reinforcement and resin together in the mold and then consolidate and impregnate simultaneously.

- *Impregnation after shaping.* In film stacking technology, layers of reinforcement fiber are interleaved with layers of film and the whole is consolidated into simple shapes. The principal advantage of this approach, especially for drapable products such as co-woven or hybridized fiber is the potential for forming a complex shape by hand and subsequently consolidating. The most obvious disadvantage of this approach is that it requires a relatively protracted molding cycle in order to carry out the impregnation stage and that impregnation stage contributes to product quality in two ways: by defining the microstructure of the molding, and by establishing the interface between the resin and the fiber on which the service properties of the composite depend.

- *Preimpregnated product forms*. Each product form has its own particular advantages: impregnated woven fabrics are most appropriate for thin skinned structures; woven single tow tapes, including tied uniaxial materials, offer modest drapability and can be made as very broad products. Uniaxial tapes are the most versatile product especially suitable for designing anisotropic composites.

Table 8.3 shows a comparison of strengths of samples made from the two types of products. It appears that the products relying on a post impregnation technology do not generate the full potential strength of the fiber reinforcement. Cogswell [3] attributed this difference to fiber attrition during the molding impregnation stage where the forces necessary to squeeze the resin into the spaces between the fibers also force the fibers together. By contrast, in preimpregnated products each fiber is lubricated by a protective coating of viscous polymer.

4.1.1. Heat Transfer-Melting

As discussed in Chapter 1, in order to make a composite, the resin has to wet the fibers. To do this, the resin has to have low viscosity (in liquid

TABLE 8.3 Properties as a Function of Impregnation Route [3].

	Axial Flexural Strength (MPa)	Short Beam Shear Strength (MPa)	Impact Energy 2 mm Sheet (J)
Preimpregnated products			
Cross plied uniaxial	907	76	23
Woven single tow	929	68	23
Woven fabric	1052	80	29
Impregnation after shaping			
Co-woven fibers	782	60	13
Film stacked	680	67	9
Powder coated fabric	545	54	

form). For PMC in the form of tapes, the resin has already wetted the fibers. As such the need for the resin to be liquid during the final fabrication stage is not essential as far as fiber wetting is concerned (liquefaction is still necessary to fill up all empty space inside the material to prevent voids).

For PMC in other forms, melting of the resin is necessary to wet the fibers. In order to determine the temperature and the energy required for melting the resin, differential scanning calorimeter (DSC) tests can be used. Figure 8.12 shows the DSC curve for amorphous PEEK. Figure 8.13 shows a typical DSC scan for a thermoplastic composite (PEKK).

The DSC curve contains a few features.

- First the ordinate shows the energy, positive is up and negative is down. Negative (endothermic) means that energy is entered into the material system, and positive (exothermic) means energy is released from the system. Along the curve, the positive and negative peaks can be interpreted along this way.
- Figure 8.13 contains two curves. The lower curve represents the heating and the upper curve represents the subsequent cooling.

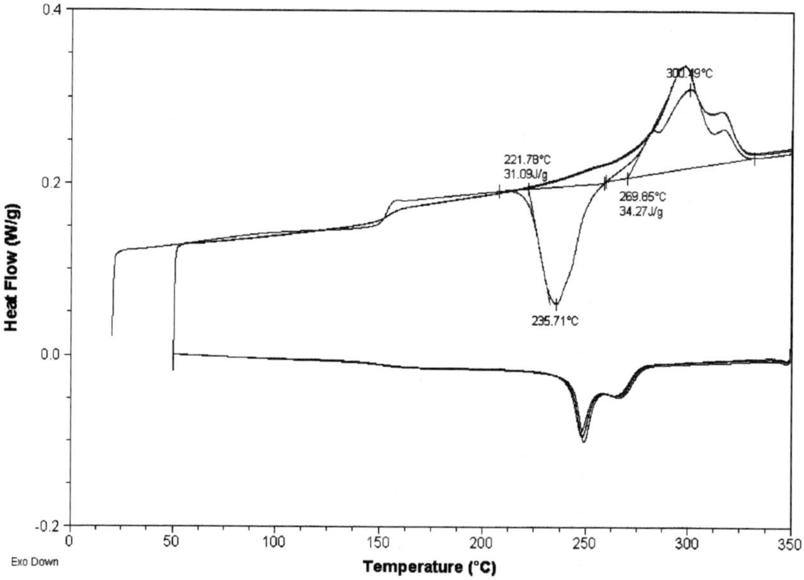

FIGURE 8.12 DSC curve for PEEK (DS) resin.

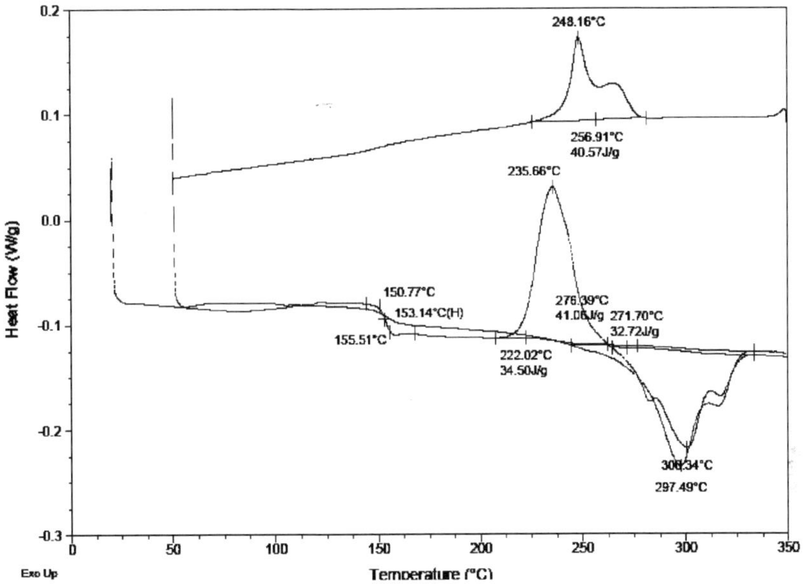

FIGURE 8.13 Typical DSC scan for carbon/PEKK composite.

- For the heating curve, as the temperature of the test is increased, there is some change in the slope of the curve indicating that energy is needed to be entered into the system to increase the temperature of the material. This is due to the heat capacity of the material according to the equation: $Q = mC_p\Delta T$ where m is the mass of the material, C_p is the heat capacity of the material and ΔT is the temperature change.
- There is a jump in the curve and a change in the slope of the curve at about 150°C. This change indicates a change in microstructure of the materials. The jump indicates that additional energy was entered into the system to make the change. The change in slope indicates that the heat capacity of the material has changed at that temperature. This temperature is called the glass transition temperature T_g, because upon heating past that temperature, the material changes from a glassy state to a viscous state. Each of these states has a different heat capacity. In this case, the glass transition temperature is determined to be 153°C. The material becomes soft above the glass transition temperature, and it is usually recommended to use the material at a temperature of about T_g –30°C.

- There is an exothermic peak beginning at about 222°C. Heat is coming out of the system. The reason for this emission of heat is due to crystallization of the material. When a material is crystallized, its microstructure becomes ordered. Some segments of the molecular structures of the thermoplastic polymer are ordered. Figure 8.14 shows the partially ordered molecular configuration. Ordering of the structure requires some degree of connection, similar to chemical bonding discussed in Chapter 2. When a bond or some arrangement is formed, heat is coming out of the system and we have an exothermic situation. The area under the curve corresponds to the heat of crystallization.
- There is an endothermic peak at about 300°C. This corresponds to the heat of melting. Energy is required to enter into the system to break the bonds between the molecules for melting to occur.
- The upper part of the curve represents the cooling of the sample after first heating. There is an exothermic peak at 245°C, which indicates that the material crystallizes around this temperature. Slow cooling facilitates crystallization.

It is noted that crystallization occurs during both heating and cooling. The reason for crystallization to occur during heating is because when the molecules are loose, they can be rearranged when the temperature is not too high. In order to bring the temperature of the material to the melting range, heat is normally used and conduction is the normal process for heat transfer.

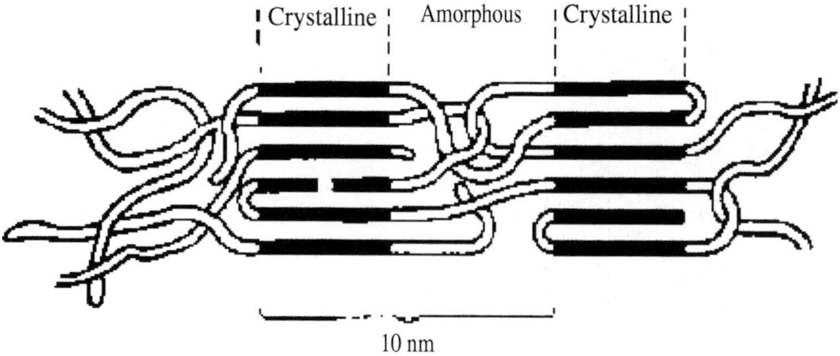

FIGURE 8.14 Configuration of partially ordered molecule.

4.1.1.1. *Heat Transfer During Molding*

Heat transfer is perhaps the most critical parameter in the processing of thermoplastic composites. Temperature distribution affects the quality and microstructure of the product in that flow, compaction, melting, and crystallization are thermally controlled processes.

Before consolidation, thermoplastic composites must be heated in order to melt the matrix resin and bond the surfaces. If the composite does not contain volatiles and solvents in its matrix, then rapid heating is preferred for short processing cycles. Several methods can be used to melt thermoplastic rapidly. Hot shoes, infrared heating, and focused laser heating are preferred in processes such as filament winding. Because of the high energy flux of a laser, the exposed surface can be melted in a few seconds, but it can be overheated and degraded if accurate control devices are not used. Resistance heating that utilizes the conductivity of carbon fibers, infrared radiation heating, and conduction heating by contacting hot rollers and shoes can be used to provide an adequate melting level for processing the composite.

The equation governing the heat transfer in a composite can be written as:

$$\rho c_p \frac{\partial T}{\partial t} = \frac{\partial}{\partial x}\left(K_{xx}\frac{\partial T}{\partial x}\right) + \frac{\partial}{\partial y}\left(K_{yy}\frac{\partial T}{\partial y}\right) + \frac{\partial}{\partial z}\left(K_{zz}\frac{\partial T}{\partial z}\right) + H \qquad (8.2)$$

where,

ρ = the density of the composite material
C_p = the specific heat of the composite material
T = the temperature
t = the time
K_{ii} = coefficients of thermal conductivity
H = the heat source

For the vast majority of commercial thermoplastic materials, no chemical reaction is involved during processing. Assume that the total heat of crystallization (H) of the polymer is negligible when compared to the heat energy released by the heating platens. Also assume that heat loss or gain through the thickness of the composite by convection and radiation is small since the thickness is much less than the length and width of the composite laminate. This reduces the problem to that of a one-dimensional heat conduction problem across the thickness of the composite as:

$$K_z \frac{\partial^2 T}{\partial z^2} = \rho C_p \frac{\partial T}{\partial t} \qquad (8.3)$$

or

$$\frac{\partial^2 T}{\partial z^2} = \frac{1}{\alpha} \frac{\partial T}{\partial t} \qquad (8.4)$$

where

K_z = the heat conductivity along the thickness direction of the composite
ρ = the density of the composite
C_p = the heat capacity
α = heat diffusivity (= $K/\rho C_p$)
T = the temperature
t = the time

The thermal properties of PEEK and carbon fiber are shown in Table 8.4 [1]. Because heat is assumed to be conducted perpendicularly across the fibers, the value for transverse heat conductivity is shown. The axial conductivity is approximately 10 times as great as the transverse conductivity.

Thermal properties of the composite are obtained by averaging the values for the neat resin and carbon fiber as:

$$\frac{1}{K_c} = \frac{V_f}{K_f} + \frac{V_m}{K_m} \qquad (8.5)$$

where

Kc = the thermal conductivity of the composite.
K_f = the thermal conductivity of the fiber.
K_m = the thermal conductivity of the matrix.
V_f, V_m = volume fraction of the fiber and matrix, respectively.

Similarly, the heat capacity of a composite can be obtained by mass averaged heat capacity of the matrix and fiber, as:

$$C_{pc} = C_{pf} m_f + C_{pm} m_m + C_{pf} m_f + C_{pm} (1 - m_f) \qquad (8.6)$$

where

C_{pc} = the heat capacity of the composite.
C_{pm} = the heat capacity of the matrix.
C_{pf} = the heat capacity of the fiber.
m_f, m_m = mass fraction of the fiber and matrix, respectively.

TABLE 8.4 Physical Properties of PEEK, Carbon Fiber (transverse) and APC2 [1].

Material	Density (g/cm^3)	Resin Weight (%)	Resin (vol.%)	Heat Capacity J/(g·K)	Thermal Conductivity J/(m·K·s)	Thermal Diffusivity cm^2/sec
PEEK	1.26	100	100	1.339	0.251	1.49×10^{-3}
Carbon fiber (transverse)	1.79			1.255	0.427	1.9×10^{-3}
APC2	1.53	40	48	1.297	0.318	1.60×10^{-3}

Assuming a void-free composite, the density of a composite can be derived as the volume average of the respective densities of the matrix and fiber, as:

$$\rho_c = \rho_f V_f + \rho_m (1 - V_f) \tag{8.7}$$

where ρ_c, ρ_f, ρ_m are density of composite, fiber, and matrix, respectively.

Finally, the thermal diffusivity of the composite can be calculated from the heat conductivity, heat capacity and density of the composite as:

$$\alpha_c = \frac{K_c}{\rho_c C_{pc}} \tag{8.8}$$

Consider the case that a flat plate of composite is to be made. Both faces of the fiber beds are in contact with the surfaces of the mold. It is reasonable to assume that the temperatures at both surfaces of the sample are the same at all times. This can be considered as the case of a plate of thickness $2L$ with initial temperature T_o and is subject to a surface temperature of T_∞ for time $t > 0$. Figure 8.15 shows the schematic of the arrangement.

The temperature as a function of time and position for the situation where the plate's surfaces are heated at a constant rate, and the origin is

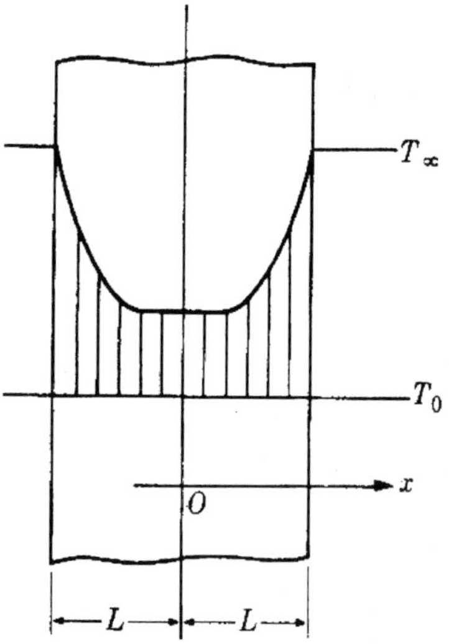

FIGURE 8.15 Schematic for the heat transfer of a flat plate.

defined at the lower surface of the plate with thickness $2L$ can be written as [5]

$$\frac{T(x,t)-T_\infty}{T_o-T_\infty}=4\sum_{n=0}^{\infty}\frac{(-1)^n}{\pi(2n+1)}\exp\left(-\frac{\alpha(2n+1)^2\pi^2 t}{4L^2}\right)\cos\left(\frac{(2n+1)\pi x}{2L}\right) \tag{8.9}$$

Equation (8.5) can be rearranged to be:

$$T(x,t)=$$

$$T_\infty+4(T_o-T_\infty)\sum_{n=0}^{\infty}\frac{(-1)^n}{\pi(2n+1)}\exp\left(-\frac{\alpha(2n+1)^2\pi^2 t}{4L^2}\right)\cos\left(\frac{(2n+1)\pi x}{2L}\right) \tag{8.10}$$

In a situation where the temperature of the surrounding environment is a function of time, for example, the case where the surrounding liquid is being cooled, $T\infty$ can be replaced by this temperature. For the case where $T_\infty = T_o + K_t$ where K is the heating rate, Equation (8.10) can be written as:

$$T(x,t)=$$

$$T_o+Kt-4Kt\sum_{n=0}^{\infty}\frac{(-1)^n}{\pi(2n+1)}\exp\left(-\frac{\alpha(2n+1)^2\pi^2 t}{4L^2}\right)\cos\left(\frac{(2n+1)\pi x}{2L}\right)$$

or

$$T(x,t)=$$

$$T_o-Kt\left\{4\sum_{n=0}^{\infty}\frac{(-1)^n}{\pi(2n+1)}\exp\left[-\frac{\alpha(2n+1)^2\pi^2 t}{4L^2}\right]\cos\left[\frac{(2n+1)\pi x}{2L}\right]+1\right\} \tag{8.11}$$

where

K = heating rate (°C/s)
L = plate half thickness (cm)
t = time (s)
T = temperature (°C)
T_i = initial temperature (°C)
x = position along the thickness direction (cm)
α = thermal diffusivity (cm^2/s)

At the center of the plate, $x = 0$ and the above equation simplifies to:

$$T(x,t) = T_o + Kt\left\{4\sum_{n=0}^{\infty}\frac{(-1)^n}{\pi(2n+1)}\exp\left[-\frac{\alpha(2n+1)^2\pi^2 t}{4L^2}\right]+1\right\} \quad (8.12)$$

The value of the thermal diffusivity of thermoplastic composites is about 1.5×10^{-3} cm²/sec (Table 8.4). Figure 8.16 shows the variation of temperature as a function of space and time. Figure 8.17 shows the variation over time of the temperature at the center of the plate, for different laminate thicknesses.

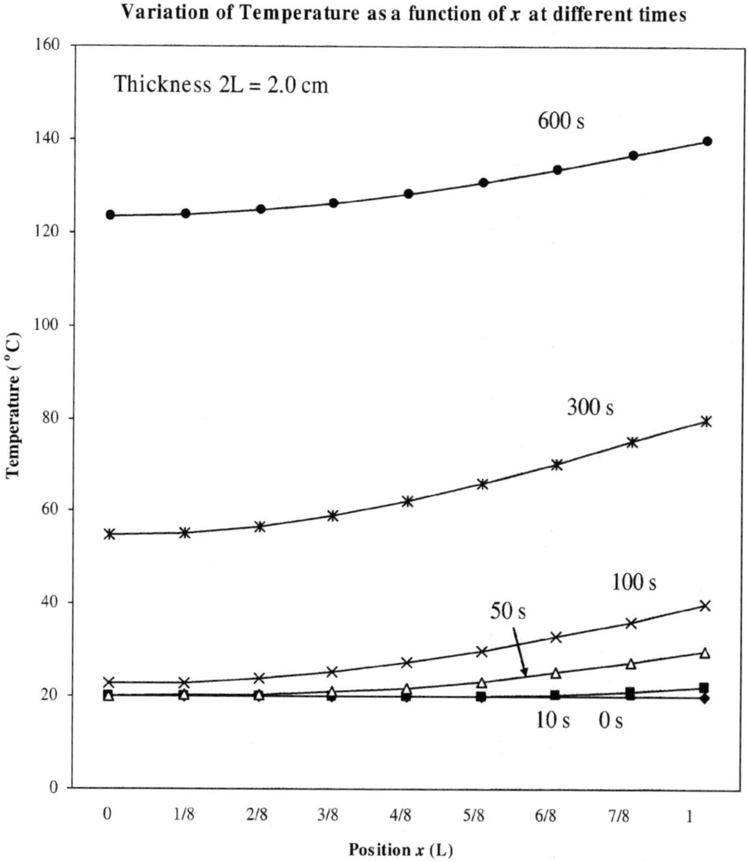

FIGURE 8.16 Variation of temperature as a function of space and time, ($T_i = 20$°C, $K =$ 0.2°C/second, $\alpha = 1.5 \times 10^{-3}$ cm²/sec).

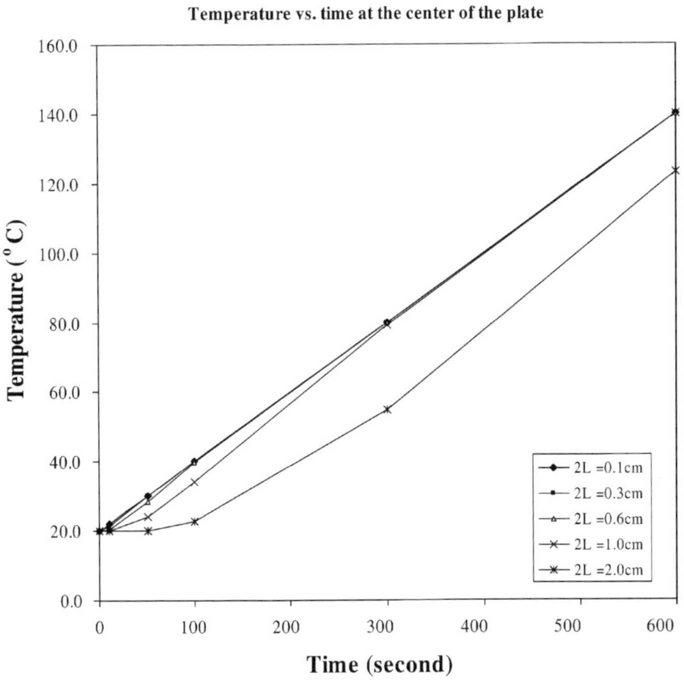

FIGURE 8.17 Variation versus time of temperature at the center of the plate for different plate thicknesses ($T_i = 20°C$, $K = 0.2°C/second$, $\alpha = 1.5 \times 10^{-3}$ cm^2/sec).

For thin plates, it is reasonable to assume that the temperature in the plate is independent of the space variable and the temperature can be expressed as:

$$T = T_i + Kt \tag{8.13}$$

It can be seen from Figure 8.17 that as the plate becomes thinner, the curve approaches the relation shown in Equation (8.13).

4.1.2. Bulk Consolidation

The consolidation of the fiber beds can be considered to consist of eliminating the empty space between the layers and eliminating the waviness of the fibers. Bulk consolidation is the phenomenon during which the individual layers in the plate are brought into intimate contact and the free space inside the composite is eliminated. The mechanism of bulk consolidation is different for each type of PCM. For tapes, the PCM is already fully impregnated, and during consolidation, spatial gaps only occur between plies because their surfaces are rough or uneven. An equation that can be used to model the bulk consolidation and analyze experimental data is:

$$P = B_y \ln\left(\frac{h_f}{h_i}\right) \tag{8.14}$$

where,

P = applied pressure
B_y = bulk modulus
h_i = composite height at beginning of resin permeation
h_o = final composite thickness

The range of bulk consolidation modulus was determined to be from approximately 1.21–8.58 MPa for APC2. The values depend on the number of layers, pressure, and temperature. The bulk consolidation of flexible towpregs is different from tapes. First, a flexible towpreg is not perfectly impregnated with the matrix. The flexibility of the towpreg makes intimate contact between the prepregs easy and immediate. Even though intimate contact is accomplished in this towpreg, the incomplete impregnation of the matrix plays a major role in bulk consolidation. Therefore, the bulk consolidation of flexible towpregs needs to include resin impregnation rather than intimate contact and can be explained by the resin flow model. One might say that bulk consolidation is the key concern when processing flexible towpregs and commingled tows.

During consolidation of the composite, the applied pressure decreases the thickness of the composite. Heat is applied, either simultaneously with the increase in pressure or shortly after bulk consolidation, to melt the resin. Resin melts and flows within the fiber network while consolidation continues through the second stage where fibers are straightened. Figure 8.18 illustrates the decrease in thickness as a function of consolidated time and indicates that several phenomena occur.

When the pressure is applied at the onset of consolidation, it is concentrated on developing intimate contact between the plies. Figure 8.19(a) shows the consolidation of a number of layers of a composite. Figure 8.19(b) shows the definition of A_{ic} (the blackened region) and how it increases as consolidation progresses. As consolidation continues, complete intimate contact is developed and A_{ic} approaches A_T. Eventually the resin is squeezed out of the plies and the fiber network carries a larger portion of the pressure.

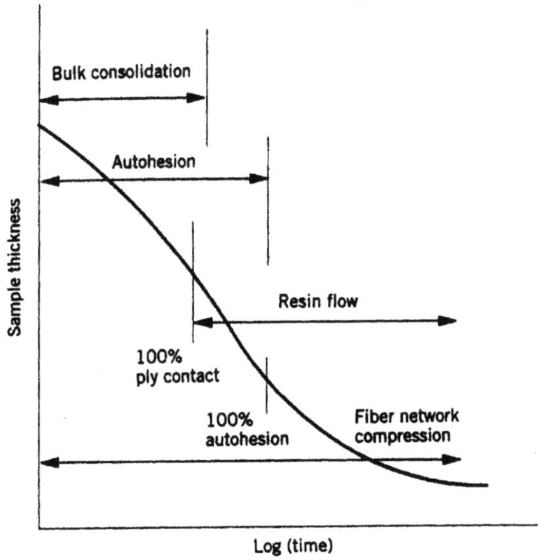

FIGURE 8.18 Decrease in thickness of a composite during consolidation (reproduced from "The processing science of thermoplastic composites," by J. D. Muzzy and J. S. Colton, in *Advanced Composites Manufacturing,* T.G. Gutowski ed., with permission from Wiley Interscience).

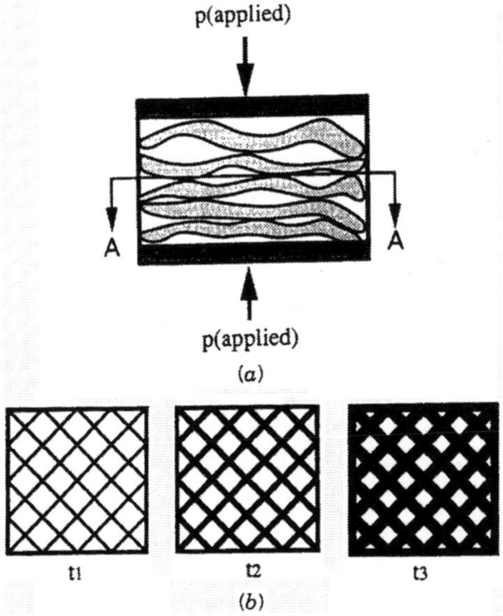

FIGURE 8.19 Different stages of fiber contact during compression (reproduced from "The processing science of thermoplastic composites," by J. D. Muzzy and J. S. Colton, in *Advanced Composites Manufacturing* T.G. Gutowski ed., with permission from Wiley Interscience).

4.1.2.1. Elasticity of Fiber Networks

During consolidation, the fiber volume fraction increases and the fiber network begins to take up an increasing amount of applied load. In the extreme case, the fiber network carries all the load and the resin pressure drops to zero. Therefore, the elastic behavior of the network must be taken into account for a correct prediction of consolidation. The relation between the compression pressure and volume fraction was presented in Chapter 3, repeated here as:

$$\sigma = \frac{3\pi E}{\beta^4} \frac{1-\sqrt{\dfrac{V_f}{V_o}}}{\left(\sqrt{\dfrac{V_a}{V_f}}-1\right)^4} \tag{8.15}$$

where

E = Young's modulus of the fiber.
β = the geometric parameter (= L/a in Figure 3.19).
V_a, V_o, V_f = maximum allowable fiber volume fraction, initial fiber volume fraction, and current fiber volume fraction, respectively.

When a thermoplastic PMC lay-up is compressed at above its melting temperature, the pressure brings the PMCs into intimate contact and eliminates any free space between the plies. This is accomplished by the lateral squeezing flow of the resin on the prepreg material and is completed when the extent of spatial gaps between the plies is minimized.

4.1.3. Resin Flow

Four basic flow patterns have been identified during the processing of PCMs into product forms. These are: resin percolation through and along the fibers, transverse flow, intraply shearing flow along the fiber direction, and interply slip cooperative flow.

Percolation. Figure 8.20 shows the schematic for percolation. Percolation of resin through the reinforcing fibers plays an important role in all processing operations. Such resin flow heals any flaws in the structure and allows the bonding of different layers of preimpregnated tape.

As an approximate description of resin flow, a composite material is considered to be a porous medium and the resin flow can be described by

Darcy's law, which states that the flow rate per unit area in a certain direction q is directly proportional to the resin pressure gradient in that direction.

The permeability of the fiber network depends on the fiber diameter, the porosity, and the network geometry, as given by the Kozeny-Carman equation. Permeability is not a physical property and depends on fiber network geometry, fiber shape, and fiber properties. Therefore, the permeability is an empirical constant that is determined by measuring the pressure drop and the flow rate through the fiber network. The Kozeny-Carman relation, which is one of the most widely accepted derivations for calculating permeability as a function of fiber volume fraction, treats the porous medium as a bundle of parallel tubes. The Kozeny-Carman relation has an important limitation in its application to composite processing. It assumes that the porous medium is macroscopically isotropic and that the pore size and the fiber distribution are uniform. Of course, this is not true for aligned fiber composites. For example, any irregular packing of the fibers results in an increased flow rate and a decrease in the Kozeny constant in the fiber direction.

The nonuniform packing of the fibers results in favored, enlarged flow paths and increases the flow rate of the fiber beds. For high fiber volume fractions, there is only a small possibility that the fibers will be distributed in an irregular manner, while for low fiber volume fractions, nonuniform packing of the fiber bed is prevalent.

Transverse flow. One flow process that allows movement of the fibers is a simple transverse flow (Figure 8. 21). This flow mechanism allows

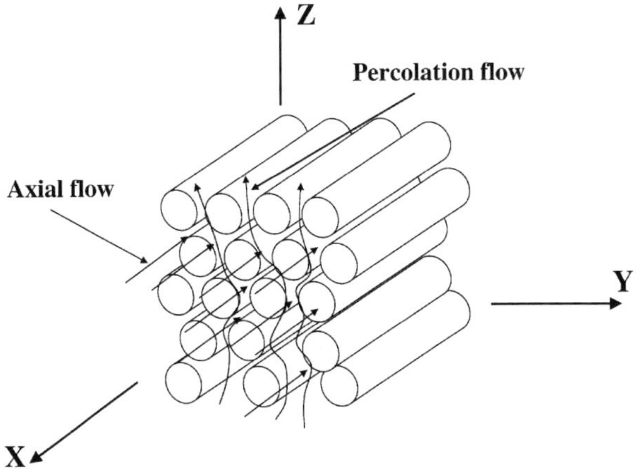

FIGURE 8.20 Flow by percolation.

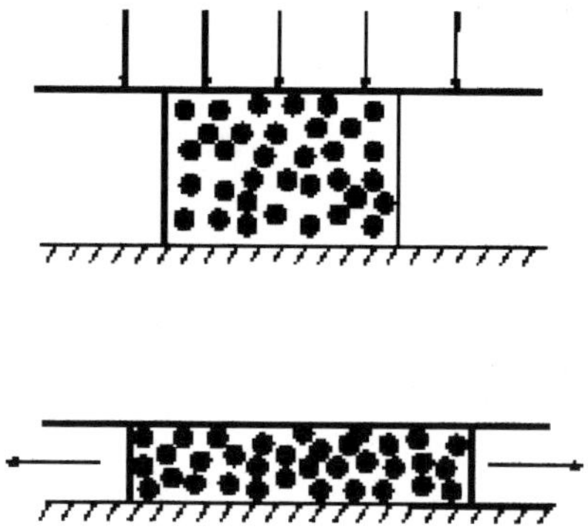

FIGURE 8.21 Transverse flow.

for healing of small gaps in the material, for example, those that occur in structures woven from impregnated single tows.

Axial shearing flow. In this flow, shearing along the fiber direction is observed in any flow process that alters the orientation of one layer of fibers relative to another.

Interply slip. Interply slip is the situation where layers of the fibers slide relative to each other. Figure 8.22 show the schematic of the two types of shearing flow.

Figure 8.23 shows the areas where the flows are applicable.

4.1.3.1. Resin Flow for Manufacturing Using Flexible Tows

For tapes, the resin has already wetted the fibers before compression molding takes place. The flow of resin during compression molding is required mainly to fill up the space between the tapes. For the case of flexi-

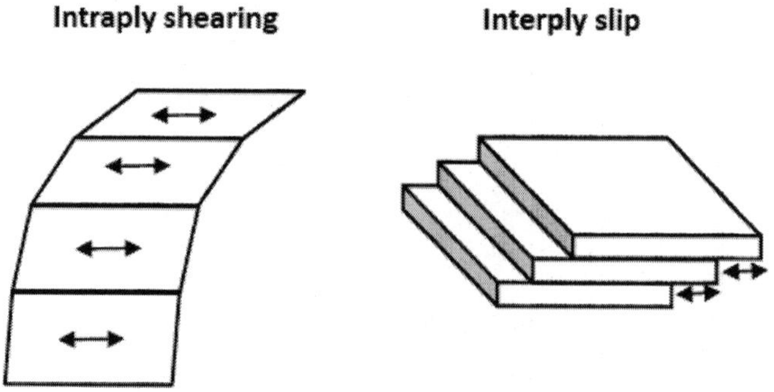

FIGURE 8.22 Shearing flows.

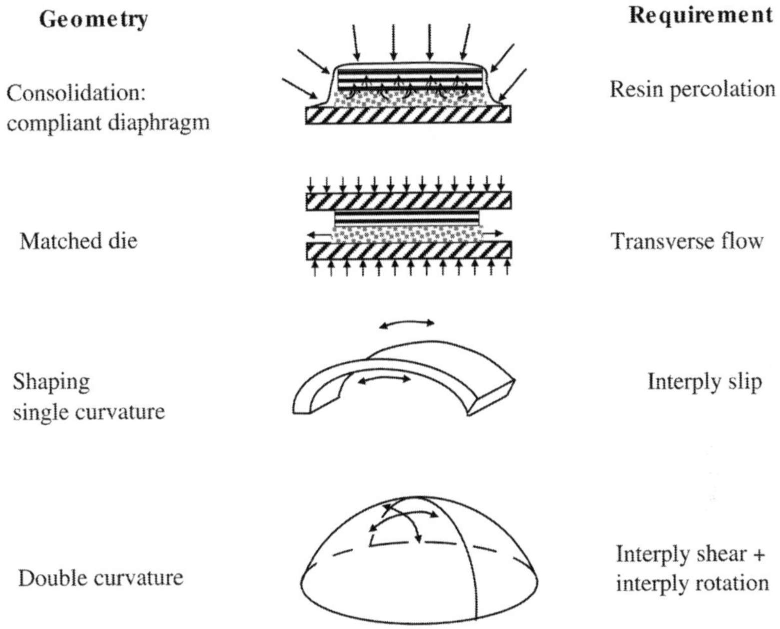

FIGURE 8.23 Deformation processes [4].

ble tows such as tows with clinging powder, commingled tows, or with fabric/film sandwich, apart from filling up the space between the tows, resin is also required to wet the fibers since the fibers are not wetted in the PMC form. An idealized configuration of a cross section of a commingled tow is shown in Figure 8.24. It can be seen that once the resin melts, it has to flow around the fibers to wet them, apart from the need to fill the empty space.

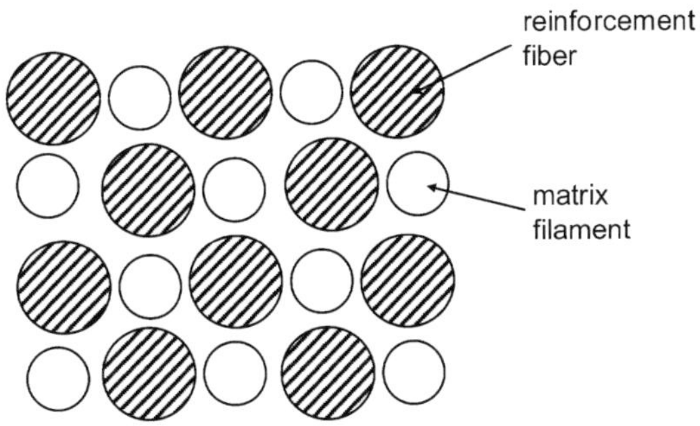

FIGURE 8.24 An ideal cross section of a commingled tow.

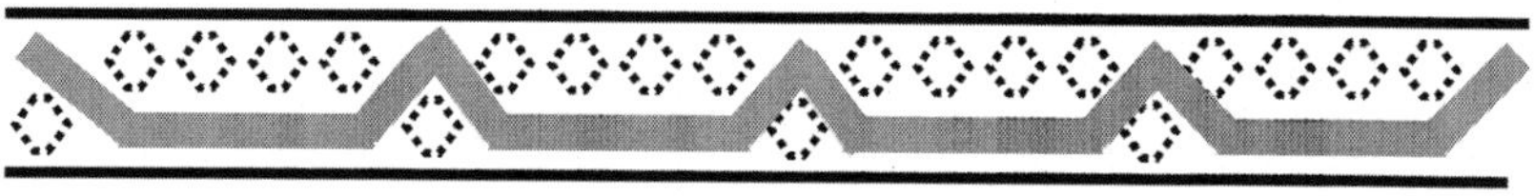

FIGURE 8.25 Cross section of a fabric/film sandwich.

Figure 8.6 shows an arrangement for the case of fiber tows with clinging powder. Again, the melted resin has to flow along the length of the fibers to wet them, in addition to the need to fill the empty space.

Figure 8.25 shows the cross section of a fabric/film sandwich. The arrangement in this figure is for the case of 5-harness satin, but the requirement for the melted resin to flow into the empty space applies to other types of weave also.

4.1.3.2. Resin Flow Analysis Model

For situations involving tapes, Loos and coworkers [6] developed an interplay intimate contact model by combining a prepreg surface roughness model with a resin flow analysis. A surface topology machine was used to measure the waviness or roughness of the resin rich prepreg surfaces. The measurements were taken across the width of the prepreg sheet, perpendicular to the fibers. From the topology measurements it appears that the surface waviness (roughness) of the prepreg is somewhat sinusoidal. For modeling purposes, the prepreg surface topology was represented by a series of rectangular elements of height h_o, with b_o and spacing w_o, as shown in Figure 8.26. The magnitudes of b_o and w_o are equal.

Application of the consolidation pressure resulted in deformation of the rectangular elements, the deformation of the rectangular elements

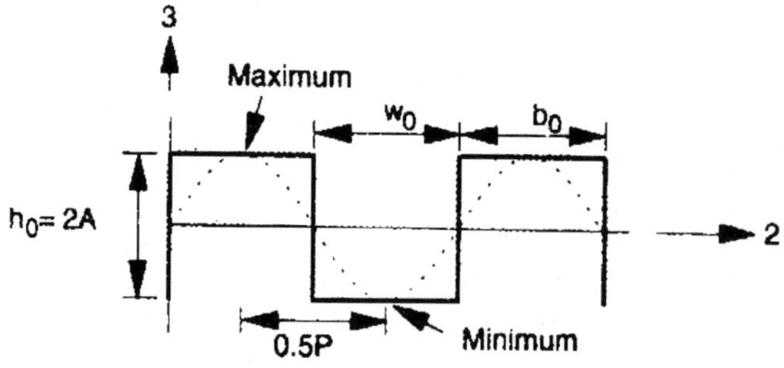

FIGURE 8.26 Prepreg surface roughness model. *A* and *P* are the amplitude and the period of the sinusoidal wave [6].

represented deformation and flow of the resin rich prepreg surfaces and was modeled as a squeezing flow between two rigid parallel plates. A formulation using the zero-shear-rate viscosity of the neat resin, which is a function of temperature, was derived to predict the interplay intimate contact achievement.

The degree of intimate contact (DIC) is defined as the ratio of the noncontact area over the total interface as:

$$DIC = \frac{b}{w+b} \tag{8.16}$$

The samples were processed at different combinations of temperature, pressure and time. Optical microscopy was used to take the degree of intimate contact from the unidirectional samples.

4.1.4. Autohesion [9]

After the prepregs coalesce, matrix interfaces can adhere to one another by the molecular diffusion of the matrix. This healing process is called autohesion. Autohesion is defined as the diffusion of chain segments across an interface, leading to the elimination of the interfaces. Conversely, adhesion is due to the chemical bonding of two dissimilar materials at an interface.

After the spatial gaps in the laminate are removed, the diffusion of polymer chains takes place at the contact surface. When two thermoplastic parts are brought in contact above the glass transition temperature of the resin, interdiffusion of polymer chains takes place at the contact surface that progressively heals the interface. The motion of a chain in an amorphous material has been modeled by reptation theory, which was developed by De Gennes [7]. This theory postulates that the motion of the polymer chains can be considered to be constrained in a tube that represents the steric effects applied by the other chains in close proximity. De Gennes made the following assumptions:

- The chain is moving in a fixed isothermal network and therefore is not allowed to cross any obstacles.
- The chain is able to move between obstacles by a snake-like motion.

Consequently the chain is confined in a tube of length *L*, which represents the steric constraints exerted by the network. As the chain moves in the tube because of Brownian motion, its extremities exit the tube. The chain ends are then free to move. The reptation theory predicts that the

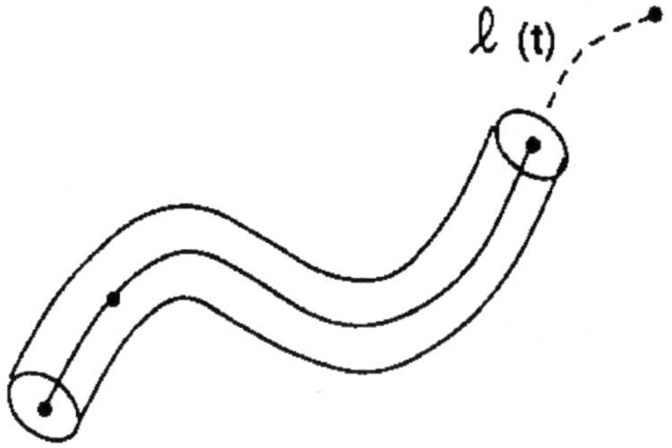

FIGURE 8.27 Minor chain length.

length, l, of the chain ends (called minor chains) varies with the square root of time (Figure 8.27). The reptation time T_r, is defined as the time at which the chain has totally exited its original tube ($l = L$). Therefore, the following relation holds:

$$\frac{l}{L} = \left[\frac{t}{T_r}\right]^{0.5} \tag{8.17}$$

If one is interested in the interdiffusion of chain across an interface, it is possible to define the average interpenetration distance of the chain across the interface, y, which varies as the square root of the minor chain length (Figure 8.28). For the reptation time, $x = x\infty$ which is equal to the radius of gyration of the chain. The following relation is derived:

$$\frac{x}{x_\infty} = \left[\frac{t}{T_r}\right]^{0.25} \tag{8.18}$$

As the chain motion grows, the penetration length of the chain across the interface increases and the fracture toughness and the tensile strength of the interface are enhanced. Kim and Wool [8] showed that mechanical properties such as stress, modulus and impact energy, are related to time,

FIGURE 8.28 Interpenetration distance of a chain across the interface.

temperature, and molecular weight, and developed the following relations:

$$\frac{S}{S_\infty} = \left(\frac{t}{T_r}\right)^{1/4} \tag{8.19}$$

$$\frac{G}{G_\infty} = \left(\frac{t}{T_r}\right)^{1/2} \tag{8.20}$$

where,

S_∞ and G_∞ = the tensile strength and fracture energy of the fully bonded interface.

T_r = the time to achieve a cohesive surface.

These relations are valid for the isothermal case and for a process time t less than T_r.

Dara and Loos defined the degree of autohesion by comparing the bonding strength of an interface to the strength of a completely bonded interface as:

$$D_{Au} = \frac{S}{S_\infty} \tag{8.21}$$

where D_{Au} is the degree of autohesion. Once D_{Au} reaches unity, there is no increase in the bond strength because any failure is entirely cohesive at this stage. For a reference temperature of 210°C and a zero shear rate viscosity of about 30 MPa·s, the contact time with $D_{Au} = 1$ can be estimated to be:

$$t_c = 22.9\eta \tag{8.22}$$

where η is viscosity in terms of MPa·s. Therefore, the contact time for perfect autohesion is directly proportional to the viscosity of the polymer.

The above equation shows that the required contact time for good autohesion is very short for PEEK at a normal processing temperature. The melt viscosity of PEEK above 360°C is much less than 1 kPa·s and the corresponding contact time is less than 1 second.

The autohesion time can be ignored for slow processes such as compression molding because the processing time is much greater than the autohesion time. For faster processes, such as filament winding, fiber placement, and pultrusion, the time of contact at high temperature and pressure is much shorter, and may be on the order of magnitude of the time of complete autohesion. In this case, autohesion must be taken into account.

4.1.5. Cooling

After the consolidation or formation of a part, a thermoplastic matrix composite has to be cooled and solidified. The heat transfer during the cooling period can be examined by considering the thermoplastic composite plate as a plate of thickness $2L$ having a uniform initial temperature T_o (Figure 8.29). At time $t = 0$, the plate is plunged suddenly into a bath at the constant temperature T_∞ (similar to the case of quenching). Assuming that the surface temperature of the plaque remains at T_∞ during the whole process of cooling. If one assumes that the amount of heat generated due to crystallization during the cooling process is negligible, then the governing equation for heat transfer can be written as:

$$\frac{\partial \theta}{\partial t} = \alpha \frac{\partial^2 \theta}{\partial z^2} \tag{8.23}$$

With the initial and boundary conditions:

$$\theta(x,0) = \theta_o = T_o - T_\infty$$
$$\frac{\partial \theta(0,t)}{\partial x} = 0 \qquad \theta(L,t) = 0 \tag{8.24}$$

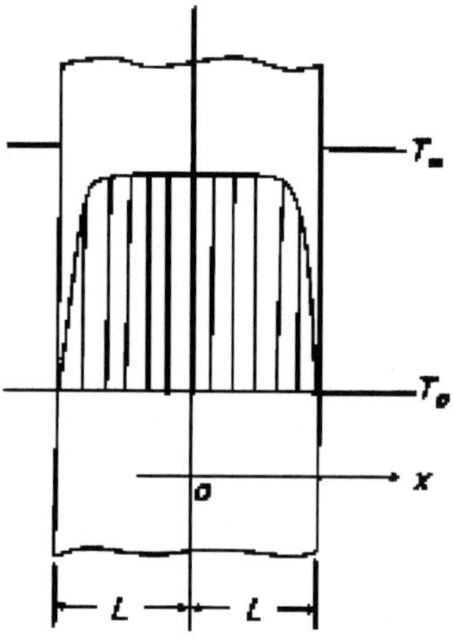

FIGURE 8.29 Plate exposed to constant surface temperature.

where

$$\theta = T - T_\infty \tag{8.25}$$

By using the method of separation of variables, it can be shown that the solution for the above problem can be written as [5]:

$$\frac{T(x,t) - T_\infty}{T_o - T_\infty} = 4\sum_{n=0}^{\infty} \frac{(-1)^n}{\pi(2n+1)} \exp\left(-\frac{\alpha(2n+1)^2\pi^2 t}{4L^2}\right)\cos\left(\frac{(2n+1)\pi x}{2L}\right) \tag{8.26}$$

Equation (8.26) can be rearranged to be:

$$T(x,t) = T_\infty + 4(T_o - T_\infty)\sum_{n=0}^{\infty} \frac{(-1)^n}{\pi(2n+1)} \exp\left(-\frac{\alpha(2n+1)^2\pi^2 t}{4L^2}\right)\cos\left(\frac{(2n+1)\pi x}{2L}\right) \tag{8.27}$$

In a situation where the temperature of the surrounding environment is a function of time, for example, the case where the surrounding liquid is being cooled, T_∞ can be replaced by this temperature. For the case where $T_\infty = T_o - Kt$ where K is the cooling rate, Equation (8.23) can be written as:

$$T(x,t) = T_o - Kt + 4Kt\sum_{n=0}^{\infty} \frac{(-1)^n}{\pi(2n+1)} \exp\left(-\frac{\alpha(2n+1)^2\pi^2 t}{4L^2}\right)\cos\left(\frac{(2n+1)\pi x}{2L}\right)$$

or

$$T(x,t) = T_o + Kt\left\{4\sum_{n=0}^{\infty} \frac{(-1)^n}{\pi(2n+1)} \exp\left[-\frac{\alpha(2n+1)^2\pi^2 t}{4L^2}\right]\cos\left[\frac{(2n+1)\pi x}{2L}\right] - 1\right\} \tag{8.28}$$

Equations (8.26) to (8.28) are the solution for the case where the surface temperature of the plate is kept at a constant temperature T_∞ throughout the cooling process. For the case where there is a convective coefficient h between the surface and the cooling liquid, the initial and boundary conditions Equations (8.24) are modified to be:

$$\theta(x,0) = \theta_o = T_o - T_\infty$$

$$\frac{\partial \theta(0,t)}{\partial x} = 0 \qquad \frac{\partial q(L,t)}{\partial z} + \frac{h}{k}\theta(L,t) = 0 \tag{8.29}$$

The solution for the above problem can be shown to be [5]:

$$\frac{T(z,t) - T_\infty}{T_0 - T_\infty} = 2\sum_{n=1}^{\infty}\left(\frac{\sin \lambda_n L}{\lambda_n L + \sin \lambda_n L \cos \lambda_n L}\right)\exp(-\alpha\lambda_n^2 t)\cos \lambda_n z \tag{8.30}$$

where the zeros of the equation $(\lambda_n L)\sin(\lambda_n L) = (hL/K)\cos(\lambda_n L)$ are the characteristic values.

During the solidification, the maintenance of pressure is required until the temperature of the composite is below its matrix glass transition temperature. This restricts the nucleation of voids within the resin, suppresses the recovery of the fiber network, and enables the composite to maintain the desired dimensions.

Example 8.2

A thermoplastic composite plate mm thick was processed at 350°C. It is quenched in water at 2°C. Assume that the temperature at the surface of the plate remains at 20°C during the course of the cooling process. Plot the variation of the temperature in the plate across its thickness at $t = 5$ seconds, 10 seconds, 50 seconds, 100 and 150 seconds. $\alpha = 1.5 \times 10^{-3}$ cm²/sec.

Equation (8.26) is used to determine the temperature as a function of time and space.

Figure 8.30 shows the temperature variations across the thickness of the laminate at different times for a thin plate (0.1 cm thick). Figure 8.31 shows the similar graphs for a thicker plate (2 cm thick). It can be seen that for a thicker plate, the temperature on the surface drops more quickly than at the center of the plate, even for a slow cooling rate of –0.2°C/second. Figure 8.32 shows the variation of temperature at the center of the plates of different thicknesses. Figures 8.33–8.35 shows similar curves for a cooling rate of –100°C/sec. It can be seen that at this fast cooling rate, the temperature drops quickly and there is a sharp temperature gradient across the thickness for thicker laminates.

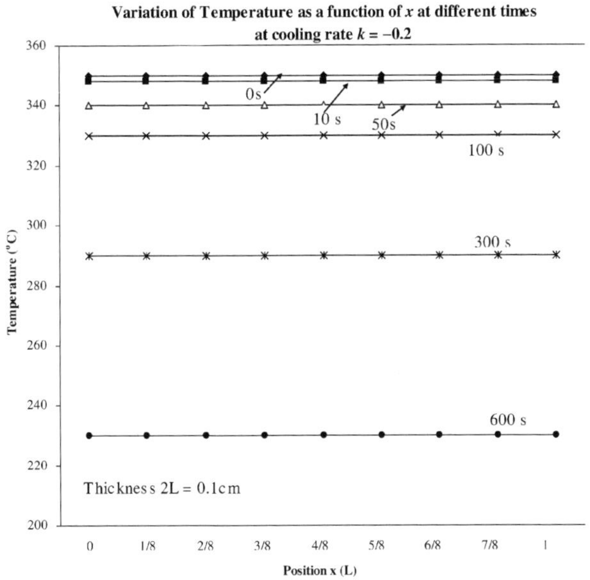

FIGURE 8.30 Temperature variation of a plate subject to cooling rate of −0.2°C/sec, 0.1 cm thick.

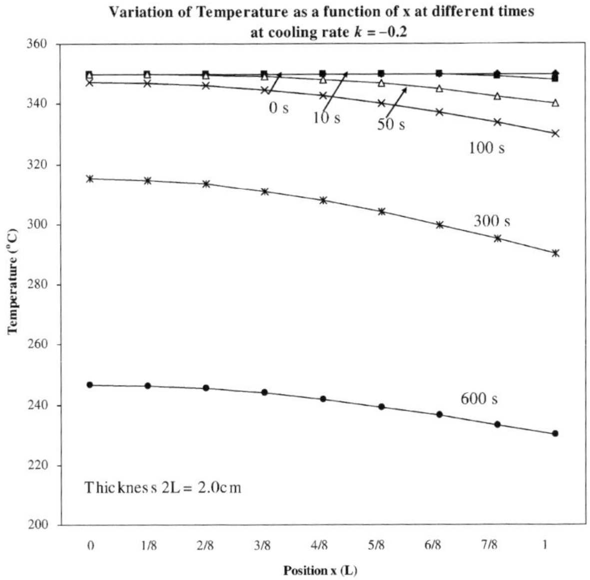

FIGURE 8.31 Temperature variation of a plate subject to cooling rate of −0.2°C/sec, 2.0 cm thick.

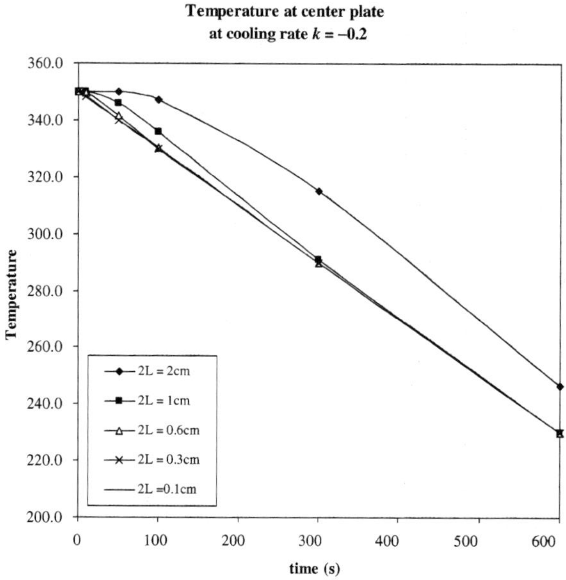

FIGURE 8.32 Temperature at center of laminate as a function of time for plates of different thicknesses. Cooling rate –0.2°C/sec.

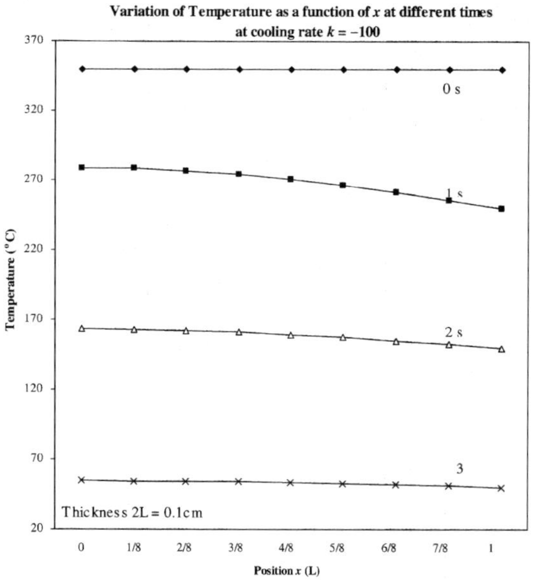

FIGURE 8.33 Temperature variation of a plate subject to cooling rate of –100°C/sec, 0.1 cm thick.

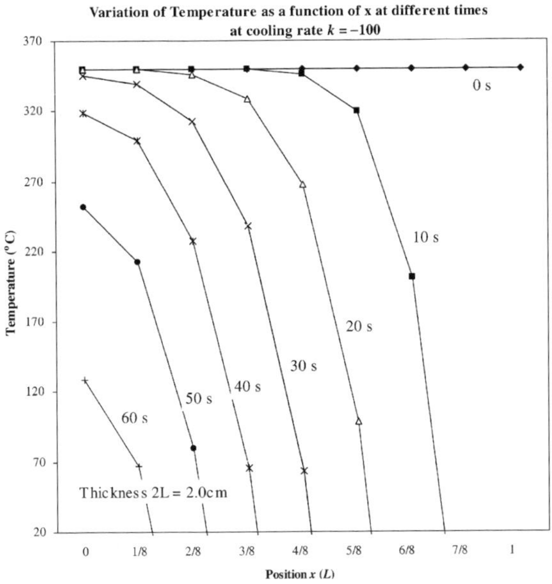

FIGURE 8.34 Temperature variation of a plate subject to cooling rate of –100°C/sec, 2.0 cm thick.

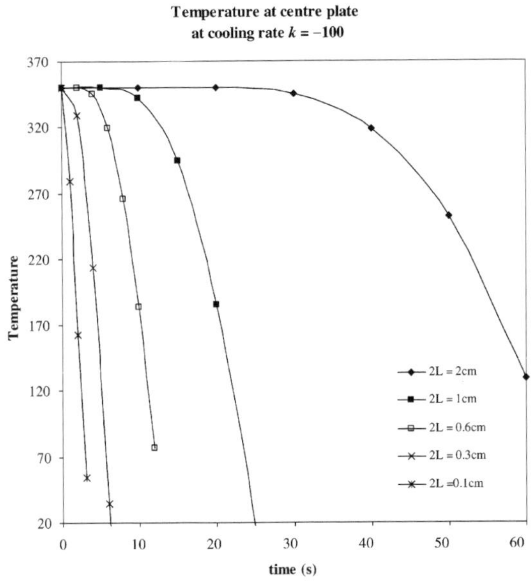

FIGURE 8.35 Temperature at center of laminate as a function of time for plates of different thicknesses. Cooling rate –100°C/sec.

4.1.6. Crystallization

A semi-crystalline thermoplastic crystallizes when an appropriate cooling rate is applied. This crystallization has a significant effect on the mechanical properties and solvent resistance of the composites and results in crystallization shrinkage. In addition, the difference in thermal expansion coefficients between the fiber and the resin can lead to residual stresses and warping.

Fortunately, the residual stresses can be relieved partly during processing if the thermoplastic matrix is maintained near its glass transition temperature (annealed) or the cooling rate is controlled to provide an even temperature profile throughout the part. The degree of crystallinity drops as the cooling rate is increased. Figure 8.36 shows the crystallinity versus cooling rate using the spherulite growth model.

4.1.7. Solidification

If the applied pressure is reduced when the resin is still molten, the composite exhibits a significant elastic recovery. Figure 8.37 illustrates this elastic recovery in terms of thickness change. This recovery cannot be explained by the resin flow model alone. Therefore, it is necessary to introduce the elastic deformation of the fiber network in the model. In

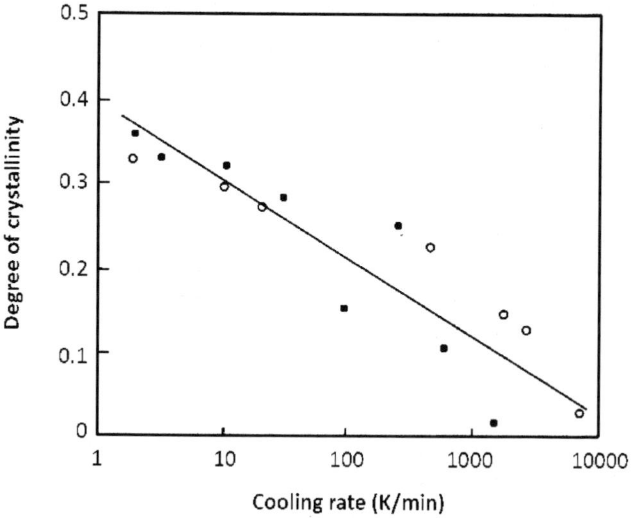

FIGURE 8.36 Variation of degree of crystallinity versus cooling rate for PEEK and APC (carbon/PEEK).

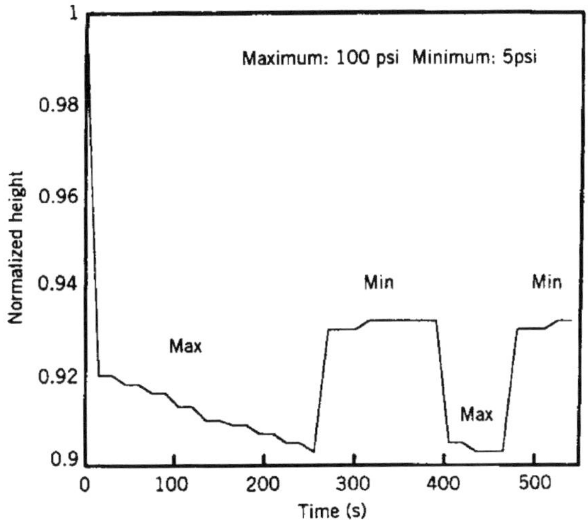

FIGURE 8.37 Elastic recovery of fiber network (reproduced from "The processing science of thermoplastic composites," by J. D. Muzzy and J. S. Colton, in *Advanced Composites Manufacturing* T.G. Gutowski ed., with permission from Wiley Interscience).

many practical situations, the fiber network is compressed to the point that it starts to take up a significant portion of the applied load. This elastic behavior may be more significant if the fiber forms a woven mat or a flexible tow. From the reasoning above, the applied pressure during consolidation $P_{applied}$ can be divided into two components.

$$P_{applied} = P_f v_f \frac{A_{ic}}{A_T} + P_r v_r \frac{A_{ic}}{A_T} \tag{8.31}$$

where

P_f = the pressure supported by the fiber network
V_f = the volume fraction of fibers
A_{ic} = the current area of intimate contact
A_T = the total area of the part
P_r = the pressure supported by the resin
V_r= -thevolume fraction of the resin

4.1.8. Variation of Degree of Crystallinity Across the Thickness of the Part in the Case of Fast Cooling (Quenching)

It can be seen from the above discussion that the cooling rate has an effect on the crystallinity of the material. Faster cooling rate results in smaller crystallinity. In cases where fast cooling rate is used (quenching), the temperature of the surface of the sample can go from about 350°C to room temperature (20°C) in the order of a few seconds. Assuming that the time involved is 3 secs, this gives a cooling rate of about 100°C/sec. Even though the temperature on the outer surface can be cooled quickly, due to the low thermal conductivity of the composite material, the inside of the material cools more slowly. The variation of the temperature as a function of location and time for a plate made of PEEK is shown in Figure 8.38.

The variation of temperature for a similar plate made of carbon/PEEK is shown in Figure 8.39.

The difference in cooling rate at different positions along the thickness of the plate gives rise to different degrees of crystallinity at these positions. Figure 8.40 shows the calculated volume fraction crystallinity for plates made of pure PEEK and carbon/PEEK for a 5-mm-thick plate cooled at 114°C/second.

Thermoplastic resins with higher crystallinity have higher stiffness. The variation in crystallinity along the thickness direction gives rise to changes in stiffness along this direction. Also the outer skin cools first

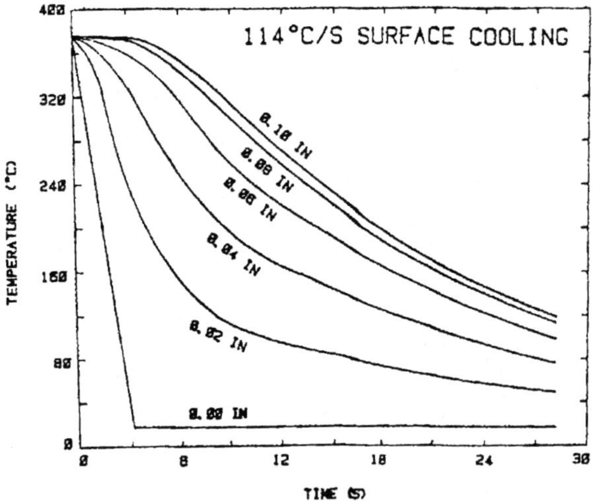

FIGURE 8.38 Predicted temperature as a function of time and position from the surface, for a 5-mm-thick neat PEEK plate with surface cooled at 114°C/second (reproduced from Velisaris C. N., and Seferis J. C. "Heat transfer effects on the processing-structure relationships of polyetheretherketone (PEEK) based composites," *J. Science and Engineering of Composite Materials,* Vol. 1, No. 1, 1988, pp. 13–22, with permission from Freund Publishing Ltd.).

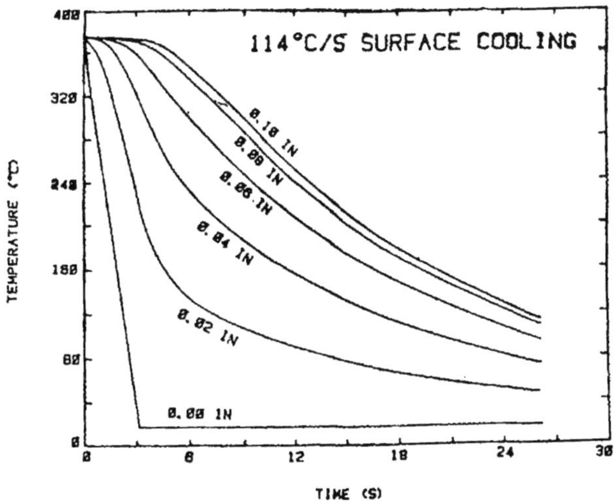

FIGURE 8.39 Predicted temperature as a function of time and position from the surface, for a 5-mm-thick neat carbon/PEEK plate with surface cooled at 114°C/second (reproduced from Velisaris C. N., and Seferis J. C. "Heat transfer effects on the processing-structure relationships of polyetheretherketone (PEEK) based composites," *J. Science and Engineering of Composite Materials,* Vol. 1, No. 1, 1988, pp. 13–22, with permission from Freund Publishing Ltd.).

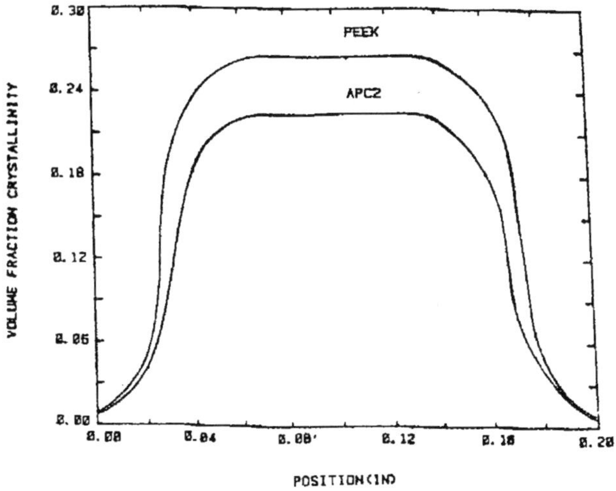

FIGURE 8.40 Calculated volume fraction crystallinity at different positions along the thickness for a 5-mm-thick neat PEEK and carbon/PEEK plates, with surfaces cooled at 114°C/second (reproduced from Chapman T. J., Gillespie J. W., Pipes R. B., Manson J. A. E, and Seferis J. C. "Prediction of process-induced residual stresses in thermoplastic composites," *J. Composite Materials,* Vol. 24, June 1990, pp. 616–643, with permission from Sage Publications).

and therefore becomes solid first. The inner core cools later. Cooling also gives rise to shrinkage. The interaction between the shrinking core and the rigid outer skin gives rise to residual stresses. Figure 8.41 shows the influence of cooling rate on the transverse residual stresses in a 40-ply carbon/PEEK unidirectional laminate.

4.2. Fiber Placement Process

Figure 8.42 shows a schematic of the fiber placement process. The concept for fiber placement is a process that was started in the 1980s. Recently, with the development of new equipment with good control, the fiber placement process can be used to make good quality thermoplastic composite parts.

The process is similar to filament winding except that the preimpregnated tapes are pushed toward the mandrel. At the nip point, a heat source is directed toward the fiber to heat and melt the tape. A roller is used to apply pressure at the nip point to spread the fiber and also to apply compaction. As the material point is moved away from the nip point, the material should cool down at an appropriate rate to solidify.

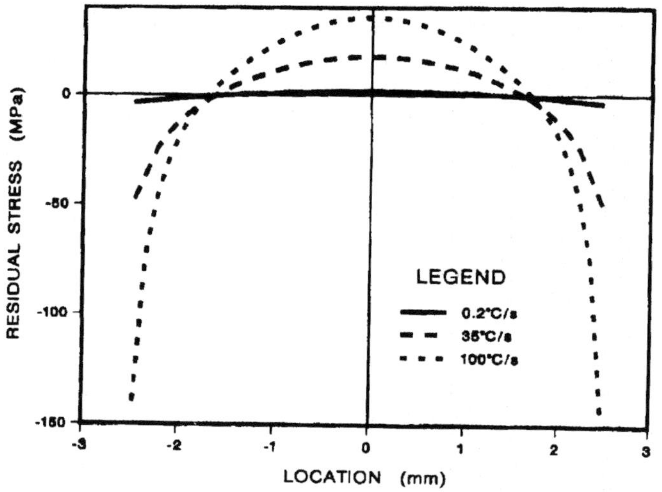

FIGURE 8.41 Predicted temperature as a function of time, and position from the surface, for a 5 mm (0.2 in) thick neat carbon/PEEK plaque with surface cooled at 114°C/second. Reproduced from Velisaris C.N., and Seferis J.C. "Heat transfer effects on the processing-structure relationships of polyetheretherketone (PEEK) based composites", *J. Science and Engineering of Composite Materials,* Vol. 1, No. 1, 1988, pp. 13–22, with permission from Freund Publishing Ltd.

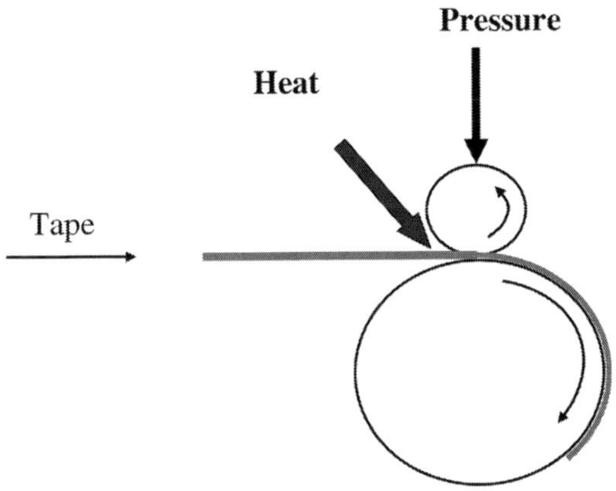

FIGURE 8.42 Schematic of the fiber placement process.

5. REFERENCES

1. Verrey J. et al. "Manufacturing cost comparison of thermoplastic and thermoset RTM for an automotive floor pan," *Composites. Part A,* 2006, Vol. 37, pp. 9–22.
2. Beland S.High performance thermoplastic resins and their composites, Noyes Data Corporation, 1990.
3. Cogswell F. N. "The processing science of thermoplastic structural composites," *International Journal of Polymer Processing,* 1987, Vol. 1, No. 4, pp. 157–165.
4. Lebel L. L. et al. "Processing and properties of carbon/nylon thermoplastic composites made by commingled tows and micro-braided tows," *Proc. of 5th Canada-Japan Workshop on Composites, Yonezawa, Japan, September 2004,* pp. 161–170.
5. Arpaci V. S., *Conduction Heat Transfer,* Addison Wesley, 1966.
6. Li M. C. and Loos A. C. "Modeling the consolidation process of thermoplastic composites," MD-Vol. 74, *Advanced Materials: Development, Characterization, Processing, and Mechanical Behaviour* (book of abstracts), ASME, 1996.
7. De Gennes P. G. *J. Chem. Phys.,* 1971, Vol. 55, p. 572.
8. Kim Y. H. and Wool R. P., *Macromolecules,* 1983, 16, p. 1115.
9. Muzzy J. D. and J. S. Colton, "The processing science of thermoplastic composites," in *Advanced Composites Manufacturing*, F. G. Gutowski ed., John Wiley and Sons, 1997.

Index

About the Author

DR. SUONG VAN HOA is a professor at the Department of Mechanical and Industrial Engineering of Concordia University, Montreal, Quebec, Canada. He obtained his B.Sc. from California State University San Luis Obispo, Master of Applied Science and Ph.D. from University of Toronto, Canada.

He became a professor at the Department of Mechanical Engineering of Concordia University in 1977. He was chair of the department from 1994 till 2000 and then again from 2003 till 2006. He has been working on Polymer Composites since 1979. He established the Concordia Center for Composites in 1993 and has been Director of the Center since. He formed the Canadian Association for Composite Structures and Materials (CACMA) in 1988. He co-established together with a Japanese colleague the series of Canada-Japan workshop on composites since 1996.

Dr. Hoa is a fellow of the American Society of Mechanical Engineers, Canadian Society for Mechanical Engineering and Engineering Institute of Canada. He is a recipient of the Society of Automotive Engineers Ralph Teetor award, the Canadian Society for Mechanical Engineering G.H. Duggan medal, the Synergy award from the Natural Sciences and Engineering Research Council of Canada, the Association des Directeurs de Recherche Industrielle du Quebec prize, the NanoQuebec Nano Academia prize, and he is a research fellow of Pratt & Whitney Canada Ltd.

Dr. Hoa has been carrying research and teaching on polymer composites since 1979 and polymer nanocomposites since 2000. He has been collaborating with many companies in composites in Canada, USA and internationally.